FIELD
SOLUTIONS
ON
COMPUTERS

FIELD
SOLUTIONS
ON
COMPUTERS

Stanley Humphries, Jr.

CRC Press
Taylor & Francis Group
Boca Raton London New York

CRC Press is an imprint of the
Taylor & Francis Group, an **informa** business

CRC Press
Taylor & Francis Group
6000 Broken Sound Parkway NW, Suite 300
Boca Raton, FL 33487-2742

© 1998 by Taylor & Francis Group, LLC
CRC Press is an imprint of Taylor & Francis Group, an Informa business

First issued in paperback 2019

No claim to original U.S. Government works

ISBN 13: 978-0-367-44799-1 (pbk)
ISBN 13: 978-0-8493-1668-5 (hbk)

Visit the Taylor & Francis Web site at
http://www.taylorandfrancis.com

and the CRC Press Web site at
http://www.crcpress.com

Preface

There is probably no hell for authors in the next world — they suffer so much from critics and publishers in this.

Bovee

The last two decades have seen explosive growth in computer methods for field solutions. Recent publications on mesh generation and applications of discrete mathematics could fill a hockey arena. There are several advanced references that summarize the full scope of research. In writing this book, my goal was more modest. I wanted to create an introductory work useful for beginning students and applied scientists. The book reflects my priorities — I am an experimentalist rather than an expert on numerics. As such, answers take priority over methods. In my experience, the main efforts of a design analysis are setting the physical constraints of the model and understanding the limitations of the output. Grinding numbers through the computer is the most straightforward part — settings of Puree, Frappe, or Liquefy give about the same results.

My involvement in numerical solutions began inauspiciously. Several years ago, Carl Ekdahl of Los Alamos National Laboratory asked me to assemble a personal computer version of the Poisson codes. The two-dimensional static field programs, written in 1966 by Ronald Holsinger, had proved to be remarkably long-lived and useful. I compiled the source code, added some color graphics, and assumed that was the end of it. But Carl developed an uncanny ability to crash the program and to develop long lists of essential new features. There came a point when there was no choice but actually to look at the code. From here, there was no turning back. Dissecting Poisson was the equivalent of a crash course in numerical methods, with clever ideas hidden around every corner. My work was greatly aided by the Poisson technical reference prepared by John Warren of LANL. Considerable material in this book follows from the code and reference, and I am happy for the opportunity to make the information widely available.

To me, the most enjoyable part of numerical work has been the chance to work with researchers and to learn something of diverse applications ranging from nuclear weapons physics to electrosurgery. I would like to thank the following people for their support and patient explanations: Carl Ekdahl and Daniel Rees of Los Alamos National Laboratories, Robert Platt and Thomas Ryan of Valleylab Incorporated, Ralph Genuario of Virginia Accelerators Corporation, John Petillo of Science Applications International Corporation, Richard Adler of Northstar Research Corporation, Joseph Heremans of

General Motors Research and Development Center, Garnett Bryant of the National Institute of Standards and Technology, Thomas Lockner of QM Technologies, and William Herrmannsfeldt of the Stanford Linear Accelerator Center.

I would also like to thank my loyal software customers for their insights and forbearance. Notable among them are Jean Miscopein, Yuval Carmel (University of Maryland), Richard Slattery (MIT Lincoln Laboratories), John Sethian (Naval Research Laboratory), Michael Baron and Phillip Kithil (Advanced Safety Concepts), Don Klein (Alpha Magnetics Corporation), Bernhard Kulke (BK Consulting), Craig Burkhart (First Point Scientific), Vladimir Derenchuk (University of Indiana Cyclotron), Derek Mous (High Voltage Engineering Europa), Michael Read (Physical Sciences Incorporated), Donald Voss (Voss Scientific), Duc Trinh (Joslyn Manufacturing), John Leopold (RAFAEL), Paul Luke (Lawrence Berkeley Laboratory), Steven Meddaugh (Schlumberger), Steven Morton (Oxford Micro Devices), Evan Pugh (Science Research Laboratories), Eiji Tanabe (AET Associates), and John Simonetti (EMR Photoelectric). I want to take this opportunity to assure these people that whatever the problem, I am working on a fix.

Much of the work preparing this book was performed on a sabbatical leave from the Department of Electrical Engineering at the University of New Mexico. I would like thank John Gahl for organizing new research programs during my absence. I received generous support from Sandia National Laboratories to develop a course on numerical field solutions for the US Particle Accelerator School. This effort was organized by Craig Olson and Thomas Sanford. I would like to thank Richard Cooper of Los Alamos National Laboratory for his advice over the years and the application of his legendary proofreading talents to this book. Finally, I want to express my gratitude to my wife Kristi for her love, support, and skills as a canal boat pilot.

Author

Stanley Humphries, Jr. is an experimentalist who has worked in the fields of controlled fusion, pulsed power applications, charged particle acceleration, and high-pressure hydrodynamics. His most notable contribution is the creation of methods to generate and to transport intense pulsed ion beams. He is the author of over 130 journal publications and the textbooks *Principles of Charged Particle Acceleration* (Wiley, 1986) and *Charged Particle Beams* (Wiley, 1990). He received a B.S. in Physics from the Massachusetts Institute of Technology and a Ph.D. in Nuclear Engineering from the University of California at Berkeley. He is currently a professor in the Department of Electrical and Computer Engineering at the University of New Mexico. He is also owner of Field Precision, an engineering software company. Prof. Humphries is a Fellow of the American Physical Society and the Institute of Electrical and Electronics Engineers.

Contents

Chapter 1 Introduction

Chapter 2 Finite-Element Electrostatic Solutions

Chapter 3 Minimum-Energy Principles in Electrostatics

Chapter 4 Finite-Difference Solutions and Regular Meshes

Chapter 5 Techniques for Numerical Field Solutions

Chapter 6 Matrix Methods for Field Solutions

Chapter 7 Analyzing Numerical Solutions

Chapter 8   Nonlinear and Anisotropic Materials

Chapter 9   Finite-Element Magnetostatic Solutions

Chapter 10   Static Field Analysis and Applications

Chapter 11 Low-Frequency Electric and Magnetic Fields

Chapter 12 Thermal Transport and Magnetic Field Diffusion

Chapter 13 Electromagnetic Fields in One Dimension

Chapter 14 Two- and Three-Dimensional Electromagnetic Simulations

Complementary Educational Software

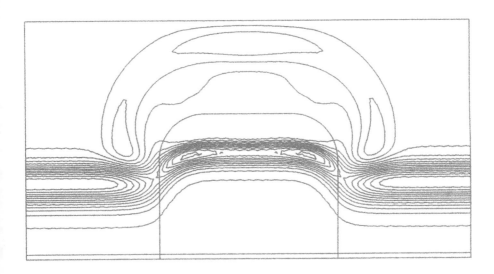

Turn-key programs to perform most of the two-dimensional calculations discussed in this book are available on the Internet. The software is provided to purchasers of the book at no charge by Field Precision. The programs run on IBM compatible computers under DOS, Windows 3.1, Windows 95 and Windows NT. The minimum requirements are a VGA screen and 4 MB of memory. The educational version of the TriComp System (Version 1.0) will handle up to 10,000 elements. The programs give solutions in electrostatics, magnetostatics (with non-linear materials), and time-domain and frequency-domain electromagnetics. An automatic mesh generator and interactive post-processors are included. Registered users can download executable programs and instruction manuals. For information, see the Field Precision site at

<http://www.rt66.com/~fieldp>.

From the home page, transfer to *Users' Groups/Educational Software*.

1

Introduction

Delightful task! to rear the tender Thought,
To teach the young Idea how to shoot,
To pour the fresh Instruction o'er the Mind,
To breathe the enlivening Spirit, and to fix
The generous Purpose in the glowing breast.

James Thompson

1.1 Overview

Numerical techniques often give straightforward answers to problems that are difficult or impossible to solve with analytic methods. This book is a toolbox of practical numerical methods for scientists and engineers who deal with electric and magnetic fields. It is designed for self-study or as a one-term course for advanced undergraduates. The emphasis is on representing physical problems by numerical simulations and understanding the results. We shall concentrate on finite-element techniques that can handle complex geometries and materials. The methods apply to a broad range of systems, including biological media, plasma processing reactors, accelerators, solid-state devices, and rotating electromagnetic equipment.

The material can help you write your own programs, either for direct field solutions or analyses of results from other programs. It can also help you apply packaged software more effectively. Because hands-on experience is an essential to understand the methods, educational software keyed to the text is available. The two-dimensional programs handle the full spectrum of physical systems that we shall study.

The goal of this book is to act as a bridge between introductory texts on electromagnetism and the growing list of advanced references on numerical field techniques. Standard texts emphasize analytic techniques and usually include exercises that are solvable on hand calculators. Despite the growing importance of computers in electromagnetic design, limits on time and hardware preclude detailed coverage of numerical techniques in introductory courses. This book addresses the problem partly by coordinating the material with integrated software for personal computers. Furthermore, electromagnetic theory is cast in a form aligned to numerical techniques. This approach has an added advantage — it can help enhance your intuitive grasp of field theory. Computer solutions are concrete representations of the abstract concepts of vector calculus. The viewpoint of fields as the interactions of simple elements is particularly valuable to students familiar with circuit theory. Another feature of the book is an emphasis on interdisciplinary applications. We shall touch on related areas of physics and engineering, including gas dynamics, thermal transport, and charged-particle optics. One motivation is to demonstrate the versatility of the numerical methods, which extend to a remarkable spectrum of applications. A second goal is to encourage a broad viewpoint that can be helpful working with electromagnetic devices. For example, in the design of a magnet we can apply similar techniques to calculate field strength, magnetic forces, strain components and cooling requirements.

The literature on numerical techniques for electromagnetism has grown considerably in the past decade. With few exceptions, available books are comprehensive reviews of advanced work aimed at experienced readers. A consequence is that it is often difficult to find practical guidelines for applying the methods. In contrast, this book has the modest goals of summarizing underlying physics and describing methods that are easy to use. To achieve a manageable length, many topics are not covered, including moment methods and high order finite-elements. Nonetheless, the book can serve as an introduction to the literature when your application demands more advanced techniques. The criterion for choosing topics was that solutions could be accomplished on standard personal computers. For example, although we shall study conformal triangular meshes in two dimensions, the discussion of three-dimensional solutions is limited to regular box elements.

1.2   The Load Ahead

The material roughly follows the order of an introductory electromagnetism course, starting with electrostatics and progressing to electromagnetic waves. The topics are arranged in a linear progression — early chapters provide a groundwork for more involved treatments that follow. Numerical techniques are introduced as needed for the increasingly challenging solutions. The book includes sections on good design practices, such as choosing run parameters, applying dimensionless variables, making effective benchmark tests, and interpreting results. There are also sections on auxiliary numerical techniques, such as interpolation and matrix inversion.

We begin with *boundary value problems* where the goal is to find static solutions in space that are consistent with conditions on surrounding boundaries. We then proceed to *initial value problems*, in which we start from a given state of a system and follow its evolution in space and time. The first eight chapters emphasize electrostatic solutions. Electrostatics has a strong intuitive appeal, and the derivations create a foundation of theory for applications that follow. Chapter 2 plunges immediately into *finite-element* solutions on arbitrary triangular meshes. This method gives two-dimensional solutions with high accuracy because a set of triangles can conform to curved and slanted boundaries of electrodes and dielectrics. To begin, the chapter reviews the differential and integral equations of electrostatics with dielectrics. The first task in a numerical solution is to convert the continuous equations into a set of difference equations suitable for digital computers. The approach in Chapter 2 is to apply Gauss's law over a volume defined by triangular elements surrounding a mesh vertex. In contrast, Chapter 3 derives the finite-element equations from the principle of minimum field energy. Although the approach is less intuitive, the formulation is more easily extended to higher-order field approximations and three-dimensional solutions.

Chapter 4 introduces electrostatic solutions on regular meshes with elements shaped like boxes. We review *finite-difference* methods. Here, the idea is to generate difference equations by direct conversion of differential equations like the electrostatic Poisson equation. The conversion is straightforward when the solution space is divided into box volumes with rectangular sides parallel to the coordinate axes. This type of mesh limits the solution accuracy but minimizes the amount of information that must be stored in memory. Finite-difference expressions play an important role in the time-dependent solutions of Chapters 12, 13, and 14.

Working from a set of difference equations, the following three chapters concentrate on techniques for solution and analysis. Chapter 5 covers the preliminary task of mesh generation. The term *mesh* refers to the way that a solution area or volume is divided. Figure 1.1 shows an example of a two-dimensional conformal mesh. The triangles constitute *elements*, the fundamental spatial unit. Material properties, like the dielectric constant, are assumed uniform over each element volume. *Vertices* are the intersections of element boundary edges. Solutions of the difference equations yield values for the electrostatic potential at the vertices. Values at intervening points and the field components can be estimated by interpolation. The chapter describes techniques to set up regular rectangular meshes in three dimensions. It also covers the more challenging task of defining a set of irregular triangular elements for two-dimensional solutions. Here, the triangle sides closely follow the boundaries of physical objects. Another task is to set up a mesh indexing system so that we can determine the elements surrounding a vertex, the neighboring vertices and the material identity of elements. The result of mesh generation and difference conversion is a large set of coupled linear equations. The final section of Chapter 5 covers solution of equation sets by relaxation methods. These methods are easy to program and run rapidly, but they may fail with difficult geometries or complex materials. As an alternative, Chapter 6 reviews direct inversion of linear equation sets by matrix methods. Initial sections cover familiar techniques like Gauss-Jordan inversion that

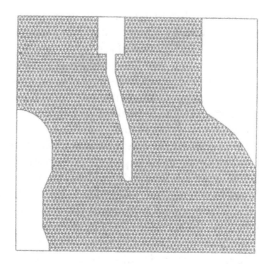

FIGURE 1.1
Conformal two-dimensional mesh. The solution area is divided into triangular elements to represent different materials. A numerical solution gives field quantities at the vertices.

apply to moderate equation sets. The chapter concludes with a discussion of block matrix methods to solve the large equation sets associated with field problems.

Chapter 7 covers the important topic of what to do with electrostatic potential values after you find them. The first sections deal with the calculation of gradients to determine electric fields. The least-squares-fit method is useful for arbitrary meshes because the number of available data points may vary. The remaining sections cover techniques for graphical display of data, including plots of boundaries, meshes, and element properties. The chapter also describes finding equipotential contours, making field line plots, and displaying potential as a three-dimensional wireframe elevation.

Chapter 8 addresses field solution techniques for materials with complex properties. The topic is an important preliminary for the magnetic solutions of the following chapter. Nonlinear materials have local properties that depend on the field quantities. Because the fields are not known in advance, we must employ a cyclic process to derive solutions. This usually involves an initial field approximation, calculation of the local material properties, correction of the fields, and so forth. The chapter addresses the convergence of cyclic calculations and interpolation methods to extract nonlinear material parameters from numerical tables. The response of anisotropic materials (such as birefringent crystals or permanent magnets) to fields depends on orientation. The final section reviews finite-element equations for these materials.

We proceed to magnetostatics in Chapter 9. The first section reviews Ampere's law in differential and integral form. The relationships lead, respectively, to finite-difference and finite-element equations. The chapter concentrates on two-dimensional solutions in planar and cylindrical geometries. In this limit, the vector potential plays a role analogous to the electrostatic potential. We can directly apply the solution methods developed in previous chapters. The chapter also discusses the properties of permanent magnets, materials with both nonlinear and anisotropic properties. The final section covers cyclic methods to find self-consistent operating points in permanent magnet devices.

Chapter 10 reviews several applications for static field solutions. Initial sections deal with volume and surface integrals of electric and magnetic field quantities over the regions of a triangular mesh. The integrals yield useful quantities like field energy and forces on structures. For electrostatic solutions we can find induced surface charge to determine self and mutual capacitance. Integrals over magnetic solutions yield information on inductance, force, and torque. The remaining sections review three applications. Sections 6 and 7 cover charged-particle devices. The first section addresses relativistic equations of motion and time-centered solutions for ordinary differential equations. The next section covers advanced techniques like self-consistent space-charge forces and field-limited flow in electron guns. Section 8 reviews advanced boundary conditions in finite-element solutions with application to the design of Hall effect sensors.

Chapter 11 initiates our study of electric and magnetic fields that change in time. We first address frequency-domain solutions where steady-state fields vary harmonically. The assumption is that the frequency is low enough so that the effects of radiation are small. The approximation allows neglect of inductively generated electric fields in electrical problems and displacement currents in magnetic problems. In this limit, electric and magnetic field solutions are separable. The governing equations are similar to those for static fields and we can adapt methods of previous chapters. The models of Chapter 11 apply to a variety of applications, including eddy currents in alternating current (AC) transformers, radio frequency (RF) electric fields in biological media, and inductive coupling in microcircuits. Chapter 12 introduces the topic of initial-value solutions through a detailed study of the diffusion equation. This relationship is critical to almost all areas of physics and engineering. We use thermal transport as a framework to develop time-dependent finite-element equations. These equations are applied to simulations of pulsed magnetic fields in the presence of conducting materials. The chapter also addresses issues of numerical stability for initial value problems.

The final two chapters remove limits on frequency to model radiation effects with coupled electric and magnetic fields. The numerical techniques incorporate the full set of Maxwell's equations. Applications include microwave devices, RF shielding, and communications. Electromagnetic solutions encompass a much broader range of possibilities than the static and quasi-static results of previous chapters. Often, solutions are easy to generate but challenging to understand. We shall take a step-by-step approach, starting from an extensive discussion of one-dimensional solutions in Chapter 13.

The advantage is that we can address the basic physics issues without worrying about details of mesh generation and complex interacting waves. With confidence in the validity of the methods, we proceed in Chapter 14 to two- and three-dimensional solutions.

The first section of Chapter 13 reviews the theory of plane waves. The material is useful for constructing benchmark tests for the numerical approaches that follow. Electromagnetic solutions are divided into two main classes: time-domain and frequency-domain. Section 2 covers numerical time-domain solutions. Here, electromagnetic pulses with arbitrary time variations propagate in ideal or absorbing media. The process involves direct solution of Maxwell's equations by replacing time derivatives with time-centered difference operators. There are important limits on the time-step so that the solutions do not violate the principle of causality. We shall use the integral form of the equations to derive finite-element representations of spatial variations. The section reviews an important issue, the simulation of infinite space by a finite solution volume. For this function we shall concentrate on the method of absorbing boundary layers, physically equivalent to matched terminations on transmission lines.

Frequency-domain calculations apply to continuous excitation of systems at a single frequency, $f = 2\pi\omega$. Here, we replace time derivatives by $j\omega$ and solve the resulting complex-number difference equations to find the amplitude and phase of field quantities. It is straightforward to model generalized material losses by taking complex values for the dielectric constant ϵ and magnetic permeability μ. Frequency-domain solutions divide into two types: scattering and resonant. Scattering solutions, discussed in Section 3, apply to open systems where radiation in the form of traveling waves can escape. Section 4 covers solutions for resonators, closed structures that support a standing wave pattern. Applications include waveguides and particle acceleration structures. Although we can find valid solutions for any frequency, we are usually interested in specific values that correspond to resonant modes. In this case, waves interfere constructively giving strong fields for moderate drive power. The calculation of resonant modes involves several solutions at different frequencies to identify the interference condition.

Chapter 14 extends the numerical methods to two- and three-dimensional electromagnetics. The foundation developed in previous chapters makes this a relatively easy process. We need make only a few extensions to theory, such as optimizing absorbing boundaries for waves incident at an angle. This chapter is the reward for the effort expended in previous sections. We can create a wealth of interesting solutions with moderate effort. Initial sections concentrate on two-dimensional solutions on triangular meshes. Figure 1.2 shows an example of a scattering-type frequency-domain solution, the electric fields in the radiation pattern of an electric dipole antenna. The final topic is three-dimensional time-domain solutions. We shall apply a finite-element viewpoint on a regular mesh. In comparison to the popular finite-difference time-domain method, the resulting difference equations are easy to interpret and provide an accurate representation of material boundaries.

FIGURE 1.2
Radiation from an electric dipole antenna, electric field lines for $\lambda = 0.5$ m. The geometry is cylindrical with the axis at the bottom. One half of the solution is shown with a symmetry boundary on the left. A thin ideal absorbing boundary defines an anechoic chamber of radius 2.0 m.

1.3   Some Precautions

A prevailing myth of our age is that computer simulations of increasing complexity lead to a better understanding of the universe. In reality, it often works the other way. People with the most prodigious codes and supercomputers may not have the most acute physical insight. This is because they must spend so much time buying cutting-edge hardware, improving graphical-user-interfaces, and converting working codes to C⁺⁺. Finances also contribute to the problem: methods that circumvent complexities through insight seldom attract money. Funding agencies are more comfortable showering resources on definable trends like *massive parallel processing*. Finally, it is infinitely more pleasant to sit at a computer than to visit experimentalists in the trenches.

Computers are an undeniably powerful tool to expand our design capabilities in science and engineering if we already know the topic thoroughly. When used with a solid grasp of analytic theory, numerical solutions can amplify our understanding of complex systems. The power to make rapid parameter searches can build valuable physical intuition if we know what to look for in the results. On the other hand, the computer has proved to be the twentieth century's most profound contribution to the art of wasting time. A computer can ingest an endless supply of hours while generating nothing but an effluvium of colored view graphs. The following guidelines, the result of difficult experience, may be useful in the struggle against computer-generated sloth.

- The validity of code results depends more on the identification of the relevant physical processes and the soundness of approximations than on the sophistication of the numerical methods or the size of the computer.

- Extensive analytic benchmark testing of new or existing technical software is essential because bugs are everywhere. I can remember numerous episodes where high-powered programs I have written or used have been defeated by the simplest textbook examples. Equally important, the creation of tests is a good way to bracket the parameter space and to learn more about the physics.

- Well-organized programming is essential for any technical task. With common sense, good programs can be written in any language.

- A technical code without comprehensive documentation of the underlying equations and approximations ranges from useless to dangerous.

- Anyone who has fought the quirks of packaged routines knows that libraries are not the answer to all problems. You can often save time by writing your own routines. An added benefit is that the process may have educational value.

- Buying software to solve engineering problems is not an alternative to knowing how to solve engineering problems.

- If your simulation brings forth an amazing potentially publishable result, look for a numerical instability.

- Beware the illusion of simplicity. Codes that appear to have all the data at hand and to make all the decisions are often collections of bad habits and questionable approximations. In such programs, initial demonstrations may appear effortless but real design problems become nightmares.

- Never believe anyone over thirty whose presentation hinges on a computer-generated movie.

2

Finite-Element Electrostatic Solutions

Never learn to do anything: if you don't learn,
you'll always find someone else to do it for you.

Mark Twain

Electrostatic calculations provide a good introduction to numerical solutions because the physics is easy to understand. The goal is to find electric forces that act in volumes containing distributions of charge. The charge can be distributed in free space, on metal electrodes, or in dielectric materials. In this chapter, we concentrate on two-dimensional solutions using finite-element methods with triangular elements. Integral relationships applied over the elements lead to a set of coupled difference equations for electrostatic potential values at the vertices.

The first four sections review principles of electrostatics. Section 2.1 covers Coulomb's law, the foundation of electrostatics. The empirical relationship specifies forces between charges. The section introduces the electric field, a vector quantity defined throughout the solution volume that gives forces associated with a distribution of charge. Section 2.2 shows how to represent forces resulting from large numbers of charged particles. Here, the direct application of Coulomb's law is impractical. Instead, we approximate the charge with a continuous distribution in space and derive an alternative relationship, Gauss' law. Section 2.3 applies Gauss' law to small volumes to find differential relationships, the electrostatic Maxwell equations. The section also introduces the electrostatic potential, a scalar (single-valued) function of space from which vector fields can be calculated. Because the scalar function satisfies simple boundary conditions, it is easier to determine fields indirectly by first calculating the potential. Section 2.4 discusses the classes of charge density that arise in electrostatics, emphasizing the charge displacement in dielectric materials.

Section 2.5 initiates the discussion of finite-element techniques. The section introduces the *computational mesh*, where the solution volume is divided into elements. The set of element boundaries looks like a mesh or screen. In two

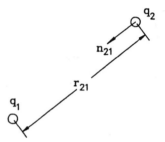

FIGURE 2.1
Coulomb's law — force between two point charges.

dimensions, the best choice of elements is a set of flexible triangles with sides aligned along the boundaries of electrodes and dielectrics. This choice gives an accurate representation of material discontinuities. The section also discusses the general strategy of numerical solutions and the importance of boundaries. Section 2.6 reviews useful relationships from analytic geometry for triangles. Section 2.7 contains the most important derivation in the book. The section shows how the application of Gauss' law in Cartesian coordinates over the elements surrounding a vertex yields a simple difference relationship between the potential at the point and its neighbors. The resulting set of coupled linear equations can be solved on a digital computer. Section 2.8 extends the derivation to three-dimensional systems with cylindrical symmetry.

2.1   Coulomb's Law

Electrostatics describes the forces between charged bodies at given positions. We denote the charge on an object, a positive or negative quantity, by the symbol q. Throughout this book, we shall use the International System of Units where the charge is given in units of *coulombs*. The charge of an electron is -1.60219×10^{-19} coulombs. The empirical relationship for the forces between stationary charges was discovered by Coulomb in 1771. Suppose there are two small objects with charges q_1 and q_2. As shown in Figure 2.1, the unit vector $\hat{n}_{21}$ points from object 2 to object 1 and the distance between the charges is r_{21}. The electric force on object 1 from object 2 is

$$F_1 = \hat{n}_{21} \frac{q_1 q_2}{4\pi\varepsilon_o\, r_{21}^2}.$$

(2.1)

Equation 2.1 is a vector equation that gives force components in three directions. It implies that (1) the force is parallel to the line between the charges, (2) the strength is proportional to the product of the charges and decreases as the inverse square of their separation, and (3) like charges repel one another while opposite charges attract. The constant ε_o with value 8.854×10^{-12} farads/m ensures that the force is in newtons if distances are measured in meters and charges in coulombs.

The *principle of superposition* is a second experimental result that allows extension of Coulomb's law to multiple charges. It states that the total force on a test charge from several charges is the vector sum of individual forces. Suppose we have a test charge q_o at location (x_o, y_o, z_o) in the presence of a set of charges q_i. The quantities $\hat{n}_i$ are unit vectors that point from the surrounding charges to the test charge and the quantities r_i are the corresponding separations. The total force on the test charge is

$$F(x_o, y_o, z_o) = q_o \sum_{i=1}^{N} \hat{n}_i \frac{q_i}{4\pi\varepsilon_o r_i^2}. \tag{2.2}$$

For a given distribution of surrounding charges, we could define a normalized force

$$\frac{F(x_o, y_o, z_o)}{q_o} = \sum_i \hat{n}_i \frac{q_i}{4\pi\varepsilon_o r_i^2}. \tag{2.3}$$

The summation on the right-hand side is a function of space that depends on the charge distribution and not on the properties of the test charge. After evaluating the summation we could find the force on any test charge at position (x_o, y_o, z_o) by multiplying the normalized force by q_o. The process can be extended by calculating the normalized force for a charge distribution at all points in space, generating a *vector field*. The field information gives the force on any test charge at any position. The normalized force distribution, called the *electric field*, is given at position $[x_o, y_o, z_o]$ by

$$E(x_o, y_o, z_o) = \sum_i \hat{n}_i \frac{q_i}{4\pi\varepsilon_o r_i^2}. \tag{2.4}$$

The electric field has units of volts per meter (V/m). Given the electric field for a charge distribution, the force on a test charge q at location $x = (x, y, z)$ is

$$F(x) = q\, E(x). \tag{2.5}$$

We usually illustrate electric fields with plots of *field lines*. These are a set of curves aligned along the local field direction with spacing inversely proportional to the field strength. The field is strong in regions where the lines are closely packed. Figure 2.2 shows the electric field lines between two charged spheres. By convention they point from regions of positive to negative charge.

FIGURE 2.2
Electric field lines between two charged spheres, positive on the left and negative on the right.

2.2   Gauss' Law and Charge Density

The direct application of Equations 2.4 and 2.5 may be useful for atomic scale problems involving interactions between a few ions or electrons. The procedure is impractical for macroscopic problems that may include more than 10^{10} charged particles. To deal with large numbers of charges, we need an alternative mathematical expression of the content of Equation 2.4. Gauss' law specifies electric field properties over the surface of a macroscopic volume in terms of the enclosed charges, independent of their position:

$$\iint_S \mathbf{E} \cdot \hat{\mathbf{n}} \ dS = \frac{\sum_i q_i}{\epsilon_o} . \tag{2.6}$$

The left-hand side is a surface integral of the normal component of electric field. Here, dS is the area of a segment of the surface and **n** is a local unit vector pointing out of the volume. The summation on the right-hand side extends over all charges inside the volume.

Gauss' law follows from the inverse square variation of electric field with distance from a charge. Figure 2.3a shows a point charge q surrounded by a spherical surface of radius R. The electric field is normal to the surface and has the uniform magnitude

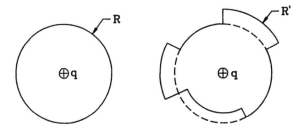

FIGURE 2.3
Derivation of Gauss' law. Left: point charge centered in a spherical surface of radius R. Right:
Point charge inside a closed surface composed of several spherical segments.

$$E_r(R) = \frac{q}{4\pi\epsilon_o R^2} . \qquad (2.7)$$

Multiplying Equation 2.7 by the area of the sphere gives the surface integral value $(q/4\pi\epsilon_o R^2)(4\pi R^2) = q/\epsilon_o$, consistent with Equation 2.5. Next, consider a test charge surrounded by an irregular volume like that shown in Figure 2.3b. The surface consists of a number of small elements that are either parallel with or normal to r, the direction pointing away from the charge. There are no contributions to the integral from portions of the surface parallel to r. Consider comparing the integral on a normal surface segment a distance R′ from the charge to that of a segment covering the same solid angle on the reference sphere of Figure 2.3a. The field is lower by the ratio $(R/R')^2$, but the surface area is higher by $(R'/R)^2$. Therefore, the segments make the same contribution to the surface integral. Integrating over all solid angle, we find that the integral for the surface in Figure 2.3b also equals q/ϵ_o. By making stair-step approximations with smaller divisions, we can extend the result to surfaces of any shape. Equivalently, the integral does not depend on the position of the charge inside the volume. Finally, the principle of superposition implies that if there are several charges inside the volume, the total surface integral is $\Sigma q_i/\epsilon_o$.

In most applications on a macroscopic scale, it is unnecessary to preserve information about the discrete nature of charges. For example, consider the field at a distance of 1 μm from two electrons with an atomic scale spacing of 1 Å. The difference in the field amplitude compared to two coincident charges is only about one part in 10^7. Usually we can approximate a large number of discrete charges as a continuous cloud represented by a charge density $\rho(x,y,z)$. The quantity ρ is a scalar field with units of coulombs/m³. It equals the sum of charge in a differential element at location (x,y,z) divided by the element volume,

$$\rho(x, y, z) \cong \frac{\sum_i q_i}{\Delta x \Delta y \Delta z} . \qquad (2.8)$$

The summation in the numerator extends over all charges in the range $x \leq x' \leq x + \Delta x$, $y \leq y' \leq y + \Delta y$ and $z \leq z' \leq z + \Delta z$.

We can rewrite Equation 2.4 in terms of a volume integral over the charge density. If the vector x gives the coordinates where the field is calculated and x' is the position of the charges, then

$$E(x) = \iiint dV \, \frac{\rho(x')}{4\pi\epsilon_o} \, \frac{x - x'}{|x - x'|^3} \,. \tag{2.9}$$

Similarly, we can rewrite Gauss' law as

$$\iint_s E \cdot n \, dS = \frac{\iiint_V dV \, \rho}{\epsilon_o} \,, \tag{2.10}$$

where the volume V is enclosed by the surface S.

Electric field lines have fluid-like geometric properties. Equation 2.10 shows that the total flux of lines out of a volume is proportional to the total charge enclosed. If there is no enclosed charge, every field line that penetrates a volume must exit. The condition that electric field amplitudes are finite implies that lines flow smoothly. They cannot cross each other or kink back on themselves (Figure 2.4a).

2.3   Differential Equations for Electrostatic Fields

In this section we derive differential relationships equivalent to Equations 2.9 and 2.10 that are useful for the finite-difference derivations of Chapter 4. An equation for the divergence of electric field follows from Equation 2.10 in the limit of small volumes. Suppose that we apply Gauss' law over the box shown in Figure 2.4b. The box is small enough to neglect variations in ρ; therefore, the volume integral on the right-hand side of Equation 2.10 is approximately $\rho(x,y,z)\Delta x\Delta y\Delta z$. We must evaluate the surface integral over the six faces. First, consider the two faces normal to the x-axis with area $\Delta y\Delta z$. The contribution to the surface integral is

$$-E_x(x,y,z) \, \Delta y\Delta z + E_x(x + \Delta x, y, z) \, \Delta y\Delta z \,. \tag{2.11}$$

If the box is small, variations of E_x in y and z are nearly the same on both faces and we can approximate the variation in x with a Taylor expansion,

$$E_x(x + \Delta x, y, z) \cong E_x(x, y, z) + \frac{\partial E_x}{\partial x} \, \Delta x \,. \tag{2.12}$$

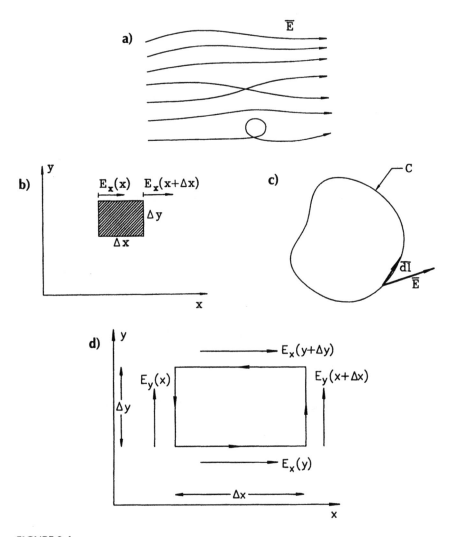

FIGURE 2.4
Derivation of the differential equations of electrostatics. (a) Geometric constraints on field lines — kinked or crossing lines give infinite field magnitude. (b) Application of Gauss' law over a differential volume to find the divergence equations. (c) Definition of a line integral around a closed curve. (d) Evaluation of a line integral around a differential volume in the x–y plane to find one component of the curl equation.

Substituting Equation 2.12 into Equation 2.11 gives

$$\frac{\partial E_x}{\partial x} \, \Delta x \Delta y \Delta z. \tag{2.13}$$

Contributions on the faces normal to the y and z axes have similar form. Setting the surface integral of field equal to the volume integral of charge density and canceling the common factor of $\Delta x \Delta y \Delta z$ gives

$$\frac{\partial E_x}{\partial x} + \frac{\partial E_y}{\partial y} + \frac{\partial E_z}{\partial z} = \frac{\rho}{\epsilon_o} . \tag{2.14}$$

We recognize the left-hand side of Equation 2.14 as the divergence of electric field. Therefore, the differential form of Gauss' law is

$$\nabla \cdot \mathbf{E} = \frac{\rho}{\epsilon_o} . \tag{2.15}$$

We can find another useful vector relationship that represents the condition that static electric field lines cannot curve back on themselves (Figure 2.4a). An equivalent statement is that the *line integral* of electric field around any closed curve in space equals zero,

$$\oint \mathbf{E} \cdot \mathbf{dl} = 0 . \tag{2.16}$$

Figure 2.4c illustrates the meaning of Equation 2.16. A closed curve is divided into a number of vector segments $\mathbf{dl}$ that point in the direction of positive rotation. The line integral is the sum of values for the dot product of segment vectors with the local electric field. Consider an integral around the small rectangle shown in Figure 2.4d normal to the z axis near the point (x,y,z).

$$E_x(x,y,z)\Delta x + E_y(x+\Delta x, y, z)\Delta y - E_x(x, y+\Delta y, z)\Delta x$$
$$- E_y(x,y,z)\Delta y = 0. \tag{2.17}$$

Applying a Taylor expansion, Equation 2.17 becomes

$$-\frac{\partial E_x}{\partial y} + \frac{\partial E_y}{\partial x} = 0 . \tag{2.18}$$

The left-hand side of Equation 2.18 is the z-component of the vector curl operator. We can derive similar expressions in planes normal to the x and y axes. Therefore, Equation 2.16 implies the differential relationship,

$$\nabla \times \mathbf{E} = 0 . \tag{2.19}$$

It is easy to prove that the curl of any vector field that can be expressed as the gradient of a scalar field is identically zero,

$$\nabla \times (\nabla \phi) = 0 . \tag{2.20}$$

An analogy to a contour map makes the physical meaning of Equation 2.20 clear. Here, elevation is the scalar field and the gradient gives lines of the slope. The lines may radiate away from a point (a peak or depression) or flow smoothly between locations, but can never curl or cross each other if we exclude cliffs and overhangs. An integral of slope around a closed curve must return to the same elevation.

Comparison of Equations 2.19 and 2.20 shows that the static electric field can be written as

$$E = -\nabla \phi. \tag{2.21}$$

The minus sign is a convention — we shall discuss the motivation shortly. The quantity ϕ (the *electrostatic potential*) is a scalar field with values defined over a region of space. It has units of volts (V). To understand the physical meaning of potential, consider the work performed moving a test charge q in an electric field from point x_1 to x_2.

$$\Delta W = -\int_{x_1}^{x_2} F \cdot dl = -q \int_{x_1}^{x_2} E \cdot dl. \tag{2.22}$$

Noting that the change in *potential energy* of the charge, ΔU, equals the work performed and substituting from Equation 2.21, we find that

$$\Delta U = q \int_{x_1}^{x_2} \nabla \phi \cdot dl = q \left[\phi(x_2) - \phi(x_1) \right]. \tag{2.23}$$

Therefore, the quantity $q\phi$ equals the relative potential energy of charged particles in a static electric field.

Combining Equations 2.15 and 2.22 gives the *Poisson equation* for the electrostatic potential,

$$\nabla \cdot (\nabla \phi) = -\frac{\rho}{\epsilon_0}. \tag{2.24}$$

Equation 2.24 gives an alternative means to find electric fields. We can solve the equation for ϕ and extract fields using Equation 2.21. In applications such as the evaluation of fields between biased metal electrodes, the method has the following advantages.

- The function ϕ is single valued while E has three components.
- The electrostatic potential varies continuously through space. In contrast, we shall see that the electric field has discontinuities at the surfaces of dielectric materials.
- The boundary condition on electrostatic potential on a metal surface is simple — it assumes a constant value.

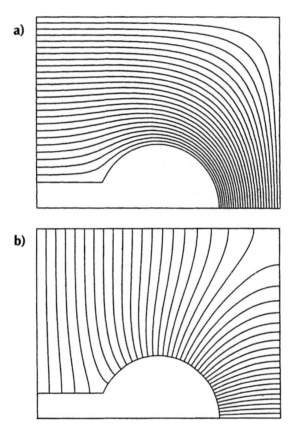

FIGURE 2.5

Electrostatic calculation of a rod and support at 50,000 V inside a grounded box. Neumann conditions apply on the symmetry boundaries on the bottom and on the left. Boundaries: x_{min} = 0.0 cm, x_{max} = 10.0 cm, y_{min} = 0.0 cm, y_{max} = 7.5 cm. (a) Equipotential lines with 2000 V intervals. (b) Electric field lines.

It is easy to prove the third point. The electric field must equal zero inside a metal; otherwise, it would drive large currents. Therefore, the integral of Equation 2.23 must be zero between any two points in the conductor implying constant potential.

Figure 2.5 shows the field distribution between a rod electrode and grounded plate to illustrate the constant potential boundary condition. The electrode and its support rod (lower left) are at potential ϕ = 50,000 V while all points on the plate and surrounding chamber (right and top boundaries) are at ϕ = 0. Figure 2.5a shows constant potential lines separated by equal increments $\Delta\phi$ = 2000 V, while -Figure 2.5b is a plot of electric field lines. Both plots give complete information about the field. In the equipotential plot the electric field direction is normal to the lines and the field magnitude is inversely proportional to the distance between the lines.

2.4 Charge Density Distributions and Dielectric Materials

In principle, we can find electric fields using the methods of the previous sections, given the charge density distribution. Sometimes we have this information. One example is a charged-particle beam with specified properties. Charge density with a predefined distribution in the solution space is called *space charge* and is denoted ρ_s. The challenge is to handle other types of charge where the spatial distribution depends on the field solution. For example, consider the charges on a metal electrode, denoted by ρ_e. Electric fields are created by moving these charges to the electrode with a power supply. The charge must reside on the electrode surface to ensure zero electric fields inside. Because we are unsure of the layer thickness, it is more convenient to refer to a surface charge density,

$$\sigma_e = \int_{\text{Layer}} \rho_e ds , \tag{2.25}$$

with units of coulombs/m^2. For complex electrode shapes it is difficult to predetermine the distribution σ_e (x,y,z) that makes the parallel component of electric field zero everywhere on the surface.

Fortunately, there is a strategy that makes this foreknowledge unnecessary. Because the fields inside electrodes equal zero, we are interested only in fields in the spaces between. The procedure is to solve the Poisson equation for ϕ in the intervening spaces where ρ_e is zero, taking the electrode surfaces as equipotential boundaries. Then, we can use the calculated values of electric fields near the electrodes to find the surface charge density from the formula

$$\sigma_e = \epsilon_o E_\perp . \tag{2.26}$$

The quantity E_Z is the amplitude of the field pointing out of the electrode. Figure 2.6a shows the origin of Equation 2.26. Gauss' law is applied to the volume shown, a thin box with cross section area dA that encompasses the charge layer on an electrode. Because the field is zero inside the electrode, the surface integral of the electric field is $E_\perp dA$ while the total enclosed charge is $\sigma_s dA$.

As an example, we shall calculate the electric field between parallel plates of area A and spacing d (Figure 2.6b). The right-hand plate is set to a voltage V_o, the left-hand plate is at ground potential, and there is no charge in the enclosed space. In the limit that d « A, we can neglect variations in y and z. The solution of Poisson's equation gives the potential variation

$$\phi(x) = V_o \left(\frac{x}{d} \right) . \tag{2.27}$$

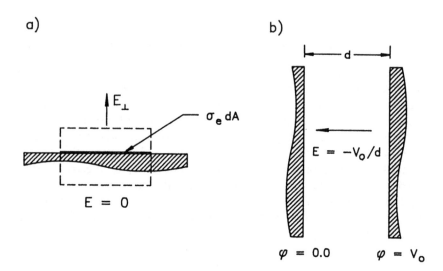

FIGURE 2.6
Electrostatic calculations using the Poisson equation. (a) Field produced by a surface charge density on an electrode. (b) Electric field between infinite parallel plates.

The electric field has the constant value $-V_o/d$. The surface charge densities are almost uniform over the inner surfaces of the left and right plates. Applying Equation 2.26, the total charges on the plates are $\pm(\epsilon_o A/d)V_o$.

The charges in dielectric materials behave differently from those on conductors. Nonetheless, we can use an analogous technique to find fields in the presence of dielectrics with no foreknowledge of the associated charge density $\rho_d(x,y,z)$. While the electrons in metals move freely through the material, the charges in gaseous, liquid, or solid dielectrics are bound to polar molecules. These molecules, shown schematically in Figure 2.7a, are electrically neutral but have a spatial separation of the positive and negative charges. Normally, thermal motions randomize the molecular orientations so there is no net charge density in the medium. An applied electric field causes a small fractional alignment of molecules. Suppose we create a field E_{xo} in the volume element shown in Figure 2.7b. The element is small enough so that the material is approximately homogeneous and the field is uniform. Inside the volume, there is no charge density because the positive and negative charges of adjacent molecules cancel. On the other hand, there is a surface charge density on the faces normal to the x-axis. The charge induced on the element and its neighbors in y and z creates an electric field with magnitude proportional to E_{xo} that points in the opposite direction. This effect implies that the total field in the dielectric, E_x, points in the same direction as E_{xo} and has a smaller magnitude,

$$E_x = \frac{E_{xo}}{\epsilon_{xr}}.$$

(2.28)

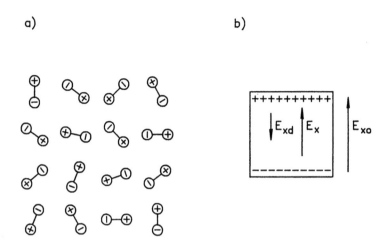

FIGURE 2.7
Dielectric materials. (a) Polar molecules with random orientation. (b) Effect of dielectric charge on the electric field inside a small volume.

The quantity ϵ_{xr} in Equation 2.28 is called the *relative dielectric constant*. It has a value greater than unity that depends on the properties of the material.

For moderate applied fields at ambient temperature, the change in potential energy associated with the orientation of a molecule is much less than its thermal energy. As a result, the fractional orientation in a material is small and ϵ_{xr} is almost independent of the field magnitude. In this case, we say that the medium is *linear* because Equation 2.28 is a linear relationship. In crystals, the relative dielectric constant may have different values along material axes. In gases, liquids, and amorphous solids (most ceramic and plastic insulators), the dielectric constant is the same in all directions. In such *isotropic* materials, the total electric field vector points in the same direction as the applied field,

$$E = \frac{E_o}{\epsilon_p}. \tag{2.29}$$

In this chapter we will concentrate on linear, isotropic dielectrics. Chapter 8 covers numerical methods for nonlinear, anisotropic dielectrics.

We can use Equation 2.29 to convert the equations we have derived to a form that automatically includes the effects of dielectric charge. For example, consider the form of Equation 2.15 in a region between electrodes that may include both dielectric and space charge,

$$\nabla \cdot \mathbf{E} = \frac{\rho_d}{\epsilon_o} + \frac{\rho_s}{\epsilon_o}. \tag{2.30}$$

Equation 2.30 explicitly shows the contribution of the dielectric charge density. By the superposition principle, we can also write an equation for the applied field that arises from space charge and surface charges on electrodes,

$$\nabla \cdot \mathbf{E}_o = \frac{\rho_s}{\epsilon_o} \tag{2.31}$$

Substituting Equation 2.29 into Equation 2.31 gives an equation that *implicitly* includes the contribution of dielectric charge,

$$\nabla \cdot \left(\epsilon_r \mathbf{E} \right) = \frac{\rho_s}{\epsilon_o} \tag{2.32}$$

We can solve Equation 2.32 to find the total electric field without advanced information on $\rho_d(x,y,z)$. It is important to remember that spatial variations of $\epsilon_r(x,y,z)$ must be included when calculating the left-hand side of Equation 2.32.

In the presence of isotropic dielectrics, Gauss' law takes the form

$$\iint_S \epsilon_r \mathbf{E} \cdot \hat{n} \, dS = \frac{\iiint_V dV \, \rho_s}{\epsilon_o}. \tag{2.33}$$

Similarly, the generalized Poisson equation is

$$\nabla \cdot \left(\epsilon_r \nabla \phi \right) = -\frac{\rho_s}{\epsilon_o}. \tag{2.34}$$

Although the electrostatic potential is continuous across a dielectric boundary, the electric field is discontinuous because of the dielectric surface charge. Changes in the field can be expressed as boundary conditions at the surface. The left-hand side of Figure 2.8 shows a thin box that encloses the interface between two media with relative dielectric constants ϵ_{r1} and ϵ_{r2}. The electric field on each side is resolved into components parallel and perpendicular to the surface. Applying Gauss' law (Equation 2.33) to the box shown gives the following relationship for the normal field component,

$$\epsilon_{r1} E_{\perp 1} = \epsilon_{r2} E_{\perp 2}. \tag{2.35}$$

The application of Equation 2.16 to a line integral around the boundary on the right-hand side of Figure 2.8 gives a relationship for the parallel field components,

$$E_{\parallel 1} = E_{\parallel 2}. \tag{2.36}$$

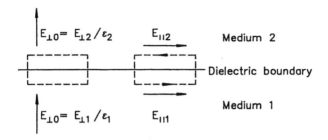

FIGURE 2.8
Boundary conditions at a dielectric interface.

The numerical solution of Figure 2.9 illustrates the nature of fields at dielectric boundaries. An isotropic dielectric slab with $\epsilon_{r2} = 10$ is suspended in a vacuum space ($\epsilon_{r1} = 1.0$) midway between parallel electrodes with $\phi = 1.0$ V at the top and $\phi = 0.0$ V on the bottom. Only half the solution is shown — a symmetry boundary is used on the left-hand side (see Section 2.8). The equipotential plot shows that the slab has an average potential midway between that of the electrodes and that there is a substantial reduction of the field inside the material. Note that the electric fields are consistent with Equations 2.35 and 2.36 on the right-hand slab boundary. Because the normal field component is relatively small inside the dielectric, the internal field lines are almost parallel to the surface.

2.5 Finite Elements

Because digital computers handle discrete numbers, it is impossible to solve the differential or integral equations of electrostatics directly. Computers cannot deal with continuous variations of charge density and electric field. The strategy we shall follow for computer field solutions is to divide the solution space into a large number of volume elements. The quantities ρ and **E** have approximately uniform values over the small elements. In this limit the field relationships transform to a large set of simple linear equations that can be handled on a computer. The result is a good approximation to the continuous solution. Because the volumes have finite size, this approach is called the *finite element* method. We must address three tasks to employ the method.

- Optimum division of the solution volume into elements
- Derivation of linear equations to represent electrostatics in the limit of small volumes
- Solution of large sets of simultaneous linear equations

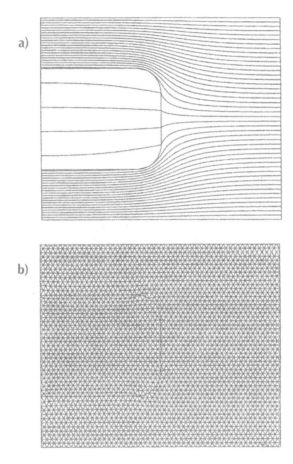

FIGURE 2.9
Electrostatic field solution with a dielectric material. Dielectric block with $\epsilon_r = 10.0$ between parallel plate electrodes, Neumann boundaries on the left and right. Boundaries: $x_{min} = 0.0$ cm, $x_{max} = 10.0$ cm, $y_{min} = 0.0$ cm, $y_{max} = 8.0$ cm. (a) Equipotential lines. (b) Computational mesh.

In this section, we will discuss some of the terminology and philosophy related to element generation. Chapter 7 covers this topic in detail. The remainder of this chapter concentrates on the second task — finite-element expressions that are equivalent to Gauss' law.

The set of elements that comprise the division of a solution volume is called the *computational mesh*. Figure 2.9b shows the two-dimensional mesh used to find the field solutions of Figure 2.9a. There are two possible two-dimensional geometries. In *planar* geometry, material properties vary in x and y with no variation in z. Therefore, the field component E_z is either zero or constant and there are nontrivial solutions for E_x and E_y. The volume element is a shape projected in the x–y plane that extends a unit distance along z. The second option is the *cylindrical* geometry with variations in r and z and symmetry in θ. The field components are E_r and E_z. A volume element is a toroidal object where a projected element shape in the r–z plane extends around the axis. To create meshes, we must decide on the best element shape for division of the x–y or r–z planes. It is difficult to handle curves with analytic geometry so we shall concentrate on polygons with straight-line boundaries. The triangle is a fundamental shape because any polygon can be decomposed into two or more triangles. In this book, we will use only triangular elements — any set of polygons can easily be converted to a pure triangular mesh.

There is no set formula to divide a solution space into elements. Nonetheless, there are two general guidelines to create good meshes. The first has to do with the *shape* of the triangles. A critical condition in the derivation of finite-element equations (Section 2.7) is that material properties are uniform in an element. Therefore, we should choose the shapes of triangles so that they are either completely inside or outside the volumes of electrodes and dielectrics. The result is that triangle sides closely follow the contours of material boundaries. The term *conformal mesh* means that we change the shape of elements so that they conform to the boundaries of objects.

The second guideline applies to the *size* of triangles. An assumption in the derivation of Section 2.7 is that the electric field is uniform in an element. Clearly, we must create several triangles to resolve a significant change in the field. A good mesh has small elements in regions of strong field changes and large elements where the field changes gradually. A mesh with regions of different average triangle sizes is called a *variable resolution mesh*. With the high speed and memory capacity of modern computers, we might be tempted to skip a careful analysis and simply use a large number of small elements. Fortunately, a feature of numerical solutions punishes such inelegance. Because of roundoff errors, the validity of solutions may actually degrade with extremely fine meshes. The art of mesh generation resides in picking appropriate element shapes and understanding how the choice affects the solution accuracy.

2.6 Coordinate Relationships for Triangles

This section reviews some mathematical relationships for triangles. The equations will be useful for the derivation of Section 2.7. First, we shall find an expression for the area of a triangle in terms of the coordinates of its vertices. Consider a triangle lying in the x–y plane with one vertex at the origin (Figure 2.10a). The triangle is defined by the vectors $x_1 = (x_1, y_1)$ and $x_2 = (x_2, y_2)$, where the vectors are ordered in a sense of positive rotation about the z axis. The enclosed area equals one half the cross product of the vectors,

$$a = \frac{x_1 y_2 - x_2 y_1}{2}. \tag{2.37}$$

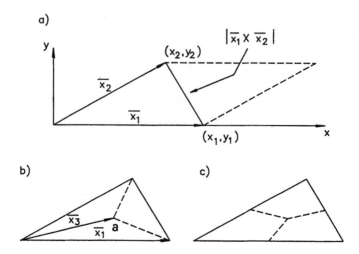

FIGURE 2.10
Relationships for triangles. (a) Calculating the area of a triangle. (b) Division of a triangle into three equal parts connecting the centroid to the vertices. (c) Division of a triangle into three equal parts connecting the centroid to the midpoints of the sides.

The formula also applies to triangles projected in the r–z plane for cylindrical problems. By convention, we shall illustrate most derivations in Cartesian coordinates and use the correspondence

$$x \rightarrow z, \ y \rightarrow r. \tag{2.38}$$

If the reference vertex is at a point (x_o, y_o), we can generalize Equation 2.37 to

$$a = \frac{(x_1 - x_o)(y_2 - y_o) - (x_2 - x_o)(y_1 - y_o)}{2}. \tag{2.39}$$

To ensure a positive value for the area, it is important to take the points 0, 1, and 2 in the order of positive rotation (counterclockwise), as in Figure 2.10.

In Section 2.7, we shall apportion the space charge contained in elements equally to their three vertices. In other words, we want to divide the area of a triangle into three equal parts. Consider the triangle of Figure 2.10b with one vertex (x_o, y_o), at the coordinate origin. The center-of-mass position is the average of the vertex coordinates,

$$x_3 = \frac{(x_o + x_1 + x_2)}{3} = \frac{(x_1 + x_2)}{3},$$

$$\tag{2.40}$$

$$y_3 = \frac{(y_o + y_1 + y_2)}{3} = \frac{(y_1 + y_2)}{3}.$$

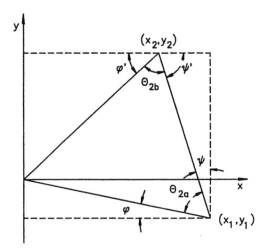

FIGURE 2.11
Definition of quantities to derive the triangle cotangent relationships.

The final form applies when $x_o = y_o = 0$. One way to divide the triangle is to draw lines from the center-of-mass point to the three vertices, as shown in Figure 2.10b. The area of triangle a is one half the cross product of the vectors from the origin to points 1 and 3,

$$a_a = \frac{|x_1 \times x_3|}{2}.$$ (2.41)

Substituting the components of x_3 from Equation 2.40 and comparing a_a with the expression of Equation 2.37 confirms that triangle a covers one third the triangle area. A second way to divide the triangle is to draw lines from the center-of-mass to the midpoints on each side. Figure 2.10c shows that these lines bisect the triangles of Figure 2.10b. Therefore, the area of triangle d equals the sum of half the areas of triangles a and b, or one third the full triangle area.

Our final task is to find formulas that relate the angles between triangle sides to the vertex coordinates. To simplify the mathematics we start with the triangle of Figure 2.11 with one vertex at the origin and the others at points (x_1, y_1) and (x_2, y_2). The quantities θ_a and θ_b are the enclosed angles opposite the main vertex. Note that the vertex points and angles are ordered with positive rotation. Figure 2.11 defines two additional angles, ξ and ψ, at vertex 1. The enclosed angle at this vertex can be written as

$$\cot \theta_a = \cot\left(\frac{\pi}{2} - \xi - \psi\right) = \tan(\xi + \psi).$$ (2.42)

A trigonometric identity gives the tangent for the sum of two angles,

$$\tan(\xi + \psi) = \frac{\tan\xi + \tan\psi}{1 - \tan\xi \tan\psi} \, . \tag{2.43}$$

We can express the ξ and ψ in terms of vertex coordinates,

$$\tan\xi = -\frac{y_1}{x_1} \, , \tag{2.44}$$

and

$$\tan\psi = \frac{(x_1 - x_2)}{(y_2 - y_1)} \, . \tag{2.45}$$

Substituting Equations 2.43, 2.44, and 2.45 into Equation 2.42 gives the desired relationship for the enclosed angle

$$\cot\theta_a = \frac{-y_1(y_2 - y_1) - x_1(x_2 - x_1)}{x_1 y_2 - x_2 y_1} \, . \tag{2.46}$$

Recognizing that the denominator of Equation 2.46 as twice the triangle area leads to the final expression

$$\cot\theta_a = \frac{-y_1(y_2 - y_1) - x_1(x_2 - x_1)}{2a} \, . \tag{2.47}$$

Figure 2.11 shows alternate angles ξ' and ψ' for vertex 2. We can use the relationship,

$$\cot\theta_b = \cot(\pi - \xi' - \psi') \, , \tag{2.48}$$

to show that

$$\cot\theta_b = \frac{y_2(y_2 - y_1) + x_2(x_2 - x_1)}{2a} \, . \tag{2.49}$$

2.7   Gauss' Law for Elements at a Vertex Point

We now have enough background to derive the finite-element equations for electrostatic fields. the following assumptions are used in the model.

- The problem has a planar geometry where ϕ is a function of x and y. The projected solution space is divided into a number of small triangular elements.
- The material properties are uniform over each element area. In electrostatic problems these quantities are the space charge density ρ and the relative dielectric constant ϵ_r.
- For a numerical solution, we seek discrete values of ϕ at the mesh vertices such that interpolations approximate the ideal values.
- Elements are the fundamental units of division, and we cannot attain a knowledge of $\phi(x,y)$ on a finer scale. Because three points determine a plane, the values of ϕ at the vertices of a triangle define a linear function of x and y with a constant gradient. Therefore, the model is based on constant values of E in each element.

Figure 2.12 shows a sample vertex in the mesh at position (x_o, y_o) surrounded by elements and neighboring vertices. We shall take the test vertex at the origin of a local coordinate system and measure X and Y relative to this point. To generalize the calculation, we can make the substitution,

$$X \rightarrow (x - x_o), \quad Y \rightarrow (y - y_o). \tag{2.50}$$

The triangles and vertices are labeled with positive rotation. Although the figure shows six triangles, the number is arbitrary. The goal is to apply Gauss' law (Equation 2.33) to a volume surrounding the vertex point. The volume is bounded by a closed curve in the X–Y plane around the reference vertex that extends vertically an arbitrary distance Δz. Figure 2.12 shows a particular boundary choice that makes it easy to find the enclosed charge. The surface consists of 12 straight line segments connecting the midpoints of the mesh lines to neighboring vertices to the element mass centers. Section 2.6 showed that this path encloses one third of the area of each triangle. We can immediately write an expression for the right-hand side of Equation 2.33,

$$\sum_{i=1}^{6} \frac{\rho_i a_i \Delta z}{3 \epsilon_o}. \tag{2.51}$$

In Equation 2.51, the quantity ρ_i is the charge density of triangle i and a_i is the element area projected in the X–Y plane.

Turning next to the surface integral on the left-hand side of Equation 2.33, we need values for the electric field in each triangle. To illustrate the calculation, consider triangle 2 in Figure 2.12. The potential in the element is a linear function of position,

$$\phi(x, y) = \phi_o + uX + vY, \tag{2.52}$$

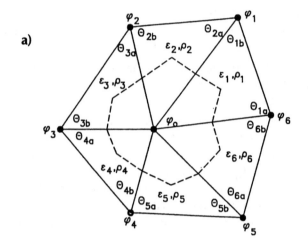

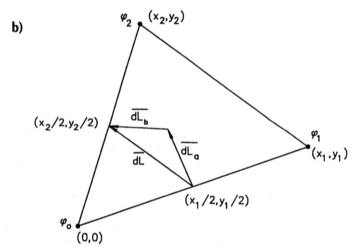

FIGURE 2.12
Gauss' law on a conformal triangular mesh. (a) Elements surrounding a vertex point, illustrating numbering conventions for elements, element angles, and vertices. (b) Detailed view of triangle 2 showing the surface integral path.

with ϕ_o is the potential at the reference vertex and u and v are constants. The electric field in triangle 2 is

$$\mathbf{E}_2 = u\hat{x} - v\hat{y}, \tag{2.53}$$

where $\hat{x}$ and $\hat{y}$ are unit vectors along the coordinate axes. The values of u and v ensure that Equation 2.52 has the correct values at the triangle vertices. Applying the condition at vertices 1 and 2 gives the two equations

$$uX_1 + vY_1 = \phi_1 - \phi_o, \tag{2.54}$$

and

$$uX_2 + vY_2 = \phi_2 - \phi_o. \tag{2.55}$$

The solution of Equations 2.54 and 2.55 gives the value of the quantity u,

$$u = \frac{-\phi_o(Y_2 - Y_1) + \phi_1 Y_2 - \phi_2 Y_1}{X_1 Y_2 - X_2 Y_1}. \tag{2.56}$$

Recognizing that the denominator is twice the area of triangle 2, Equation 2.56 becomes

$$u = \frac{-\phi_o(Y_2 - Y_1) + \phi_1 Y_2 - \phi_2 Y_1}{2a_2}. \tag{2.57}$$

Similarly, the quantity v is given by

$$v = \frac{\phi_o(X_2 - X_1) - \phi_1 X_2 + \phi_2 X_1}{2a_2}. \tag{2.58}$$

We can represent portions of the Gaussian surface by a set of vectors ordered with positive rotation. In triangle 2, the vector L_a extends from $(X_1/2,Y_1/2)$ to $([X_1+X_2]/3,[Y_1+Y_2]/3)$ and the vector L_b from the center of mass to $(X_2/2,Y_2/2)$. The Gaussian integral over the first section of surface is

$$\Delta z \left| L_A \right| \epsilon_2 \, E_2 \cdot \hat{n}_a, \tag{2.59}$$

where $\hat{n}_a$ is a unit vector pointing outward normal to L_a. The unit vector is

$$\hat{n}_a = \frac{L_a \times \hat{z}}{|L_a|}, \tag{2.60}$$

where $\hat{z}$ is a unit vector along the z axis. The cross product in the numerator gives the correct direction and the length in the denominator normalizes the vector. Applying a vector identity, we can write the contribution to the integral

$$\Delta z E_2 \cdot (L_a \times \hat{z}) = \Delta z (E_2 \times L_a) \cdot \hat{z}. \tag{2.61}$$

Extending the results to the surface section defined by L_b, we can write an expression for the total surface integral in triangle 2 as,

$$\Delta z \, \epsilon_2 \left[(E_2 \times L_a) + (E_2 \times L_b) \right] \cdot \hat{z} = \Delta z \, \epsilon_2 \left[E_2 \times (L_a + L_b) \right] \cdot \hat{z}. \tag{2.62}$$

The sum of the line vectors is a vector that points from $(X_1/2, Y_1/2)$ to $(X_2/2, Y_2/2)$,

$$L_a + L_b = \frac{X_2 - X_1}{2}\hat{x} + \frac{Y_2 - Y_1}{2}\hat{y}. \qquad (2.63)$$

Substituting from Equation 2.53 and Equation 2.63, Equation 2.62 becomes

$$\Delta z\ \epsilon_2\left(-\frac{Y_2 - Y_1}{2}u + \frac{X_2 - X_1}{2}v\right). \qquad (2.64)$$

Substituting for the quantities u and v from Equations 2.57 and 2.58, we can write the surface integral over triangle 2 in terms of the potential values at the vertices,

$$\Delta z\ \epsilon_2\ \frac{\left(+\phi_0(Y_2 - Y_1) - \phi_1 y_2 + \phi_2 y_1\right)(Y_2 - Y_1)}{4a_2}$$

$$+\frac{\left(\phi_0(X_2 - X_1) - \phi_1 X_2 + \phi_2 X_1\right)(X_2 - X_1)}{4a_2}. \qquad (2.65)$$

Finally, we can use the cotangent expressions of Equation 2.47 and 2.49 to simplify Equation 2.65 to

$$\Delta z\ \epsilon l_2\left[\phi_0\left(\cot\theta_{2b} + \cot\theta_{2a}\right) - \phi_1\cot\theta_{2b} - \phi_2\cot\theta_{2a}\right]. \qquad (2.66)$$

Similar expressions hold for the other triangles.

Summing terms for all surrounding elements to find the left-hand side of Equation 2.33 and substituting from Equation 2.51 for the right-hand side, we can write the finite-element equivalent of Gauss' law,

$$\phi_0\sum_{i=1}^{6}W_i - \sum_{i=1}^{6}\phi_i W_i = \sum_{i=1}^{6}\frac{\rho_i A_i}{3\epsilon_0}. \qquad (2.67)$$

The dimensionless numbers W_i, called the *coupling coefficients* at the vertex point, are given by

$$W_1 = \frac{\epsilon_2\cot\theta_{2b} + \epsilon_1\cot\theta_{1a}}{2},$$

$$W_2 = \frac{\epsilon_3\cot\theta_{3b} + \epsilon_2\cot\theta_{2a}}{2},$$

$$\dots \qquad (2.68)$$

$$W_6 = \frac{\epsilon_1\cot\theta_{1b} + \epsilon_6\cot\theta_{6a}}{2}.$$

The coupling constant to a neighboring vertex depends on the relative dielectric constants and geometries of the triangles adjacent to connecting line.

We can write Equation 2.67 in an alternate form that better illustrates the physical content

$$\phi_o = \frac{\displaystyle\sum_{i=1}^{6} \phi_i W_i + \sum_{i=1}^{6} \frac{\rho_i A_i}{3\epsilon_o}}{\displaystyle\sum_{i=1}^{6} W_i}. \tag{2.69}$$

In the absence of space-charge, Gauss' law implies that the potential value at a vertex is a weighted average of the potentials at neighboring vertices. The weighting factor depends on the material properties and shapes of surrounding elements. A nearby accumulation of positive space charge elevates the potential at a vertex.

2.8 Solution Procedure and Boundary Conditions

In this section we shall discuss the general strategy for applying finite-element equations to find global electrostatic field solutions. The methods will be covered in detail in following chapters. The first step in a solution is mesh generation — the solution volume is divided into a large number of elements and vertices. Next, we set up an equation similar to Equation 2.68 for each vertex point. The result is a large set of coupled linear equations, typically 50,000 for a two-dimensional problem. The variables are the potential values at the vertices.

Equation 2.67 gives the potential at a point in vacuum or dielectric internal to the solution region. To carry out electrostatic solutions, we must also deal with electrodes in the volume and vertex points on the boundaries. Representing internal electrodes is easy. We simply assign a constant value of potential at the point and remove the vertex equation from the set. This reduces both the number of variables and number of equations by one. Because there is no equation for vertices in and on electrodes, the properties of internal elements play no role.

Because finite-element solutions deal with bounded volumes, we must address the question of what to do with points on the boundary. Some of these points may represent a surface of constant potential. For example, the right and top boundaries of the solution region of Figure 2.5 are part of a grounded chamber wall. For a fixed boundary, we set specified values for the vertex potentials and eliminate the associated equations. In this case, there is no need to consider field values or material properties outside the solution volume. This type of boundary satisfies a *Dirichlet condition* where the value of the unknown function (electrostatic potential) is specified.

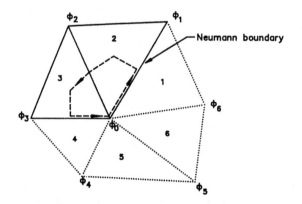

FIGURE 2.13
Gauss' law integration path at a vertex on a Neumann boundary. Elements 2 and 3 are inside the solution volume, while elements 4, 5, 6, and 1 are outside. The boundary lies on the lines from vertex 3 to 0 to 1.

To complete the process, we must consider how to handle boundary points that do not have fixed potential. The convention in finite-element solutions is to apply Equation 2.67 at the point as though there were additional vertices and elements outside the solution volume. For electrostatic solutions, we assign the following material properties to the external elements,

$$\rho_i = 0, \tag{2.70}$$

$$\epsilon_i = 0. \tag{2.71}$$

To understand the implications of the procedure, consider the finite-element equation for the boundary vertex of Figure 2.13a. The elements marked 2 and 3 are inside the solution volume, and there are four arbitrary external elements. The dark lines represent the boundary. Suppose we carry out the Gaussian volume and surface integrals described in Section 2.7 along the dashed line. Equation 2.70 implies that only the internal elements contribute to the space-charge integral. Similarly, an inspection of Equations 2.66 and 2.71 shows that only the internal elements contribute to the surface integral.

Next, consider Gaussian integrals around alternate path of Figure 2.13b. This path includes the internal elements and follows the boundary. The enclosed space-charge is the same as that of Figure 2.13a; therefore, the surfaces integrals are equal. The implication is that the surface integral of the normal component of electric field along the boundary is zero. An equivalent statement is that the potential satisfies the equation

$$\frac{\partial \phi}{\partial \hat{n}} = 0, \tag{2.72}$$

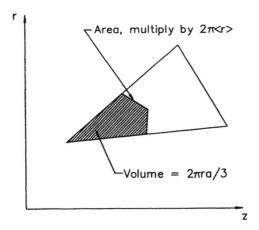

FIGURE 2.14
Modification of the volume and surface integrals in Gauss' law for cylindrical coordinates.

along the boundary. The notation on the left-hand side denotes the derivative of potential normal to local direction of the boundary. Equation 2.72 is a special case of a *Neumann boundary condition*, where a *derivative* of the unknown quantity is specified.

The procedure described ensures that unspecified boundary points in a finite-element solution automatically assume the specialized Neumann condition of Equation 2.72. An inspection of Equations 2.68 and 2.71 shows that all coupling constants to the hypothetical external vertices are zero. The condition of Equation 2.72 means that electric field lines are parallel to the boundary. The potential at a boundary vertex assumes the value that would occur if the external potential variation mirrored potential values inside the solution volume. For this reason, the specialized Neumann condition can be used to represent a reflection plane so that it is necessary to simulate only half of a symmetric system. For example, the specialized Neumann conditions applies on the left-hand boundary of Figure 2.9.

2.9 Electrostatic Equations in Cylindrical Coordinates

The derivation of Section 2.7 provides a clear picture of the physical origin of the finite-element equations. We can make some simple modifications to extend the results to cylindrical coordinates. Figure 2.14 illustrates a sample vertex with neighbors and surrounding triangles projected in the r–z plane. The dashed line shows the cross-section of the surface for the Gaussian integral. Each triangle represents a toroidal shape with approximate volume

$$2\pi\, r_i\, a_i, \tag{2.73}$$

where a_i is the cross section area and r_i is the center-of-mass radius,

$$r_i = \frac{r_{oi} + r_{1i} + r_{2i}}{3}. \tag{2.74}$$

Equation 2.73 implies that the charge contained inside the surface is approximately,

$$\sum_{i=1}^{6} \frac{2\pi r_i \rho_i a_i}{3\epsilon_o}. \tag{2.75}$$

In the discussion of planar elements in Section 2.7, the areas of the Gaussian surfaces equaled the segment lengths multiplied by Δz, an arbitrary length in z. For the cylindrical case, we must multiply by $2\pi R$, where R is the average radial position of the segment. It is usually sufficient to take the segment average radius equal to the center-of-mass element radius of Equation 2.74, $R \approx r_i$. With this convention, the contribution to the surface integral from triangle 2 is

$$2\pi\, r_2\, \epsilon_2 \left[\phi_o \left(\cot\theta_{2b} + \cot\theta_{2a} \right) - \phi_1 \cot\theta_{2b} - \phi_2 \cot\theta_{2a} \right]. \tag{2.76}$$

Equation 2.76 is similar to Equation 2.66 with the replacement $\Delta z \to 2\pi r_2$. Extending the equation to the other surrounding triangles, setting the result equal to Equation 2.75 and canceling the common factor of 2π gives the cylindrical form of Gauss' law,

$$\phi_o = \frac{\displaystyle\sum_{i=1}^{6} \phi_i\, W_i + \sum_{i=1}^{6} \frac{r_i\, \rho_i\, a_i}{3\epsilon_o}}{\displaystyle\sum_{i=1}^{6} W_i}. \tag{2.77}$$

Note the factor of r_i in the space-charge term. The coupling coefficients have the modified form

$$W_1 = \frac{\epsilon_2 r_2 \cot\theta_{2b} + \epsilon_1 r_1 \cot\theta_{1a}}{2},$$

$$W_2 = \frac{\epsilon_3 r_3 \cot\theta_{3b} + \epsilon_2 r_2 \cot\theta_{2a}}{2}, \tag{2.78}$$

$$\cdots$$

$$W_6 = \frac{\epsilon_1 r_1 \cot\theta_{1b} + \epsilon_6 r_6 \cot\theta_{6a}}{2}.$$

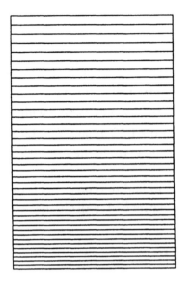

FIGURE 2.15
Benchmark test — electric field between coaxial cylinders. Bottom boundary: inner electrode with radius 0.02 m and potential 1000 V. Top: Outer electrode with radius 0.05 m and potential 0 V. Equipotential lines at 25 V intervals.

Here, the cotangents of the element angles are multiplied by the average triangle radius.

Figure 2.15 shows a benchmark test of the method, equipotential lines from a numerical solution for the field between long coaxial cylinders. Neumann boundaries on the left and right represent the infinite extent of the rod. A solution of the cylindrical Poisson equation gives the radial field as

$$E_r(r) = \frac{V_o}{r \ln\left(\dfrac{r_o}{r_i}\right)}. \tag{2.79}$$

In Equation 2.79, r_i is the radius of the inner cylinder with voltage V_o and r_o is the radius of the outer cylinder. For a choice of parameters $r_i = 0.02$ m, $r_o = 0.05$ m, and $V_o = 1000$ V, Equation 2.75 predicts a field of 54.568 kV/m on the inner electrode. A numerical solution with 80 elements in the radial direction gives a value of 54.543 kV/m, correct to 0.046%.

Chapter 2   Exercises

2.1. Find the force per unit area on each of two large parallel plates with separation d and voltage difference V_o.

(a) Show the force density is attractive with magnitude $V_o^2 \epsilon_o / 2d$.

(b) Give a value for d = 0.02 m and $V_o = 1.0 \times 10^5$ V.

(c) Compare the value in (b) to the force per unit area corresponding to 1 atm.

2.2. Use Coulomb's law to find the components of force on the test charge $q_o = 1.0 \times 10^{-6}$ C in Figure E2.1. (All charges are in the plane z = 0.)

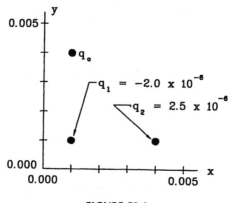

FIGURE E2.1

2.3. A line charge extending along the z-axis has linear charge density Σ coulombs/m. Use Coloumb's law to show that the electric field outside the line is $E_r(r) = \Sigma / 2\pi\epsilon_o r$. (Divide the line into a set of individual charges Σdz and integrate electric field contributions at the point $(r,0,0)$.)

2.4. An electric field is generated between coaxial cylinders of radii R_i and R_o by applying a voltage $-V_o$ to the center conductor.

(a) Find an expression for the kinetic energy of a proton of mass m_p that moves azimuthally in a circular orbit of radius $(R_i + R_o)/2$. (Balance the electric and centrifugal forces.)

(b) Find a value for the kinetic energy for $m_p = 1.67 \times 10^{-27}$ kg, $V_o = 5000$, $R_i = 0.02$ m and $R_o = 0.05$ m.

2.5. A cylindrical electron beam of radius R_o has charge density $\rho(r) = \rho_o(1 - r^2/R_o^2)$, where $R_o = 0.005$ m and $\rho_o = 1.0 \times 10^{-8}$ C/m³.

(a) Find a value for Σ, the charge per unit axial length of the beam.

(b) Use Gauss' law to find the value of the radial electric field at the beam envelope, $r = R_o$.

(c) The current of the beam is the charge that crosses a plane normal to the z-axis per unit time. If the beam velocity is v_z, show that the current equals $I = \Sigma v_z$.

(d) Give a value for the current if $v_z = 1.5 \times 10^7$ m/s.

2.6. Two concentric metal spheres have radii R_i and R_o. The inner sphere carries a charge Q_o.

(a) Apply Gauss' law to find an expression for the radial electric field between the spheres.

(b) Integrate the electric field from R_i to R_o to confirm the following expression for the voltage difference:

$$V_o - V_i = \frac{Q_o}{4\pi\epsilon_o}\left[\frac{1}{R_i} - \frac{1}{R_o}\right].$$

(c) Considering the properties of metals, how is the charge distributed on the inner sphere?

2.7. Extend the derivation of the divergence equation in Section 2.3 to cylindrical coordinates. Apply Gauss' law to a volume element with dimensions Δr, $r\Delta\theta$, and Δz to show that

$$\frac{1}{r}\frac{\partial}{\partial r}(rE_r) + \frac{1}{r}\frac{\partial E_\theta}{\partial \theta} + \frac{\partial E_z}{\partial z} = \frac{\rho}{\epsilon_o}.$$

2.8. Prove that the curl of any vector function that can be written as the gradient of a scalar function equals zero by directly evaluating the expressions in Cartesian coordinates:

$$\nabla \times (\nabla\phi) = 0.$$

2.9. A spherical plastic bead of diameter 0.001 m with density 700 kg/m³ has charge Q on its surface. The bead is in a region of vertical electric field of magnitude $E = 5 \times 10^6$ V/m. For what value of Q does the electric force balance the force of gravity?

2.10. Apply the Poisson equation to find the distribution of potential in a planar region of width d between two grounded metal plates filled with uniform space charge ρ_o. Give a value for the maximum potential when $d = 0.02$ m and $\rho_o = 2.5 \times 10^{-6}$ coulombs/m³.

2.11. A cylindrical charged-particle beam of radius R_o with uniform charge density ρ_o propagates along the axis of a grounded pipe of radius R_w.

(a) Use Gauss' law to find expressions for the electric field within the beam and in the space between the beam and the wall.

(b) By integrating E_r from $r = 0$ to R_w, show that the on-axis electrostatic potential is

$$\phi_o = \frac{\rho_o r_o^2}{4\epsilon_o}\left[1 + 2\ln\left(\frac{r_o}{r_i}\right)\right].$$

2.12. The equipotential lines in Figure 2.5a are spaced at 2000 V intervals. Use a scale to estimate the electric field amplitude at the tip of the spherical electrode and compare the results to the numerical prediction of 2.498×10^6 V/m.

2.13. A high-voltage coaxial transmission line has electrode radii $R_i = 0.075$ and $R_o = 0.150$ m. We want to reduce the electric field stress near the center conductor by using a radial distribution of dielectrics.

(a) Suppose the space between the electrodes is divided into two dielectric layers with a boundary at $(R_i + R_o)/2$. The outer layer has $\epsilon_r = 2$. What value of ϵ_r in the inner layer gives $E_r(R_i) = E_r(R_o)$?

(b) Suppose we had an ideal material that could be fabricated with any desired spatial variation of relative dielectric constant. If the value at the center conductor is ϵ_{ro}, give the functional form $\epsilon_r(r)$ that maintains a radially uniform electric field.

2.14. A parallel plate capacitor has the dielectric distribution shown in Figure E2.2. The bottom plate is grounded and the top plate is at voltage $V_o = 2000$ V. Dielectric 1 has $\epsilon_{r1} = 3.5$ and $d_1 = 0.002$ m while dielectric 2 has $\epsilon_{r2} = 5.0$ and $d_2 = 0.003$ m.

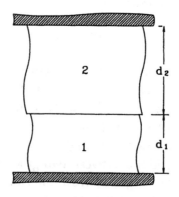

FIGURE E2.2

(a) Find the values of the electric field in each of the layers.

(b) Give values of surface charge density for the free and dielectric charges at the top electrode.

2.15. A triangle that lies in the x–y plane has the following vertex coordinates: [2.0,2.0], [4.6,3.5], [2.0,5.0].

(a) Find the area of the triangle using Equation 2.39.

(b) Show that the triangle is equilateral by computing θ_a and θ_b from Equations 2.47 and 2.49.

(c) Show that the area of an equilateral triangle with sides of length a is $A = a^2 \tan(60°)/2$. Compare the predicted value to that of part (a).

2.16. The three vertices of a triangle have coordinates [0.010,0.010], [0.034,0.014], and [0.028,0.030], where the dimensions are given in meters. The corresponding values of potential are 45.6, 57.8, and 72.0 V. Find values of E_x and E_y in the element.

2.17. Find the six coupling coefficients at the point marked *A* in Figure E2.3. Give an interpretation of the results.

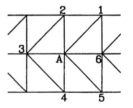

FIGURE E2.3

2.18. Figure E2.4 shows an electrostatic sextupole. The system has sixfold symmetry in the x–y plane and the electrodes have alternate potentials $\pm V_o$. Show that the fields can be calculated by simulating one twelfth of the system and applying symmetry conditions. Sketch a possible simulation volume and indicate Neumann and Dirichlet boundaries.

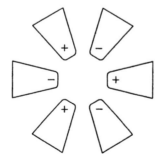

FIGURE E2.4

3

Minimum-Energy Principles in Electrostatics

So-called professional mathematicians have, in their reliance on the relative incapacity of the rest of mankind, acquired for themselves a reputation for profundity very similar to the reputation for sanctity possessed by theologians.

G. C. Lichtenberg

In this chapter we shall study an alternative derivation of the electrostatic finite element equations using the principle of minimum energy. There are several reasons to understand this approach. First, it gives insight into the physical implications of the Poisson equation. Second, the derivation is used in many advanced references on finite element methods. Third, it is easier to extend the procedure to three-dimensional calculations with a tetrahedronal mesh. Finally, the material gives a good opportunity to review concepts of electric field energy.

The potential energy stored in a distribution of charge can be written as a volume integral over a function of the electric field. Section 3.1 derives this function and shows how the contribution of dielectric charges is automatically included. Section 3.3 shows that the spatial distribution of electric field given by a solution of the Poisson equation corresponds to a state of minimum field energy integrated over the system volume. In preparation, Section 3.2 reviews results from the calculus of variations that are necessary for the derivation.

The minimum energy is applied to the fields of triangular elements in a two dimensional system in Section 3.4. The electrostatic energy in elements surrounding a vertex depends on the electrostatic potential at the vertex and at neighboring points. The finite-element relationship that represents an energy minimum is identical to the equation derived from Gauss's law in Section 2.7. Section 3.5 extends the procedure to a three-dimensional mesh consisting of contiguous tetrahedrons of arbitrary shape. To conclude, Section 3.6 summarizes the motivations for and procedures to derive finite-element equations from high-order polynomial expansions of the potential. This approach offers enhanced accuracy in applications where materials have uniform properties.

3.1   Electrostatic Field Energy

Section 2.3 showed that the amount of work necessary to move a charge from position 1 to position 2 in an electric field is

$$\Delta W = q(\phi_2 - \phi_1), \tag{3.1}$$

where ϕ is the electrostatic potential. Taking $\phi = 0$ at an infinite distance, the work to move a test charge to position x in a region of electric field is

$$\Delta W = q\phi(x). \tag{3.2}$$

Consider two point charges, Q_1 and Q_2. Applying Coulomb's law, the electric field and potential created by the second charge are given by

$$\phi(r) = \frac{Q_2}{4\pi\epsilon_o r}, \tag{3.3}$$

where r is the distance from the charge. The work to move Q_1 to a distance R from Q_2 is

$$\Delta W = Q_1 \frac{Q_2}{4\pi\epsilon_o R} = Q_1\phi_1, \tag{3.4}$$

where ϕ_1 is the potential at the position of Q_1. Conversely, we could view the process as moving charge Q_2. In this case, the energy could be written

$$\Delta W = Q_2 \frac{Q_1}{4\pi\epsilon_o R} = Q_2\phi_2. \tag{3.5}$$

The work to assemble the charge distribution can be written in a symmetric form by combining Equations 3.4 and 3.5,

$$\Delta W = \frac{Q_1\phi_1 + Q_2\phi_2}{2}. \tag{3.6}$$

In general, the work to assemble a collection of N charges Q_i at positions x_i is

$$W = \frac{1}{2} \sum_{i=1}^{N} Q_i \phi(x_i). \tag{3.7}$$

When there are large numbers of charges we can write Equation 3.7 in terms of charge density

$$W = \frac{1}{2} \iiint dV \, \rho(x)\phi(x). \tag{3.8}$$

Equation 3.8 is the potential energy of the charge distribution. Because the assembly process creates an electric field, we often refer to the potential energy as the *field energy*.

Often it is convenient to express the field energy entirely in terms of the electric field. Consider a sample volume where the charges that comprise the density ρ have entered from infinity. These charges could be placed on electrodes by a power supply or injected into the volume as a charged-particle beam. The charge density is related to the field through Equation 2.15,

$$\rho = \epsilon_o \nabla \cdot E. \tag{3.9}$$

Substituting Equation 3.9 in Equation 3.8 gives the field energy as

$$W = \frac{1}{2} \iiint dV \left(\epsilon_o \, \nabla \cdot E \right) \phi. \tag{3.10}$$

Applying the vector identity

$$\nabla \cdot (\phi E) = \phi \nabla \cdot E + E \cdot \nabla \phi, \tag{3.11}$$

Equation 3.10 becomes

$$W = \frac{1}{2} \iiint dV \, \epsilon_o \, \nabla \cdot (\phi E) - \frac{1}{2} \iiint dV \left(\epsilon_o \, E \cdot \nabla \phi \right). \tag{3.12}$$

Application of the divergence theorem to convert the first term on the right hand side of Equation 3.12 gives the relationship,

$$W = \frac{1}{2} \iint dS \, \epsilon_o \, \phi E \cdot n + \frac{1}{2} \iiint dV \, \epsilon_o \, E \cdot E. \tag{3.13}$$

The first integral can be taken over any surface enclosing the volume — we choose a sphere of large radius R. Depending on the distribution of charge, the potential must decrease at least as fast as R^{-1} and the electric field magnitude drops off as R^{-2} or faster. Therefore, the product of the potential and electric field must decrease at least as fast as R^{-3} while the surface area increases only as R^2. Therefore, the term must equal zero. The total field energy is

$$W = \iiint dV \, \frac{\epsilon_o \, \mathbf{E} \cdot \mathbf{E}}{2} . \tag{3.14}$$

The system energy of Equation 3.14 is the volume integral of the *electrostatic energy density,*

$$u_e = \frac{\epsilon_o \, E^2}{2} . \tag{3.15}$$

We must take care including the contributions of dielectric charges to the system energy. While the nondielectric charges enter from infinity, the dielectric charges are already inside the volume. The resolution is to calculate the total work to assemble the nondielectric charge distribution, accounting for the modification of ϕ by the presence of the dielectric charges. For a linear isotropic material, the main modification to the energy derivation is that the nondielectric charge density is related to the electric field by

$$\rho = \epsilon_r \epsilon_o \, \nabla \cdot \mathbf{E} , \tag{3.16}$$

rather than by Equation 3.9. More charge must be moved to create the same electric field. The result is that the generalized field energy density (for linear dielectrics) is given by

$$u_e = \frac{\epsilon_r \epsilon_o \, E^2}{2} . \tag{3.17}$$

To illustrate the basis of the field energy density expression, consider the parallel electrodes of Figure 3.1a. The plates have surface area A and are separated by a vacuum region of width d, with $d \ll A^{1/2}$. A power supply moves charge between the plates to create a voltage V_o. Following the discussion of Section 2.4, the total displaced charge is $\epsilon_o V_o A / d$. The potential energy of the charges equals the work performed by the power supply. When the electrodes are at voltage V, the work to move an increment of charge $d\rho$ is

$$dW = V d\rho = \frac{\epsilon_o A}{d} V dV . \tag{3.18}$$

Integrating Equation 3.18 from $V = 0$ to V_o gives the total work

a)

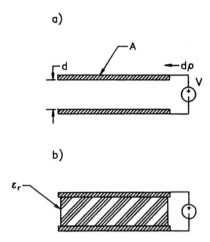

b)

FIGURE 3.1
Energy transfer to a parallel plate capacitor. (a) Air-gap capacitor. (b) Dielectric-filled capacitor.

$$W = \frac{\epsilon_o A}{d} \frac{V_o^2}{2}.$$ (3.19)

Dividing the energy of Equation 3.19 by the volume, Ad, and writing the result in terms of the electric field, $E = V_o/d$, gives an expression for the average energy density identical to Equation 3.15.

We can extend the model to clarify the role of dielectrics. Figure 3.1b shows the parallel plates with a dielectric with ϵ_r in the intervening space. Because the shifted dielectric charge does not pass through the power supply, an increment of work equals the product of an increment of *nondielectric* charge times the voltage between the plates. In the presence of the dielectric charge, more nondielectric charge must pass through the supply to create the voltage. For a linear dielectric, the flow of nondielectric charge must increase by a factor of ϵ_r. Therefore, the total system energy is

$$W = \frac{\epsilon_r \epsilon_o A}{d} \frac{V_o^2}{2}.$$ (3.20)

Equation 3.17 follows from Equation 3.20.

3.2   Elements of the Calculus of Variations

One application of differential calculus is the identification of values for one or more independent variables that give minima and maxima of functions. In contrast, the *calculus of variations* seeks unknown functions that give extreme values of functional expressions. In this book, we shall apply the calculus of variations to investigate minimization of electrostatic field energy. The energy is the volume integral of the energy density (Equation 3.17), a function of the electric fields. The electric field components are taken as unknown functions of the spatial variables (x,y,z).

For simplicity, consider a one-dimensional region of space in the range $x_1 \leq x \leq x_2$. A function $\phi(x)$ and its derivative, $\phi' = d\phi/dx$, are defined over the region along with $F(\phi,\phi',x)$, a general expression of the functions and independent variable. Consider the volume integral of the expression

$$I = \int_{x_1}^{x_2} F(\phi, \phi', x)\, dx . \tag{3.21}$$

The quantity I takes on a value for each choice of the function ϕ. It is called a *functional* of ϕ. We seek a particular form for ϕ that gives a maximum or minimum value of the functional.

Assume that $\phi(x)$ has an arbitrary variation over the range but that the end values are clamped

$$\phi(x_1) = \phi_1, \quad \phi(x_2) = \phi_2 . \tag{3.22}$$

The constraint is equivalent to fixing boundary values in an electrostatic solution. To seek a minimum, we investigate variations of the functional form of $\phi(x)$. Let $\phi_0(x)$ represent the desired minimizing function and define an arbitrary function $\eta(x)$ that satisfies the boundary conditions

$$\eta(x_1) = 0, \quad \eta(x_2) = 0 . \tag{3.23}$$

Let ϵ be a small number and add the quantity $\epsilon\eta(x)$ to define variation of $\phi(x)$ about the optimum form. The functional becomes

$$I(\epsilon) = \int_{x_1}^{x_2} F(x, \phi, \phi')\, dx = \int_{x_1}^{x_2} F(x, \phi_0 + \epsilon\eta, \phi_0' + \epsilon\eta')\, dx . \tag{3.24}$$

The condition $dI/d\epsilon = 0$ means that $\phi_0(x)$ corresponds to a minimum (or maximum) of the functional. The derivative of Equation 3.24 is

$$\frac{dI}{d\epsilon} = \int_{x_1}^{x_2} \frac{dF}{d\epsilon}\, dx = \int_{x_1}^{x_2} \left[\frac{\partial F}{\partial \phi}\eta + \frac{\partial F}{\partial \phi}\eta' \right] dx = 0 . \tag{3.25}$$

We simplify Equation 3.25 using integration by parts,

$$\int v\,du = uv - \int u\,dv.$$

(3.26)

Taking $v = \partial F/\partial\phi'$ and $du = \eta'dx$, modification of the second term in brackets in Equation 3.25 gives

$$\int_{x_1}^{x_2}\left[\frac{\partial F}{\partial\phi}\eta - \eta\frac{d}{dx}\frac{\partial F}{\partial\phi'}\right]dx + \frac{\partial F}{\partial\phi}\eta\,\bigg|_{x_1}^{x_2} = 0.$$

(3.27)

The last term on the right hand side equals zero by the condition of Equation 3.23. Therefore, the remaining bracketed term in the integral must equal zero for any choice of $\eta(x)$. The function $\phi(x)$ that gives a minimum of the functional I therefore satisfies the equation

$$\frac{\partial F}{\partial\phi} - \frac{d}{dx}\left(\frac{\partial F}{\partial\phi'}\right) = 0.$$

(3.28)

Equation 3.28 is one form of the *Euler equation*. When the function $\phi(x,y,z)$ depends on three independent variables, the three-dimensional form of the Euler equation is (see, for instance, Hildebrand, 1965)

$$\frac{\partial F}{\partial\phi} - \frac{\partial}{\partial x}\left(\frac{\partial F}{\partial(\partial\phi/\partial x)}\right) - \frac{\partial}{\partial y}\left(\frac{\partial F}{\partial(\partial\phi/\partial y)}\right) - \frac{\partial}{\partial z}\left(\frac{\partial F}{\partial(\partial\phi/\partial z)}\right) = 0.$$

(3.29)

3.3 Poisson Equation as a Condition of Minimum Energy

In this section we shall apply the Euler equation to show that Poisson's equation is equivalent to the condition of minimum electrostatic field energy. Here, the functional is the volume integral of energy density. To begin, we must carefully analyze the form of the energy function, making distinctions between the contributions of charge densities. An electrostatic solution may include charges on electrodes, charges on dielectrics, and free charges. The latter, called *space charge* and denoted as ρ_s, could represent a charged particle beam or a prescribed charge layer in a semiconductor. The field energy density expression of Equation 3.17 includes contributions from all charges. In the search for a minimum energy state, the procedure is to allow electrode and dielectric charges to shift, but to maintain the space charge in a fixed configuration. Simply minimizing the volume integral of Equation 3.17 allows all charges to shift. This process leads to the trivial conclusion that the minimum (or maximum) field energy occurs when $\rho_s = 0$.

The correct procedure is to minimize the energy integral of electrode and dielectric charges only. The associated energy density equals the field energy density minus the contribution of fixed particles, $\rho_s\phi$. Substituting $\mathbf{E} = -\nabla\phi$ in Equation 3.17 for the field energy, the energy functional is

$$I = \iiint dV \left[\frac{\epsilon_r \epsilon_0}{2} \left(\left(\frac{\partial\phi}{\partial x} \right)^2 + \left(\frac{\partial\phi}{\partial y} \right)^2 + \left(\frac{\partial\phi}{\partial z} \right)^2 \right) - \rho_s\phi \right]. \tag{3.30}$$

where $\epsilon_r(x,y,z)$ may vary in space. Noting that $\partial F/\partial\phi = \rho_s$ and $\partial F/\partial(\partial\phi/\partial x) = \epsilon\, \partial\phi/\partial x$, the Euler equation (Equation 3.29) implies that

$$\frac{\partial}{\partial x}\left(\epsilon_r \frac{\partial\phi}{\partial x} \right) + \frac{\partial}{\partial y}\left(\epsilon_r \frac{\partial\phi}{\partial y} \right) + \frac{\partial}{\partial z}\left(\epsilon_r \frac{\partial\phi}{\partial z} \right) + \frac{\rho_s}{\epsilon_0} = 0. \tag{3.31}$$

Moving the space-charge term to the right-hand side, we see that Equation 3.31 is identical to the Poisson equation (Equation 2.34).

3.4   Finite-Element Equations for Two-Dimensional Electrostatics

This section derives the finite-element difference equations for electrostatics in a two-dimensional system by seeking a minimum of adjustable field energy in a volume surrounding a mesh vertex. Here, the term *adjustable* refers to the contributions from electrode and dielectric charges as discussed in the previous section. Figure 3.2 shows the geometry. As in Chapter 2, the vertex is surrounded by six triangles of arbitrary shape with the same labeling conventions. The procedure is to construct an expression for the adjustable field energy in the surrounding elements in terms of the potentials at the test vertex (ϕ_0) and its neighbors. Fixing the neighboring potentials, we seek the value of ϕ_0 that minimizes the energy. The process leads to the same linear equation derived in Section 2.7.

Following Sections 3.1 and 3.3, the adjustable field energy for element 2 of Figure 3.2 is

$$\Delta U_2 = \Delta z a_2 \left[\frac{\epsilon_2 \epsilon_0 E_2^2}{2} - \rho_2 \frac{\phi_0 + \phi_1 + \phi_2}{3} \right]. \tag{3.32}$$

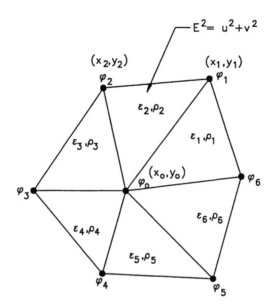

FIGURE 3.2
Calculation of electrostatic energy in triangular elements surrounding a vertex.

The quantity Δz is an arbitrary element height and a_2 is the cross section area. The quantities ϵ_2 and E_2 are the uniform values of relative dielectric constant and electric field in the element. The second term in brackets is the negative of the potential energy density of fixed charges in the triangle, the product of the uniform charge density ρ_2 times the average value of electrostatic potential. Following Equation 2.53, the square of the electric field is

$$E_2^2 = u^2 + v^2, \tag{3.33}$$

where u and v are given by Equations 2.56 and 2.57. Substitution gives an expression for the energy contribution of Equation 3.32 in terms of ϕ_o, ϕ_1, and ϕ_2 and factors that depend on the element geometry.

The total adjustable energy in the surrounding triangles is given by an expression of the form

$$U\left(\phi_o, \phi_1, \phi_2, \phi_3, \phi_4, \phi_5, \phi_6\right) = \sum_{i=1}^{6} \Delta U_i, \tag{3.34}$$

where U is a function of the vertex potential. We seek an extreme value of U by varying ϕ_o and holding the other potentials fixed,

$$\frac{\partial U}{\partial \phi_o} = 0. \tag{3.35}$$

The partial derivative of U equals the sum of derivatives for individual elements. Consider the energy expression for element 2,

$$\Delta U_2 = \frac{\Delta z a_2 \, \epsilon_r \epsilon_o}{8 a_2^2} \Big[\phi_o^2 \big(y_2^2 - y_1^2 \big)^2 - 2\phi_o \phi_1 y_2 \big(y_2 - y_1 \big)$$

$$+ 2\phi_o \phi_2 y_1 \big(y_2 - y_1 \big) + \phi_o^2 \big(x_2 - x_1 \big)^2 - 2\phi_o \phi_1 x_2 \big(x_2 - x_1 \big) \qquad (3.36)$$

$$+ 2\phi_o \phi_2 x_1 \big(x_2 - x_1 \big) + \ldots \Big] - \Delta z a_2 \rho_2 \frac{\phi_o}{3}.$$

Terms that do not explicitly contain ϕ_o have been omitted from Equation 3.36. The derivative is

$$\frac{\partial \Delta U_2}{\partial \phi_o} = \frac{\phi_o \epsilon_2 \epsilon_o \Delta z a_2}{4 a_2} \Big[y_2 \big(y_2 - y_1 \big) + x_2 \big(x_2 - x_1 \big) - y_1 \big(y_2 - y_1 \big) - x_1 \big(x_2 - x_1 \big) \Big]$$

$$- \frac{\phi_1 \epsilon_2 \epsilon_o \Delta z a_2}{4 a_2} \Big[y_2 \big(y_2 - y_1 \big) + x_2 \big(x_2 - x_1 \big) \Big] \qquad (3.37)$$

$$- \frac{\phi_2 \epsilon_2 \epsilon_o \Delta z a_2}{4 a_2} \Big[-y_1 \big(y_2 - y_1 \big) - x_1 \big(x_2 - x_1 \big) \Big] - \frac{\Delta z a_2 \rho_2}{3}.$$

Equation 3.37 can be simplified using the cotangent expressions of Section 2.6,

$$\frac{\partial \Delta U_2}{\partial \phi_o} = \frac{\phi_o \epsilon_o \Delta z}{2} \big(\epsilon_2 \cot \theta_{2b} + \epsilon_2 \cot \theta_{2a} \big)$$

$$- \frac{\phi_1 \epsilon_o \Delta z}{2} \epsilon_2 \cot \theta_{2b} - \frac{\phi_2 \epsilon_o \Delta z}{2} \epsilon_2 \cot \theta_{2a} - \frac{\Delta z A_2 \rho_2}{3}. \qquad (3.38)$$

Summing over all triangles and dividing terms by $\epsilon_o \Delta z$ gives Equation 2.67 with the same definition of coupling constants (Equations 2.68). It is not surprising that we arrive at the same result for the finite element equations because the principle of minimum energy is physically equivalent to Gauss' law.

3.5 Three-Dimensional Finite-Element Electrostatics on Arbitrary Meshes

Three-dimensional electrostatic solutions involve geometric variations along all three coordinate axes and field components E_x, E_y, and E_z. Although the

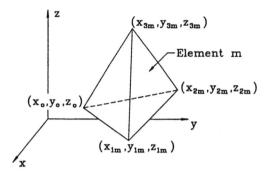

FIGURE 3.3
Labeling conventions for three-dimensional electrostatic calculations on a mesh composed of tetrahedrons.

extension of the derivation of the previous section is conceptually simple, the mathematical expressions are considerably more complex. The tetrahedron is the simplest solid body. For arbitrary meshes, the approach to three-dimensional finite-element solutions is to divide the volume into a contiguous set of tetrahedrons that may have different shapes. The choice of shapes ensures that element facets lie along the boundaries of material regions. Figure 3.3 shows one of the tetrahedrons surrounding a target vertex. We denote the vertex at position (x_o, y_o, z_o) as point 0 and mark the tetrahedron with the index m. The element is defined by three points, $x_{1m} = (x_{1m}, y_{1m}, z_{1m})$, $x_{2m} = (x_{2m}, y_{2m}, z_{2m})$, and $x_{3m} = (x_{3m}, y_{3m}, z_{3m})$. Note that the points are labeled with positive rotation looking toward the test vertex. The volume of the tetrahedron is given by evaluating the determinant

$$V_m = \frac{1}{6} \det \begin{bmatrix} 1 & x_{0m} & y_{0m} & z_{0m} \\ 1 & x_{1m} & y_{1m} & z_{1m} \\ 1 & x_{2m} & y_{2m} & z_{2m} \\ 1 & x_{3m} & y_{3m} & z_{3m} \end{bmatrix}. \tag{3.39}$$

Following the previous section, the electrostatic potential in element m can be written as a linear function of the spatial coordinates,

$$\phi_m(x, y, z) = \alpha_{0m} + \alpha_{1m}x + \alpha_{2m}y + \alpha_{3m}. \tag{3.40}$$

Equation 3.40 can also be written in vector notation as

$$\phi_m = \begin{bmatrix} 1 & x & y & z \end{bmatrix} \begin{bmatrix} \alpha_{0m} \\ \alpha_{1m} \\ \alpha_{2m} \\ \alpha_{3m} \end{bmatrix}. \tag{3.41}$$

The element electric field is a constant with the value

$$\mathbf{E}_m = \alpha_{1m}\hat{x} + \alpha_{2m}\hat{y} + \alpha_{3m}\hat{z}.$$ (3.42)

The coefficients in Equation 3.42 satisfy the equation

$$\begin{bmatrix} \phi_{0m} \\ \phi_{1m} \\ \phi_{2m} \\ \phi_{3m} \end{bmatrix} = \begin{bmatrix} 1 & x_{0m} & y_{0m} & z_{0m} \\ 1 & x_{1m} & y_{1m} & z_{1m} \\ 1 & x_{2m} & y_{2m} & z_{2m} \\ 1 & x_{3m} & y_{3m} & z_{3m} \end{bmatrix} \begin{bmatrix} \alpha_{0m} \\ \alpha_{1m} \\ \alpha_{2m} \\ \alpha_{3m} \end{bmatrix}$$ (3.43)

Solving for Equation 3.43 gives the coefficients as

$$\begin{bmatrix} \alpha_{0m} \\ \alpha_{1m} \\ \alpha_{2m} \\ \alpha_{3m} \end{bmatrix} = \begin{bmatrix} 1 & x_{0m} & y_{0m} & z_{0m} \\ 1 & x_{1m} & y_{1m} & z_{1m} \\ 1 & x_{2m} & y_{2m} & z_{2m} \\ 1 & x_{3m} & y_{3m} & z_{3m} \end{bmatrix}^{-1} \begin{bmatrix} \phi_{0m} \\ \phi_{1m} \\ \phi_{2m} \\ \phi_{3m} \end{bmatrix}$$ (3.44)

Rather than write out the involved expressions for the coefficients, we note that the inverse matrix of Equation 3.44 can be evaluated numerically using routines discussed in Section 6.1.

To apply the minimum energy principle in three dimensions, it is more effective to write Equation 3.40 in an alternate form that displays the vertex potential values explicitly,

$$\phi_m(x,y,z) = N_{0m}(x,y,z)\,\phi_{0m} + N_{1m}(x,y,z)\,\phi_{1m}$$
$$+ N_{2m}(x,y,z)\,\phi_{2m} + N_{3m}(x,y,z)\,\phi_{3m}.$$ (3.45)

The *pyramid functions*, N_{im}, are described in *Basics of the Finite Element Method* (Allaire, 1985). They are functions of position inside the element that equal unity at the corresponding vertex and go to zero at the other vertices. We can derive the pyramid functions by writing them in the form

$$N_{0m} = a_{0m} + b_{0m}x + c_{0m}y + d_{0m}z,$$
$$N_{1m} = a_{1m} + b_{1m}x + c_{1m}y + d_{1m}z,$$
$$N_{2m} = a_{2m} + b_{2m}x + c_{2m}y + d_{2m}z,$$
$$N_{3m} = a_{3m} + b_{3m}x + c_{3m}y + d_{3m}z.$$ (3.46)

Substituting in Equation 3.45, the interpolated potential can be written in matrix form as

$$\phi_m = \begin{bmatrix} 1 & x & y & z \end{bmatrix} \begin{bmatrix} a_{0m} & a_{1m} & a_{2m} & a_{3m} \\ b_{0m} & b_{1m} & b_{2m} & b_{3m} \\ c_{0m} & c_{1m} & c_{2m} & c_{3m} \\ d_{0m} & d_{1m} & d_{2m} & d_{3m} \end{bmatrix} \begin{bmatrix} \phi_{0m} \\ \phi_{1m} \\ \phi_{2m} \\ \phi_{3m} \end{bmatrix}. \tag{3.47}$$

A comparison of Equations 3.41, 3.44, and 3.47 shows that the coefficients of the pyramid functions are given in terms of the inverse of the vertex matrix

$$\begin{bmatrix} a_{0m} & a_{1m} & a_{2m} & a_{3m} \\ b_{0m} & b_{1m} & b_{2m} & b_{3m} \\ c_{0m} & c_{1m} & c_{2m} & c_{3m} \\ d_{0m} & d_{1m} & d_{2m} & d_{3m} \end{bmatrix} = \begin{bmatrix} 1 & x_{0m} & y_{0m} & z_{0m} \\ 1 & x_{1m} & y_{1m} & z_{1m} \\ 1 & x_{2m} & y_{2m} & z_{2m} \\ 1 & x_{3m} & y_{3m} & z_{3m} \end{bmatrix}^{-1}. \tag{3.48}$$

Again, the coefficients can easily be determined from the coordinates of the element vertices by a numerical inversion. The electric field components are

$$E_x = -\left(b_{0m}\phi_0 + b_{1m}\phi_{1m} + b_{2m}\phi_{2m} + b_{3m}\phi_{3m}\right),$$
$$E_y = -\left(c_{0m}\phi_0 + c_{1m}\phi_{1m} + c_{2m}\phi_{2m} + c_{3m}\phi_{3m}\right), \tag{3.49}$$
$$E_z = -\left(d_{0m}\phi_0 + d_{1m}\phi_{1m} + d_{2m}\phi_{2m} + d_{3m}\phi_{3m}\right).$$

We shall concentrate on the Poisson equation in the absence of space charge (Laplace equation). Following Section 3.3, the procedure is to form the field energy expression for the M tetrahedrons surrounding the test vertex and then to minimize the quantity with respect to ϕ_0. The resulting difference equation relating the potential to the values at nearest neighbors is

$$\phi_0 \sum_{m=1}^{M} \epsilon_m \left(b_{0m}^2 + c_{0m}^2 + d_{0m}^2\right) =$$
$$-\sum_{m=1}^{M} \epsilon_m \left[\left(b_{0m}b_{1m} + c_{0m}c_{1m} + d_{0m}d_{1m}\right)\phi_{1m} \right. \tag{3.50}$$
$$+ \left(b_{0m}b_{2m} + c_{0m}c_{2m} + d_{0m}d_{2m}\right)\phi_{2m}$$
$$\left. + \left(b_{0m}b_{3m} + c_{0m}c_{3m} + d_{0m}d_{3m}\right)\phi_{3m}\right].$$

The expressions in Equation 3.50 are involved but can be attacked step-by-step on a computer. The resulting set of equations can be cast in a form that relates the potential at each vertex point to a weighted average of the potentials at neighboring points. The large set of linear equations can then be solved using either relaxation methods (Section 5.5) or block matrix inversion (Section 6.5).

There are two main challenges in three-dimensional finite-element solutions on an arbitrary mesh. The first is setting up the computational mesh, a procedure that is considerably more challenging for three-dimensional tetrahedrons that two-dimensional triangles. Second, the storage requirements are daunting. Adding the third dimension multiplies the number of vertices by a large factor (~100). Furthermore, the number of quantities that must be stored at each vertex to represent the geometry and material characteristics is two to three times larger. As a result, except in the simplest geometries, it is difficult to perform three-dimensional solutions with free-form meshes on personal computers. Therefore, in the remainder of the book we shall apply arbitrary meshes only in two dimensions and concentrate on regular meshes for finite-difference or finite-element techniques in three-dimensional problems.

3.6   High-Order Finite-Element Formulations

The finite-element derivations of previous sections were based on an approximation of the potential by a linear function in each element. It is important to recognize that it is possible to define a hierarchy of approximations where potential variations are represented by quadratic, cubic, and higher-order expansions. In principle, the advantage is that it is possible to achieve the target accuracy with fewer elements. The price is that it is more difficult to derive the finite-element equations and to apply them in a versatile code. Furthermore, larger elements may not be advantageous if we are trying to model small-scale variations of material properties.

The mathematics is challenging even for a derivation of the second-order finite-element equations on a two-dimensional triangular mesh. It is necessary to use more abstract methods than the Gauss' law treatment of Chapter 2. Here, we shall outline the steps. A detailed description is given in *Numerical Techniques in Electromagnetics* (Sadiku, 1992). The idea is to represent the potential variation in an element by a second-order function. For two-dimensional solutions, the expression must have six terms,

$$\phi(x,y) = a_o + a_1 x + a_2 y + a_3 x^2 + a_4 xy + a_5 y^2. \tag{3.51}$$

The coefficients are determined from the potential values at six points (or *nodes*) in an element. Figure 3.4 shows the standard choice for a triangular

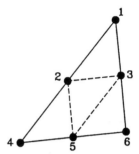

FIGURE 3.4
Nodes of a triangular element for second-order finite-element field calculation.

element: the three vertices and three midpoints on the sides. The procedure is to apply Equation 3.51 to form expressions for the electrostatic field energy in each element in terms of potential values at the nodes and then to minimize the global energy to derive equations that connect the potential at a node to the values at neighboring points. As before, this leads to a large set of linear equations that can be solved by the methods discussed in Chapters 5 and 6. The algebra involved is facilitated by matrix methods and the use of *shape functions* for the elements. These are the extensions of the pyramid functions discussed in Section 3.5.

High-order methods have advantages for certain types of problems. One example is the evaluation of mechanical strain in a homogeneous object with a complex three-dimensional shape like a crankshaft. As discussed in Section 3.5, an accurate linear solution could require a large number of elements, exceeding the memory capacity of the computer or degrading accuracy through round-off errors. On the other hand, there are reasons to avoid high-order calculations for many of the applications in the following chapters. Consider, for example, a finite-element electrostatic solution. If the dielectric constant is uniform through a macroscopic region, then there may be an advantage to large elements. On the other hand, suppose we have a nonlinear dielectric where ϵ varies with the field amplitude, $|E|$. A linear solution with small elements presents no problem — the constant element electric field defines a unique value of ϵ with good spatial resolution of variations. In contrast, Equation 3.51 implies that $|E|$ varies over the volumes of elements in a second-order treatment. Here, the choice is to determine spatial variations of ϵ within elements or to use constant values based on an average of $|E|$. Regarding the former, derivation of finite element equations by energy minimization in elements with an arbitrary nonlinear dependence of the relative dielectric constant would be extremely difficult. At the other extreme, a simple average over large elements gives a poor representation of material properties, leading to an inaccurate field solution. Generally, first-order methods are much easier to apply in solutions that depend on the resolution of self-consistent material properties. Examples that we shall study in this book include magnet fields with ferromagnetic materials and permanent magnets (Chapter 9) and thermal transport with temperature-dependent conductivity (Chapter 12). Even in solutions with simple materials, the accuracy advantage of second-order methods is not so clear-cut if we apply second-order interpolation techniques to linear solutions (Section 7.3). In the end, the choice of computational methods is similar to the choice of a tool for any job. Even though a 30-horsepower tractor mower may not be the ideal choice for the small plot of grass out front, it does make a statement to the neighbors.

Chapter 3   Exercises

3.1. Confirm Equation 3.7 by calculating the energy to assemble a distribution of three charges.

3.2. Given spherical distribution of charge of form

$$\rho(r) = \rho_o \exp\left(-\frac{r}{R}\right).$$

find an expression for the total electrostatic energy in space. Give a value for $\rho_o = 10^{-8}$ coulombs/m^3 and $R = 0.02$ m.

3.3. A nonlinear isotropic dielectric has a relative dielectric constant that varies with the amplitude of the electric field, $\epsilon_r(E)$. Show that the field energy density in the medium is given by

$$u_e = \frac{1}{2}\,\epsilon_o \int\limits_0^E E'\, d\big(E'\epsilon_r(E')\big).$$

3.4. Show that the distance between two points in the x–y plane is given by the expression

$$D = \int\limits_{x_1}^{x_2} dx\,\sqrt{1+(dy/dx)^2}.$$

Use the Euler equation to prove mathematically that the shortest distance between the points is a straight line.

3.5. A geodesic on a surface is a curve where the distance between two points is a minimum. The geodesic on a plane is a straight line. Find the equation of a geodesic on a right circular cylinder. Take $ds^2 = a^2d\theta^2 + dz^2$ and minimize the functional

$$D = \int dz\,\sqrt{1+a^2(d\theta/dz)^2}.$$

3.6. Confirm that minimizing the field energy functional

$$I = \iiint dV \; \frac{\epsilon_0 \epsilon_r E^2}{2},$$

leads to the Laplace equation for electrostatics in the absence of space-charge,

$$\nabla \cdot \left(\epsilon_r \nabla \phi \right) = 0.$$

3.7. Extend the derivation of Section 3.4 to cylindrical coordinates. Minimize the adjustable energy for toroidal elements that extend 360° around the axis and show the derivation leads to Equations 2.77 and 2.78.

3.8. Consider a one-dimensional planar gap with applied voltage $V(0) = 0.0$ and $V(1.0) = 1.0$. The lower half of the gap is filled with a dielectric with $\epsilon_r = 2.5$.

   (a) Find an expression for the potential as a function of position in the gap.

   (b) Set up a spreadsheet with a column of the analytic potential values at 21 positions (0.00, 0.05, 0.10, ..., 1.00). Set up a second column with the field energy per unit length calculated from Equation 3.17 for the 20 elements .

   (c) Set up an interactive cell that sums the field energy contributions. Experiment and show that any variation of potential from the analytic values results in an increase in field energy.

3.9. Find expressions for the pyramid functions of the right isosceles triangles with the following vertices.

   (a) $x_1 = (0,0)$, $x_2 = (d,0)$, and $x_3 = (d,d)$

   (b) $x_1 = (0,0)$, $x_2 = (d,d)$, and $x_3 = (0,d)$

3.10. Find the three pyramid functions in a two-dimensional system for a triangle with the following vertices: $x_1 = (0,0)$, $x_2 = (3,0)$, and $x_3 = (3,5)$. Give values for E_x and E_y in the element if the potentials at the three vertices are $\phi_1 = 30.0$, $\phi_2 = 35.0$, and $\phi_3 = 42.0$ V.

3.11. A two-dimensional mesh consists of square elements with sides of length d. Use the results of problem 3.10 to find an expression for the field energy of an element in terms of potential values at the four vertices by dividing the space into two right triangles.

4

Finite-Difference Solutions and Regular Meshes

A new scientific truth does not triumph by convincing its opponents and making them see the light, but rather because its opponents eventually die, and a new generation grows up that is familiar with it.

Max Planck

The finite-difference method is an alternative approach to derive difference equations for physical systems. It is based on the direct conversion of the governing differential equations for a physical system by the substitution of *difference operators*. These operators act over small divisions of space ($\Delta x, \Delta y, \Delta z$) or time ($\Delta t$). There are several ways to make the conversion. The solutions approach the exact results in the limit that Δx, Δy, Δz, $\Delta t \to 0$.

Both finite-element and finite-difference techniques seek a set of difference equations to represent a continuous physical system in terms of discrete quantities. In the finite-element method these equations are derived by application of conservation principles over a *volume*. In contrast, finite-difference equations follow from the application of physical laws at a *point*. The differential equations underlying the finite-difference method are usually referenced to Cartesian, cylindrical, or spherical coordinates. Therefore, difference operators are usually defined on a regular mesh. The term *regular mesh* means that space is divided into rectangles or boxes determined by sets of vertex coordinates along each axis. Figure 4.1 shows an example of a two-dimensional regular mesh with vertices at the set of coordinates (x_i, y_j). In contrast to the conformal meshes of Chapter 2, regular meshes require less storage of geometrical information. Given the coordinate sets x_i, y_j, and z_k, we can immediately find the location of any vertex point (x_i, y_j, z_k) from the indices (i,j,k). The disadvantage is clear in Figure 4.1. A regular mesh often gives a poor representation of slanted or curved boundaries.

Finite-difference techniques are difficult to apply on arbitrary meshes. On regular meshes we can choose finite-element or finite-difference methods. We shall see in this chapter that it is usually better to adopt the finite-element

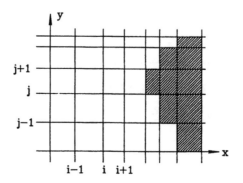

FIGURE 4.1
Two-dimensional regular mesh with indices for finite-element calculations.

viewpoint for spatial differencing. The method ensures unambiguous identification of material boundaries and avoids nonphysical divergences at the axes of cylindrical or spherical systems. On the other hand, difference operators are essential for time-dependent problems. In contrast to spatial problems, nature dictates that we must advance in one direction parallel to the time axis. In the remainder of this book, we shall represent spatial variations with finite-element equations and advance in time with finite-difference operators. The material introduced in this chapter will be important for the diffusion problems of Chapter 12 and electromagnetic pulse solutions of Chapter 13.

Section 4.1 covers methods to convert differential equations to a discrete set of linear equations on a regular mesh. We shall see how the choice of difference operator can affect the accuracy and stability of the solution. Section 4.2 illustrates the procedure with the application to initial value solutions of ordinary differential equations. Section 4.3 covers the differencing techniques for a boundary value problem, the one-dimensional Poisson equation. The simple geometry provides a good illustration of contrasts between the finite-difference and finite-element approaches. Section 4.4 shows how to solve the set of linear equations for the one-dimensional Poisson equation through the method of back-substitution. The discussion also reviews the implementation of boundary conditions. Section 4.5 extends the difference equations to two-dimensional field variations on regular meshes, both in Cartesian and cylindrical coordinates. Finally, Section 4.6 derives finite-difference and finite-element equations for three-dimensional electrostatics.

4.1 Difference Operators

Figure 4.2a shows a continuous function $f(x)$ defined over a spatial region. The function has an infinite set of values. In numerical calculations we must approximate its behavior with a finite set of quantities. The standard approach is to

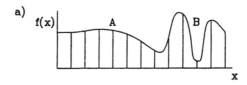

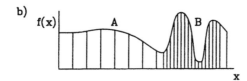

FIGURE 4.2
Discrete representations of continuous functions. (a) One-dimensional function f(x) defined on a uniform mesh. (b) Nonuniform mesh to improve the function resolution in regions of fine scale variations.

divide the space into intervals and to store discrete function values at the interval boundaries. We denote the boundary positions as x_i and the corresponding function values as f_i. The one-dimensional mesh in Figure 4.2a has 10 intervals with a total of $I = 11$ boundaries. If the function varies smoothly, we can estimate intervening values with linear or higher-order interpolations (Section 8.3). The procedure is accurate when the scale length for variations of the function extends over many intervals as in the region marked A in Figure 4.2a. On the other hand, a numerical calculation gives poor results in region B. We could improve the approximation by using finer spacing as in Figure 4.2b.

We can determine a criterion for the validity of discrete approximations from the theory of Fourier series. We can represent a continuous function in the region $x_1 \leq x \leq x_I$ as

$$f(x) = A_o + \sum_{n=1}^{\infty} \left(A_n \cos\frac{2\pi n x}{L} + B_n \sin\frac{2\pi n x}{L} \right), \tag{4.1}$$

where $L = x_I - x_1$. The coefficients in Equation 4.1 are given by

$$A_o = \frac{1}{L} \int_{x_1}^{x_I} f(x)\, dx,$$

$$A_n = \frac{2}{L} \int_{x_1}^{x_I} f(x) \cos\frac{2\pi n x}{L}\, dx, \tag{4.2}$$

$$B_n = \frac{2}{L} \int_{x_1}^{x_I} f(x) \sin\frac{2\pi n x}{L}\, dx.$$

The wave number of a Fourier component is $k_n = 2\pi n/L$. The summation in Equation 4.1 extends to infinity; therefore, the series can represent arbitrary functions with short scale variations. The content of Equations 4.1 and 4.2 can be written succinctly in complex number notation as

$$f(x) = \sum_{n=1}^{\infty} g_n \exp\left(\frac{2\pi j n x}{L}\right), \tag{4.3}$$

where

$$j = \sqrt{-1}. \tag{4.4}$$

The complex Fourier coefficients are given by

$$g_n = \frac{1}{L} \int_{x_0}^{x_1} f(x) \exp\left(\frac{-2\pi j n x}{L}\right) dx. \tag{4.5}$$

Note that the summation of Equation 4.3 extends over negative values of n. The negative components are a mathematical requirement to cancel the imaginary parts of terms.

We can derive analogous expressions for a discrete data set that approximates a continuous function. Consider a set of I values f_i at locations x_i separated by the uniform interval Δ, where $(I-1)\Delta = L$. For convenience, we shift the values of x so that they lie in the range $-L/2 \leq x \leq L/2$. In this case, $x = n\Delta$. The function values constitute a vector with I components. A familiar result from linear algebra is that such a set can be represented by an expansion in terms of I orthogonal functions. The following form of the discrete Fourier series is analogous to that of the continuous series (Equation 4.3),

$$f_i = \sum_{n=-I/2}^{I/2} g_n \exp\left(\frac{2\pi j n i}{I}\right), \tag{4.6}$$

where

$$g_n = \frac{1}{I} \sum_{i=1}^{I} f_i \exp\left(\frac{-2\pi j n i}{I}\right). \tag{4.7}$$

Although Equation 4.6 appears to have $I + 1$ components there are actually only I independent entries. Applying the periodic properties of the exponential function, we can show that $g_{-I/2} = g_{I/2}$.

Equation 4.7 follows from the orthogonality condition for the base functions,

$$\sum_{i=1}^{I} \exp\left(\frac{2\pi jni}{I}\right) \exp\left(\frac{-2\pi jn'i}{I}\right) = I\delta_{nn'}. \tag{4.8}$$

The discrete delta function on the right-hand side of Equation 4.8 has the properties

$$\delta_{nn'} = 1, \ (n = n') \ \delta_{nn'} = 0, \ (n \neq n'). \tag{4.9}$$

We can confirm Equation 4.8 by writing the summation on the left-hand side as

$$\exp\left(\frac{2\pi j(n-n')}{I}\right) \sum_{i=0}^{I-1} \left[\exp\left(\frac{2\pi j(n-n')}{I}\right)\right]^{i} = $$

$$\exp\left(\frac{2\pi j(n-n')}{I}\right) \left[\frac{1-\exp(2\pi j(n-n'))}{1-\exp\left(\frac{2\pi j(n-n')}{I}\right)}\right]. \tag{4.10}$$

The conversion of the bracketed term in Equation 4.10 follows from the following formula for a geometric progression,

$$\sum_{j=0}^{N} x^{j} = \frac{1-x^{N+1}}{1-x}. \tag{4.11}$$

Applying Euler's formula, exp(jωt) = cos(ωt) + j sin(ωt), the right-hand side of Equation 4.10 becomes

$$\frac{1}{I} \exp\left(\frac{\pi j(I+1)(n-n')}{I}\right) \frac{\sin[\pi(n-n')]}{\sin[\pi(n-n')/I]}. \tag{4.12}$$

The expression of Equation 4.12 satisfies the delta function properties of Equation 4.9.

To conclude the proof, we can use Equation 4.8 to verify Equation 4.7. Multiplying both sides of Equation 4.6 by exp(–2πjn'i/I) and summing over the data values gives the equation

$$\sum_{i=1}^{I} f_i \exp\left(\frac{-2\pi j n'i}{I}\right) = \sum_{i=1}^{I} \sum_{n=-I/2}^{I/2} g_n \exp\left(\frac{2\pi j n i}{I}\right) \exp\left(\frac{-2\pi j n'i}{I}\right)$$

$$= \sum_{n=-I/2}^{I/2} g_n \sum_{i=1}^{I} \exp\left(\frac{2\pi j n i}{I}\right) \exp\left(\frac{-2\pi n'j}{J}\right). \tag{4.13}$$

Applying the delta function relationship of Equation 4.8 to the final expression in Equation 4.13 yields the relationship,

$$\sum_{i=1}^{I} f_j \exp\left(\frac{-2j\pi n'i}{I}\right) = \sum_{n=-I/2}^{I/2} g_n \, \delta_{nn'} = g_{n'}. \tag{4.14}$$

which is identical to Equation 4.7.

An important implication of Equation 4.6 is that there is a maximum value of the spatial wave number in a discrete Fourier series,

$$k_{max} = \frac{\pi L}{2\Delta}. \tag{4.15}$$

The quantity k_{max} is called the *Nyquist wave number*. The corresponding minimum wavelength is $\lambda_{min} = 2\Delta$. Figure 4.3a shows the physical interpretation of Equation 4.15. The Fourier component at the Nyquist wavelength varies between positive and negative values at sequential points. The data set cannot represent the small scale variations shown in Figure 4.3b.

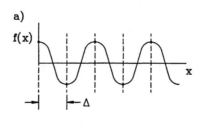

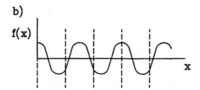

FIGURE 4.3
Discrete representations of continuous functions on a uniform mesh. (a) Fourier mode of a one-dimensional function at the Nyquist wavelength. (b) Function that cannot be represented with the given mesh.

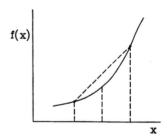

FIGURE 4.4
Centered difference operator for the derivative of the function f(x) at x_i.

Next, we shall discuss difference expressions for derivatives of a function f. First, consider approximations for the first derivative at the point x_i. As shown in Figure 4.4, a good choice is

$$\Delta'_x \, f_i = \frac{f_{i+1} - f_{i-1}}{2\Delta}. \tag{4.16}$$

The symbol Δ'_x stands for the difference operator that corresponds to the first derivative along x. The expression of Equation 4.16 gives an approximation for the derivative that is centered at the point of interest. Here, we say that the difference operator is *space-centered*. A centered operator clearly gives better accuracy than a skewed expression like $(f_{i+1} - f_i)/\Delta$. We can use Fourier analysis to check the accuracy of Equation 4.16. Let u(x) be a Fourier mode with wave number k of a known function f(x),

$$u(x) = g_k \, \exp(jkx). \tag{4.17}$$

The exact derivative of the mode at x_i is

$$\frac{du}{dx} = jkg_k \, \exp(jkx_i). \tag{4.18}$$

The difference formula of Equation 4.16 gives the expression

$$\Delta'_x u = \frac{g_k \, \exp(jkx_{i+1}) - g_k \, \exp(jkx_{i-1})}{2\Delta}. \tag{4.19}$$

Noting that $x_{i+1} = x_i + \Delta$ and $x_{i-1} = x_i - \Delta$, we can write Equation 4.19 as

$$\Delta'_x u = \frac{g_k}{2\Delta}\left[\exp(jkx_i + j\Delta) - \exp(jkx_i - j\Delta)\right]$$

$$= \frac{g_k}{\Delta} \exp(jkx_i)\frac{1}{2}\left[\exp(jk\Delta) - \exp(-jk\Delta)\right]. \tag{4.20}$$

Applying Euler's formula, the last form becomes

$$\Delta'_x u = ju \ \frac{\sin(k\Delta)}{\Delta} . \tag{4.21}$$

Replacing the sine function with a Taylor expansion gives the final result

$$\Delta'_x u = jku \left(1 - \frac{k^2\Delta^2}{6} + ... \right). \tag{4.22}$$

The expression of Equation 4.22 differs from the exact derivative by an error of about $k^2\Delta^2/6$. Because the error scales as Δ^2, the space-centered expression of Equation 4.16 is said to be *second-order accurate*.

The second derivative at x_i equals the change in the derivative per length along x. From the previous discussion, we expect that the difference in the derivatives at $x + \Delta/2$ and $x - \Delta/2$ divided by Δ would give a space-centered estimate with good accuracy,

$$\Delta''_x y_i = \frac{(y_{i+1} - y_i)/\Delta - (y_i - y_{i-1})/\Delta}{\Delta} = \frac{y_{i+1} - 2y_i + y_{i-1}}{\Delta^2} . \tag{4.23}$$

We can analyze the effect of the operator on a Fourier mode and compare the result to the exact second derivative, $d^2u/dx^2 = -k^2u$. The error is approximately $(k\Delta)^2/12$.

The expressions for difference operators are more complex on a nonuniform mesh. Suppose the mesh spacings near the point x_i are $\Delta x_i = x_i - x_{i-1}$ and $\Delta x_{i+1} = x_{i+1} - x_i$. Taylor expansions give approximations for function values at neighboring mesh points,

$$f_{i+1} \cong f_i + \left[\frac{df}{dx} \right]_{x_i} \Delta x_{i+1} + \left[\frac{d^2f}{dx^2} \right]_{x_i} \frac{\Delta x_{i+1}^2}{2} ,$$

$$f_{i-1} \cong f_i - \left[\frac{df}{dx} \right]_{x_i} \Delta x_i + \left[\frac{d^2f}{dx^2} \right]_{x_i} \frac{\Delta x_i^2}{2} . \tag{4.24}$$

Solving Equation 4.24 for the derivatives gives the following expressions for the difference operators,

$$\Delta'_x f_i = \frac{(f_{i+1} - f_i) \Delta x_i^2 + (f_i - f_{i-1}) \Delta x_{i+1}^2}{\Delta x_i \Delta x_{i+1} (\Delta x_i + \Delta x_{i+1})} , \tag{4.25}$$

and

$$\Delta_x'' f_i = 2\frac{\left(f_{i+1} - f_i\right)\Delta x_{i+1} - \left(f_i - f_{i-1}\right)\Delta x_i}{\Delta x_i\,\Delta x_{i+1}\left(\Delta x_i + \Delta x_{i+1}\right)}. \tag{4.26}$$

Equation 4.26 can be rewritten as

$$\Delta_x'' f_i = \frac{\left(f_{i+1} - f_i\right)/\Delta x_{i+1} - \left(f_i - f_{i-1}\right)/\Delta x_{i+1}}{\left(\Delta_{i+1} + \Delta_i\right)/2}. \tag{4.27}$$

The numerator is the difference of first derivatives estimated at the midpoints of adjoining intervals and the denominator is the distance between the midpoints.

4.2 Initial Value Solutions of Ordinary Differential Equations

In the following chapters, we shall apply difference operators to calculate the time evolution of physical systems from a known state. In preparation, this section covers numerical methods to solve ordinary differential equations. Here, time is the single independent variable. To begin, consider a first-order differential equation with one dependent variable.

$$\frac{dx}{dt} = f(x,t). \tag{4.28}$$

A numerical solution gives a set of values for x at points in time. For simplicity, we shall advance the solution with uniform interval Δt. Applying Equation 4.16, the *time-centered* difference form of Equation 4.28 is

$$\frac{x_{n+1} - x_{n-1}}{2\Delta t} = f(x_n, t_n). \tag{4.29}$$

where x_n is the value of x at time $t_n = n\Delta t$. Equation 4.29 gives an algorithm to advance the dependent variable,

$$x_{n+1} = x_{n-1} + 2\,\Delta t\, f(x_n, t_n). \tag{4.30}$$

Figure 4.5 shows how values of the function f advance the quantity x in increments of $2\Delta t$. The scheme is called the *leapfrog method*. For accuracy, the function should not change significantly over the interval, or

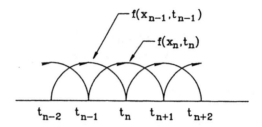

FIGURE 4.5
Advancing a function with the leapfrog method.

$$\Delta t < \frac{\Delta x_n}{\Delta f_n} \cong \frac{1}{\left. \partial f / \partial x \right|_n} \; . \tag{4.31}$$

Equation 4.31 is also a sufficient condition for numerical stability for oscilla-tory solutions (see, for instance, Potter, 1973).

Although Equation 4.30 is easy to apply and gives second-order accuracy, there are two drawbacks. First, the algorithm may exhibit numerical instabil-ity for certain types of growing or damped solutions. Second, a good solution demands an accurate knowledge of the system state at the initial time t^0 and the advanced time t^1. An alternate method must be applied to find x^1.

In most cases, it is more convenient to use a solution method that is stable for all physical problems and only involves a knowledge of quantities at the initial time. The simplest approach to estimate changes in a function over an interval t_n to t_{n+1} is the *Euler method*. The algorithm uses the known value of $f(x_n, t_n)$ to estimate x_{n+1} and $f(x_{n+1}, t_{n+1})$,

$$x_{n+1} = x_n + f(x_n, t_n) \Delta t . \tag{4.32}$$

The Euler method is clearly not time-centered. We can get an idea of the accu-racy by comparing Equation 4.32 to the exact solution of the differential equation,

$$x_{n+1} = x_n + \int_{t_n}^{t_{n+1}} f(x, t) \, dt . \tag{4.33}$$

The Euler method approximates the integral of f over the interval as $f(x_n, t_n) \Delta t$. Inspection of Figure 4.6a shows that the approximation has a first-order error in Δt. Therefore, it is necessary to use very small intervals and a large number of steps to achieve good accuracy.

Figure 4.6b suggests that we could get a much better estimate of the inte-gral if we knew the value of the function in the middle of the interval,

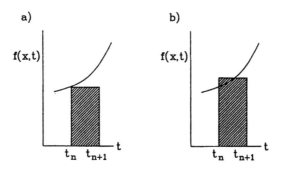

FIGURE 4.6
Numerical solutions of a first-order differential equation. (a) Euler method approximation of the integral of f over the interval Δt. (b) Approximation of the integral using an estimate of f at the interval center.

$f(x_{n+\frac{1}{2}}, t_{n+\frac{1}{2}})$. The *explicit two-step method* achieves second-order accuracy in Δt by first using the Euler method for an initial estimate of f at t_n + Δt/2. The mathematical expression is

$$x_{n+\frac{1}{2}} = x_n + f(x_n, t_n)\,(\Delta t/2) \qquad \text{(Step 1)},$$

$$x_{n+1} = x_n + f\!\left(x_{n+\frac{1}{2}}, t_n + \Delta t/2\right)\Delta t \quad \text{(Step 2)}. \qquad (4.34)$$

The condition for accuracy and stability of Equation 4.31 also applies to the two-step method. To illustrate the advantage in accuracy over the Euler method, consider numerical solutions of the exponential equation

$$\frac{dx}{dt} = x. \qquad (4.35)$$

The exact solution is x = exp(t) for x(0) = 1, giving a value of x = 20.0855 at t = 3. A two-step integration with 10 intervals (Δt = 0.3) gives x(3) = 19.3741, an error of 3.5%. In contrast, the Eulerian integration gives x(3) = 13.7858, a 31% error.

The two-step method is a specific instance of the second-order Runge-Kutta procedure. To understand this method, consider a Taylor expansion of x(t) in time about the value x_n,

$$x_{n+1} = x_n + f_n\,\Delta t + \left[\frac{\partial f_n}{\partial t} + \frac{\partial f_n}{\partial x} f_n\right]\frac{\Delta t^2}{2} + \dots . \qquad (4.36)$$

In Equation 4.36, the quantity f_n corresponds to $f(x_n, t_n)$. The goal is to find an approximation for x(t) of the form

$$x_{n+1} = x_n + \lambda_1 \, \Delta t \, f_n + \lambda_2 \, \Delta t \, f\big(t_n + \mu_1 \Delta t, \; x_n + \mu_2 \, f_n \Delta t\big). \qquad (4.37)$$

We seek parameters λ_1, λ_2, μ_1, and μ_2 such that Equation 4.37 is consistent with the Taylor expansion of Equation 4.36. The third term on the right-hand side of Equation 4.37 can be approximated with a first-order Taylor expansion of $f(x,t)$,

$$f\big(t_n + \mu_1 \Delta t, \; x_n + \mu_2 \, f_n \Delta t\big) \cong f(t_n, x_n) + \mu_1 \, \frac{\partial f_n}{\partial t} \, \Delta t + \mu_2 \, \frac{\partial f_n}{\partial x} \, f_n \, \Delta t. \quad (4.38)$$

Substitution in Equation 4.37 gives

$$x_{n+1} \cong x_n + (\lambda_1 + \lambda_2) \, f_n \, \Delta t + \lambda_2 \left(\mu_1 \, \frac{\partial f_n}{\partial t} + \mu_2 \, \frac{\partial f_n}{\partial x} \, f_n \right) \Delta t^2. \qquad (4.39)$$

There is an infinite set of parameters that will make Equations 4.36 and 4.39 consistent. The choice $\lambda_1 = \lambda_2 = \frac{1}{2}$ and $\mu_1 = \mu_2 = 1$ gives the standard symmetric form of the second-order Runge-Kutta integration,

$$x_{n+1} \cong x_n + \frac{\Delta t}{2} \left[f(t_n, x_n) + f\big(t_n + \Delta t, \; x_n + f_n \Delta t\big) \right]. \qquad (4.40)$$

The choice $\lambda_1 = 0$, $\lambda_2 = 1$ and $\mu_1 = \mu_2 = \frac{1}{2}$ corresponds to the two-step method of Equation 4.34.

Extensions of the above derivation lead to higher-order Runge-Kutta schemes. A popular choice for a high-accuracy solution with a known function $f(x,t)$ is the fourth-order method defined by the following equations.

$$x_{n+1} \cong x_n + \frac{1}{6} \big(\beta_1 + 2\beta_2 + 2\beta_3 + \beta_4 \big), \qquad (4.41)$$

where

$$\begin{aligned}
\beta_1 &= f(x_n, t_n) \, \Delta t, \\
\beta_2 &= f(x_n + \beta_1/2, \; t_n + \Delta t/2) \, \Delta t, \\
\beta_3 &= f(x_n + \beta_2/2, \; t_n + \Delta t/2) \, \Delta t, \\
\beta_4 &= f(x_n + \beta_3, \; t_n + \Delta t) \, \Delta t.
\end{aligned} \qquad (4.42)$$

Powerful tools for the solution of ordinary differential equations can be created by combining high-order integration schemes with routines to optimize

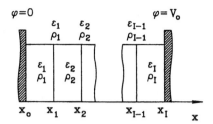

FIGURE 4.7
Numerical electric field calculation in a one dimensional gap — index conventions for finite-difference and finite-element treatments.

At continuously. Methods to implement *adaptive step size* are described in *Numerical Recipes in FORTRAN*, (Press et al., 1992).

4.3 One-Dimensional Poisson Equation

This section reviews difference equations for the one-dimensional Poisson equation,

$$\frac{\partial}{\partial x} \epsilon_r \frac{\partial \phi}{\partial x} = -\frac{\rho}{\epsilon_o}. \tag{4.43}$$

We shall compare finite-difference and finite-element viewpoints using the index conventions of Figure 4.7. To begin, we shall find a discrete form for Equation 4.43 using difference operators applied at points. Suppose that ϵ_r and ρ are continuous functions in space and have the following set of values at mesh positions:

$$\epsilon_i = \epsilon_r(x_i), \quad \rho_i = \rho(x_i). \tag{4.44}$$

An approximation for the left-hand side of Equation 4.43 at position x_i follows from the difference between centered estimates of $\epsilon \, d\phi/dx$ in adjacent intervals divided by $(x_{i+1} - x_i)/2$. The expression to the left of point x_i is

$$\epsilon \frac{d\phi}{dx} \cong \frac{\epsilon_i + \epsilon_{i-1}}{2} \frac{\phi_i - \phi_{i-1}}{x_i - x_{i-1}}. \tag{4.45}$$

The first term on the right-hand side is the average value of the dielectric constant at $(x_i + x_{i+1}/2)$. Applying Equation 4.45 to find the second derivative and setting the result equal to ρ_i/ϵ_o gives the equation

$$\frac{\dfrac{\epsilon_{i+1}+\epsilon_i}{2}\dfrac{\phi_{i+1}-\phi_i}{x_{i+1}-x_i}-\dfrac{\epsilon_i+\epsilon_{i-1}}{2}\dfrac{\phi_i-\phi_{i-1}}{x_i-x_{i-1}}}{\dfrac{x_{i+1}-x_{i-1}}{2}}=-\frac{\rho_i}{\epsilon_o}. \tag{4.46}$$

Solving Equation 4.46 for ϕ_i leads to a form reminiscent of Equation 2.66,

$$\phi_i = \frac{W_{i+1}\phi_{i+1}+W_{i-1}\phi_{i-1}+\dfrac{\rho_i\left(x_{i+1}-x_{i-1}\right)}{\epsilon_o}}{W_{i+1}+W_{i-1}}. \tag{4.47}$$

The coupling coefficients are

$$W_{i+1} = \frac{\epsilon_{i+1}+\epsilon_i}{2}\frac{1}{x_{i+1}-x_i}, \quad W_{i-1} = \frac{\epsilon_i+\epsilon_{i-1}}{2}\frac{1}{x_i-x_{i-1}}. \tag{4.48}$$

Equation 4.47 is an acceptable difference form if ρ and ϵ_r vary continuously in space. The method is ill-suited to electrostatic solutions with sharp boundaries between materials of different ϵ_r. We shall see in the next section that Equation 4.47 gives a smooth change of electric field at a discontinuity, leaving the boundary position ambiguous. The finite-element viewpoint gives better results. We take ϵ_r and ρ as element properties. The revised form of Equation 4.45 is

$$\epsilon\frac{d\phi}{dx} \cong \epsilon_i\frac{\phi_i-\phi_{i-1}}{x_i-x_{i-1}}. \tag{4.49}$$

We shall use the weighted average of space-charge density in the adjacent elements on the right-hand side of the Poisson equation,

$$\frac{\rho_i\left(x_i-x_{i-1}\right)+\rho_{i+1}\left(x_{i+1}-x_i\right)}{x_{i+1}-x_{i-1}}. \tag{4.50}$$

Substituting and solving for ϕ_i gives

$$\phi_i = \frac{W_{i+1}\phi_{i+1}+W_{i-1}\phi_{i-1}+\dfrac{\rho_{i+1}\left(x_{i+1}-x_i\right)+\rho_i\left(x_i-x_{i-1}\right)}{2\epsilon_o}}{W_{i+1}+W_{i-1}}. \tag{4.51}$$

In this case, the coupling coefficients are

$$W_{i+1} = \frac{\epsilon_{i+1}}{x_{i+1} - x_i} \quad W_{i-1} = \frac{\epsilon_i}{x_i - x_{i-1}}. \tag{4.52}$$

Although Equations 4.51 and 4.52 are similar to the finite-difference results of Equations 4.47 and 4.48, they lead to considerably different predictions of electric fields near dielectric boundaries.

4.4   Solution of the Poisson Equation by Back-Substitution

The difference equations derived in Section 4.3 for the one-dimensional Poisson equation have the general form

$$\alpha_i \phi_{i-1} + \beta_i \phi_i + \gamma_i \phi_{i+1} = Q_i. \tag{4.53}$$

Equation 4.53 represents a set of coupled linear equations. If the mesh has I vertices with fixed potentials on each boundary, there are $(I - 2)$ equations for the $(I - 2)$ unknown potential values at the intermediate points. If we postulate the relationship,

$$\phi_{i+1} = X_i \phi_i + Y_i, \tag{4.54}$$

then the solution depends on finding a set of coefficients X_i and Y_i consistent with Equation 4.53 and the boundary conditions. Substitution of Equation 4.54 into Equation 4.53 gives

$$\phi_i = -\frac{\alpha_i}{\beta_i + \gamma_i X_i} \phi_{i-1} + \frac{Q_i - \gamma_i Y_i}{\beta_i + \gamma_i X_i} \tag{4.55}$$

Equation 4.55 is consistent with the form of Equation 4.54 if

$$X_{i-1} = -\frac{\alpha_i}{\beta_i + \gamma_i X_i}, \tag{4.56}$$

and

$$Y_{i-1} = \frac{Q_i - \gamma_i Y_i}{\beta_i + \gamma_i X_i}. \tag{4.57}$$

Equations 4.56 and 4.57 are recursion relationships. To solve the set of equations represented by Equation 4.53, we first scan downward to find the set of X_i and Y_i, and then apply Equation 4.54 in an upward scan to find the values of ϕ_i.

To illustrate the procedure, consider the solution of the finite-element equations (Equation 4.51) with fixed potentials at each boundary. There are I mesh points with $\phi_1 = 0$ and $\phi_I = V_o$. In this case,

$$\alpha_i = -W_{i-1}, \quad \beta_i = W_{i-1} + W_{i+1}, \quad \gamma_i = -W_{i+1},$$

$$Q_i = \frac{\rho_{i+1}\left(x_{i+1} - x_i\right) + \rho_i\left(x_i - x_{i-1}\right)}{2\epsilon_o}. \tag{4.58}$$

The coupling coefficients W_i are given by Equations 4.52. The first quantities to find are X_{I-1} and Y_{I-1}. Because ϕ_I has a fixed value that does not depend on ϕ_{I-1}, inspection of Equation 4.54 shows that $X_{I-1} = 0$ and $Y_{I-1} = V_o$. The next step is to apply Equations 4.56 and 4.57 to find the set of coefficients down to X_1 and Y_1. Finally, Equation 4.54 is applied to find the set of ϕ_i starting at $\phi_1 = 0$.

We can also apply back-substitution to problems with mixed Dirichlet and Neumann boundary conditions. Suppose that $\phi(x_1) = 0$ and $d\phi(x_I)/dx = 0.0$. The latter condition means that the plane at x_I is a symmetry boundary, equivalent to the presence of a potential $\phi_{I+1} = \phi_{I-1}$ at position $x_I + (x_I - x_{I-1})$. Therefore, the form of the Equation 4.54 at the Neumann boundary is

$$\alpha_I \, \phi_{I-1} + \beta_I \, \phi_I + \gamma_I \, \phi_{I+1} = Q_I. \tag{4.59}$$

Substituting for ϕ_{I+1} and solving for ϕ_I gives

$$\phi_I = -\frac{\alpha_I + \gamma_I}{\beta_I} \phi_{I-1} + \frac{Q_I}{\beta_I}. \tag{4.60}$$

Comparing Equation 4.60 with Equation 4.54 gives the values of the initial coefficients as

$$X_{I-1} = -\frac{\alpha_I + \gamma_I}{\beta_I}, \quad Y_{I-1} = \frac{Q_I}{\beta_I}. \tag{4.61}$$

The recursion formulas of back-substitution are easy to implement in a spreadsheet program. Figure 4.8 shows results from a spreadsheet calculation to demonstrate the problem of dielectric boundaries in finite difference solutions. The solution covers a 1 cm planar gap with plate voltages of 0.0 and –1.0 V. The right half of the gap is filled with a dielectric ($\epsilon_r = 2.5$) while the left half is vacuum. The solution uses 11 mesh points and 10 intervals.

Theoretically, the electrical field has constant values of 0.57142857 and 1.42857143 V/cm in the dielectric and air regions, respectively. Figure 4.8a shows finite-difference results where the relative dielectric constant is defined at vertices. The transition in E_z spreads over several mesh points. The field values far from the boundary have about a 3% error. Figure 4.8b plots the field values from a finite-element solution. Elements to the right of $x = 0.5$ have $\epsilon_r = 1.0$ while elements to the left have $\epsilon_r = 2.5$. The fields are constant in each region and exhibit the correct transition at the boundary. The numerical values are within 0.0002% of the prediction.

4.5 Two-Dimensional Electrostatic Solutions on a Regular Mesh

In this section, we shall derive difference equations to represent the Poisson equation and Gauss' law on a regular mesh over two-dimensional areas. As in Section 4.4 we shall cover both finite-element method and finite-difference formulations. To begin, consider finite-difference equations for a region described by Cartesian coordinates. Figure 4.9a shows the mesh index conventions. The continuous quantities ϕ, ϵ_r, and ρ are approximated by interpolating quantities at mesh locations (ϕ_{ij}, ϵ_{rij}, and ρ_{ij}). Adding partial derivatives in the y direction to the method of Section 4.4 gives

$$\phi_{ij} = \frac{W_{i+1,j}\,\phi_{i+1,j} + W_{i-1,j}\,\phi_{i-1,j} + W_{i,j+1}\,\phi_{i,j+1} + W_{i,j-1}\,\phi_{i,j-1} + \dfrac{\rho_{ij}}{\epsilon_o}}{W_{i+1,j} + W_{i-1,j} + W_{i,j+1} + W_{i,j-1}} \tag{4.62}$$

where

$$W_{i+1,j} = \frac{\epsilon_{i+1,j} + \epsilon_{i,j}}{2} \frac{1}{\left(x_{i+1} - x_i\right)\left(x_{i+1} - x_{i-1}\right)},$$

$$W_{i-1,j} = \frac{\epsilon_{i-1,j} + \epsilon_{i,j}}{2} \frac{1}{\left(x_i - x_{i-1}\right)\left(x_{i+1} - x_{i-1}\right)},$$

$$W_{i,j+1} = \frac{\epsilon_{i,j+1} + \epsilon_{i,j}}{2} \frac{1}{\left(y_{j+1} - y_j\right)\left(y_{j+1} - y_{j-1}\right)}, \tag{4.63}$$

$$W_{i,j=1} = \frac{\epsilon_{i,j-1} + \epsilon_{i,j}}{2} \frac{1}{\left(y_j - y_{j-1}\right)\left(y_{j+1} - y_{j-1}\right)}.$$

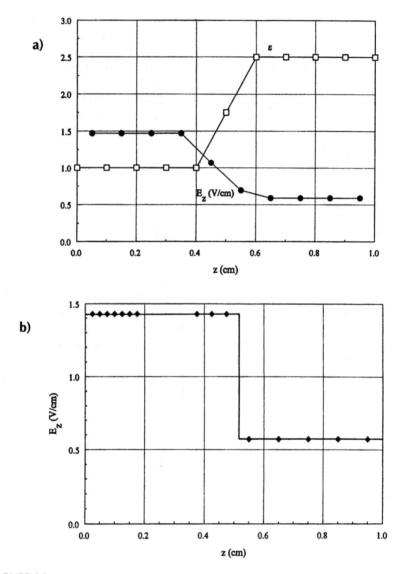

FIGURE 4.8
Numerical solution for the electric field in a one-dimensional gap. Dielectric with $\epsilon_r = 2.5$ extends from $x = 0.5$ to $x = 1.0$ cm. Number of mesh points: $I = 11$. (a) Finite-difference results. (b) Finite element results.

In the limit of a uniform mesh and a medium with no space charge, Equation 4.63 implies that potential values should equal the average of their four neighbors,

$$\phi_{i,j} = \frac{\phi_{i+1,j} + \phi_{i-1,j} + \phi_{i,j+1} + \phi_{i,j-1}}{4} . \tag{4.64}$$

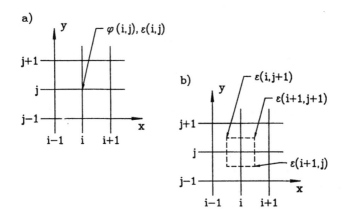

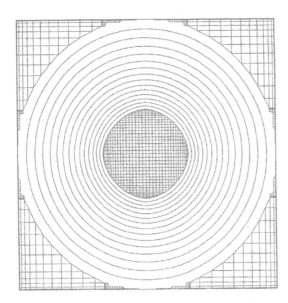

FIGURE 4.9
Two-dimensional electrostatic calculations on a regular mesh. a) Index conventions for a finite-difference treatment. b) Index conventions and Gaussian volume for a finite-element treatment.

To solve Equation 4.63 we fix the potential of certain boundary and internal points to represent the electrodes and solve the remaining set of difference equations to find the potential values ϕ_{ij}. Figure 4.10 shows a solution for a cylindrical electrode inside a grounded box. The figure illustrates the drawback of regular meshes — the vertex locations give a poor representation of the curved electrode. As a result, field estimates near the electrode may have errors.

FIGURE 4.10
Equipotential lines for a two-dimensional finite-element solution of a cylindrical electrode inside a grounded box. Regular mesh with variable resolution along the axes.

To find two-dimensional finite-element equations we can apply Gauss' law (Equation 2.33) to the rectangular volume shown as a dashed line in Figure 4.9b. The volume surrounds a test point with potential ϕ_0. The mesh intervals along the x and y axes above and below the test point are h_{xu}, h_{yu}, h_{xd}, and h_{yd}, respectively. The right-hand side of Equation 2.32 equals

$$\frac{\left(\rho_{uu}h_{xu}h_{yu} + \rho_{du}h_{xd}h_{yu} + \rho_{dd}h_{xd}h_{yd} + \rho_{ud}h_{xu}h_{yd}\right)\Delta z}{4\epsilon_0}, \qquad (4.65)$$

where Δz is an element arbitrary height. The surface integral is evaluated over the eight segments denoted a through h in Figure 4.9b. For example, the integral on segment a is

$$\epsilon_{uu}\frac{\phi_0 - \phi_{xu}}{h_{xu}}\frac{h_{yu}}{2}\Delta z. \qquad (4.66)$$

Adding contributions from all segments and setting the result equal to the quantity of Equation 4.65 gives the result,

$$\phi_0 =$$

$$\frac{W_{xu}\phi_{xu} + W_{yu}\phi_{yu} + W_{xd}\phi_{xd} + W_{yd}\phi_{yd} + \frac{\rho_{uu}h_{xu}h_{yu} + \rho_{du}h_{xd}h_{yu} + \rho_{dd}h_{xd}h_{yd}\,\rho_{ud}h_{xu}h_{yd}}{4\epsilon_0}}{W_{xu} + W_{yu} + W_{xd} + W_{yd}}. \qquad (4.67)$$

The coupling coefficients are

$$W_{xu} = \frac{h_{yd}\epsilon_{ud} + h_{yu}\epsilon_{uu}}{2h_{xu}},$$

$$W_{yu} = \frac{h_{xu}\epsilon_{uu} + h_{xd}\epsilon_{du}}{2h_{yu}},$$

$$W_{xd} = \frac{h_{yu}\epsilon_{du} + h_{yd}\epsilon_{dd}}{2h_{xd}}, \qquad (4.68)$$

$$W_{yd} = \frac{h_{xd}\epsilon_{dd} + h_{xu}\epsilon_{ud}}{2h_{yd}}.$$

To this point, we have covered three methods for numerical electrostatic solutions: (1) finite-difference on a regular mesh, (2) finite-element on a regular mesh, and (3) finite-element on a conformal mesh. In two dimensions, there is no doubt which method to use. Finite-elements on a conformal mesh have overwhelming advantages. Regular meshes are useful in three-dimensional calculations because they require less data storage. Again, the finite-element formulation is preferable because of the ease in representing material and Neumann boundaries. Because there is little motivation for finite-differencing in space, we shall concentrate on finite-element methods in the remainder of this book.

4.6 Three-Dimensional Electrostatic Solutions on a Regular Mesh

Generating conformal meshes in three dimensions is a challenging task — extended geometric information must be stored for each vertex. On the other hand, it is relatively easy to implement three-dimensional solutions with regular meshes. In this section, we shall review the associated three-dimensional finite-element equations and discuss techniques to improve accuracy. Figure 4.11 defines mesh parameters near a test point. We include the option for variable mesh spacing along each axis. Each mesh point has six neighboring vertices. Extending the notation of Section 4.5, the potentials are denoted ϕ_{xu}, ϕ_{xd}, ϕ_{yu}, ϕ_{yd}, ϕ_{zu}, and ϕ_{zd}. The distances to the neighboring vertices are h_{xu}, h_{xd}, h_{yu}, h_{yd}, h_{zu}, and h_{zd}. There are eight surrounding elements with relative dielectric constant and space charge density indicated by the notation ϵ_{udu} and ρ_{ddu}. Here, the index order refers to the x, y, and z directions, respectively. We shall take a Gaussian surface integral over a box that extends parallel to the axes halfway to each neighbor. The enclosed space charge is

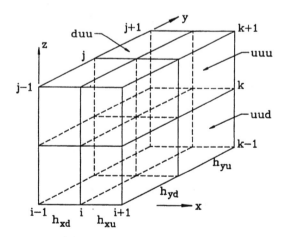

FIGURE 4.11
Mesh conventions and Gaussian volume near a test point for a three-dimensional finite-element electrostatic solution.

$$Q =$$

$$\frac{1}{8\epsilon_o}\left[h_{xu}h_{yu}h_{zu}\rho_{uuu} + h_{xd}h_{yu}h_{zu}\rho_{duu} + h_{xu}h_{yd}h_{zu}\rho_{udu} + h_{xd}h_{yd}h_{zu}\rho_{ddu}\right. \quad (4.69)$$

$$\left. + h_{xu}h_{yu}h_{zd}\rho_{uud} + h_{xd}h_{yu}h_{zd}\rho_{dud} + h_{xu}h_{yd}h_{zd}\rho_{udd} + h_{xd}h_{yd}h_{zd}\rho_{ddd}\right].$$

There are 24 facets that contribute to the surface integral. Consider the facets of the four elements in the positive x direction. Although the dielectric constant may differ in each element, Equation 2.35 implies that the electric field parallel to surfaces between dielectric regions is continuous. Therefore, E_x has approximately the same value on the four surface segments,

$$E_x\left(x + h_{xu}/2,\ y,\ z\right) \cong \frac{\phi_o - \phi_{xu}}{h_{xu}}. \qquad (4.70)$$

The contribution to the surface integral from the face is

$$\frac{\phi_o - \phi_{xu}}{4h_{xu}}\left[\epsilon_{uuu}h_{yu}h_{zu} + \epsilon_{udu}h_{yd}h_{zu} + \epsilon_{uud}h_{yu}h_{zd} + \epsilon_{udd}h_{yd}h_{zd}\right]. \quad (4.71)$$

Adding contributions from the six faces and setting the result equal to the sum of Equation 4.69 gives the finite-element difference equation for three-dimensional electrostatics,

$$\phi_o = \frac{W_{xu}\phi_{xu} + W_{xd}\phi_{xd} + W_{yu}\phi_{yu} + W_{yd}\phi_{yd} + W_{zu}\phi_{zu} + W_{zd}\phi_{zd} + Q}{W_{xu} + W_{xd} + W_{yu} + W_{yd} + W_{zu} + W_{zd}}. \quad (4.72)$$

The coupling constants follow from an inspection of Equation 4.71,

$$W_{xu} = \frac{1}{4h_{xu}}\left[\epsilon_{uuu}h_{yu}h_{zu} + \epsilon_{udu}h_{yd}h_{zu} + \epsilon_{uud}h_{yu}h_{zd} + \epsilon_{udd}h_{yd}h_{zd}\right],$$

$$(4.73)$$

$$W_{zd} = \frac{1}{4h_{zd}}\left[\epsilon_{uud}h_{xu}h_{yu} + \epsilon_{dud}h_{xd}h_{yu} + \epsilon_{udd}h_{xu}h_{yd} + \epsilon_{ddd}h_{xd}h_{yd}\right].$$

Three-dimensional solutions may require over a million vertices. Transferring large amounts of data to disk slows program operation; therefore, there is a considerable advantage to storage in random-access memory. Consider a solution with indices in the range $0 \le i \le I$, $0 \le j \le J$, $0 \le k \le K$. One option to store potential values is to define a three-dimensional array $\phi(i,j,k)$ with dimensions I, J, and K. This is a poor choice if we want a versatile program for a variety of geometries. For example, a solution in long thin volume may require only a small value of K so that a large block of memory would be wasted. A better approach is to store the values in a one-dimensional array of length $N = (I+1)(J+1)(K+1)$ with a global index n. In this case, we can use the full available memory for any consistent values of I, J, and K. The program must include subroutines for the conversions $(i,j,k) \Leftrightarrow (n)$. For example, one choice of the global index in a three-dimensional calculation is

$$n = N_{plane}k + N_{row} \, j + i. \tag{4.74}$$

Redundant calculations are eliminated by initially setting variables

$$N_{row} = I+1, \quad N_{plane} = (J+1)(I+1). \tag{4.75}$$

The inverse transformations are

$$i = \left[\left[\left(n + N_{plane} \right) \bmod N_{plane} \right] + N_{row} \right] \bmod N_{plane},$$

$$j = \frac{\left(n + N_{plane} \right) \bmod N_{plane}}{N_{row}}, \tag{4.76}$$

$$k = \frac{n}{N_{plane}}.$$

The operations in Equation 4.76 are modulus and integer division. The time penalty for index conversions is usually small compared with the time required for floating point operations.

Chapter 4   Exercises

4.1. Verify that the complex form of the Fourier series (Equations 4.3 and 4.5) is equivalent to the harmonic form of Equations 4.1 and 4.2.

4.2. Use a spreadsheet to find the complex Fourier coefficients g_n for the following data set.

n	1	2	3	4	5	6	7	8	9	10
fn	0.0	0.4	0.6	0.8	1.0	1.0	0.75	0.50	0.25	0.20

Verify the validity of the expansion by substitution in Equation 4.6. (If the spreadsheet does not handle complex numbers, you can find the real and imaginary parts of g_n explicitly using harmonic functions.)

4.3. Use a spreadsheet to confirm the orthogonality relationship of Equation 4.8 for I = 5.

4.4. Investigate the accuracy of the difference approximation $(f_{i+1} - f_i)/\Delta$ for the derivative of a function f_i defined on a uniform mesh in x. Apply the method described in Section 4.1 based on a Fourier mode $f_k = g_k \exp(ikx)$. Show that the expression has first-order accuracy as an approximation to the derivative at x_i but second-order accuracy for the position $x_i + \Delta/2$.

4.5. Verify that the error in Equation 4.23 is approximately $(k\Delta)^2/12$.

4.6. Find a centered difference expression for d^4f/dx^4 at x_i and estimate the error for a Fourier mode with wavenumber k.

4.7. Show that Equations 4.25 and 4.26 approach Equations 4.16 and 4.23, respectively, in the limit of a uniform mesh, $\Delta x_i = \Delta x_{i+1}$.

4.8. In the limit $\Delta x_i \ll \Delta x_{i+1}$, show that the first derivative expression of Equation 4.25 approaches

$$\frac{y_i - y_{i-1}}{\Delta x_i}.$$

This result implies that the expression approaches the skewed estimate based on values to the left of the point when there is a large distance to the point on the right hand side.

4.9. Use a spreadsheet to solve the equation set $dx/dt = v$ and $dv/dt = -x$ in the range $0 \le t \le 2\pi$. The initial conditions are $x(0) = 1$ and $v(0) = 0$. Use the Euler and two-step methods for $\Delta t = 2\pi/20$ and compare final values to the analytic prediction $x = \cos(2\pi) = 1$.

4.10. Use a spreadsheet to derive a numerical solution to the equation

$$\frac{dx}{dt} = x$$

over the interval $0 \le t \le 3$. Divide the range into 10 steps and take $x(0) = 1$.

(a) Use the two step method and compare the results to those quoted in Section 4.2.

(b) Apply the leapfrog method and check the stability of the solution.

4.11. Apply the fourth-order Runge-Kutta method on a spreadsheet to find the solution of the equation $dx/dt = \sin(t) - t^2/20$ in the range $0 \le t \le 5$. The initial condition is $x(0) = 0$.

(a) Find x(at t = 5) as a function of the number of steps for n = 1, 2, 5, and 10.

(b) By plotting x(at t = 5) vs. 1/n, can you estimate the value for an infinite number of steps (1/n = 0)?

(c) Compare this result to the analytic prediction x(5) = 1 – cos(5) – $5^3/60 = -1.3670$.

4.12. Program the recursion formulas of the back-substitution method to solve the following one-dimension electrostatic problem. Parallel plates separated by distance d = 0.05 m have applied potentials $\phi(0)$ = 0 and $\phi(d)$ = 100. The relative dielectric constant varies as $\epsilon_r = 1 + 3 \sin(\pi x/d)$. Use 20 elements.

4.13. The Poisson equation has the following form in cylindrical coordinates.

$$\frac{1}{r}\frac{\partial}{\partial r}\,\epsilon\,\frac{\partial \phi}{\partial r} + \frac{\partial^2 \phi}{\partial z^2} = \frac{\rho}{\epsilon_0}.$$

Find the finite difference form of the equation on uniform regular mesh by replacing the differential operators with difference operators (Section 4.3). Compare the result to Equations 4.62 and 4.63.

4.14. Find a finite-element equation to represent electrostatics in cylindrical coordinates by applying Gauss' law to the annular volume shown in Figure E4.1. Assume a uniform mesh with spacings Δr and Δz. Compare the results to those of Exercise 4.9 and Equations 4.67 and 4.68.

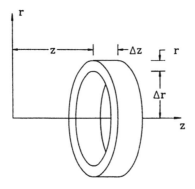

FIGURE E4.1

4.15. Derive Equations 4.76 for the inverse transformations from a global index given by Equation 4.74.

4.16. Write simplified forms of Equations 4.72 and 4.73 for the following cases.

(a) A uniform cubic mesh with spacing Δ.

(b) A region of uniform dielectric constant.

5

Techniques for Numerical Field Solutions

An ounce of action is worth a ton of theory.

Friedrich Engels

In previous chapters we covered several ways to represent the equations of electrostatics in terms of coupled linear-equation sets. This chapter addresses the mechanics of the solution: setting up a mesh, assigning material properties, dealing with boundaries, and solving the equations to find the electrostatic potential at the vertex points. When these processes are complete, we can interpolate the potential to estimate the electric field at any point in the solution volume.

Sections 5.1, 5.2, and 5.3 cover the creation of meshes to represent arbitrary geometries. In contrast to the well-defined procedures of previous chapters, mesh generation is a craft. There is no ideal approach; each numericist can rest assured that his or her generator is imperceptibly better than others. We cannot attempt to cover the full spectrum of methods in an introductory work. Instead, we shall concentrate on two straightforward methods that handle arbitrary geometries and are straightforward to program. The simple logic of connections in the resulting meshes makes it easy to locate elements and to solve the equations by matrix inversion.

Section 5.1 covers the creation of regular meshes in three dimensions. Here, the elements are boxes with rectangular sides. We shall use the method of volumes to assign vertex and element characteristics. Here, complex three-dimensional shapes are constructed in terms of positive and negative elementary volumes. The next two sections deal with conformal two-dimensional meshes. Section 5.2 describes a method to create flexible triangular meshes. The critical requirement is an indexing scheme for quick identification of vertices and elements surrounding a vertex and the vertices that bound an element. Section 5.3 covers techniques to adjust the mesh to match the boundaries of arbitrary electrodes and dielectrics. The section also discusses assignment of vertex and material properties and the determination of the coupling constants introduced in Section 2.7.

Section 5.4 discusses applications of special Neumann boundaries in finite-element solutions. In electrostatic solutions the normal derivative of potential equals zero along the boundary. Equivalently, the electric field is parallel to the boundary. The ability to generate slanted or curved symmetry boundaries is essential for solutions in applications such as resistive flow or gas dynamics. To conclude, Section 5.5 introduces a technique to solve the large sets of linear equations. The method of *successive over-relaxation* is an iterative process. It involves continual small corrections of vertex potentials to bring them into conformance with the difference relationships. The method is fast and easy to program. An important advantage for calculations on personal computers is that memory requirements are modest. The drawback is that there is no rigid criterion to pick optimum relaxation parameters. Chapter 6 discusses alternative direct solution methods based on matrix algebra.

5.1   Regular Meshes in Three Dimensions

In this section we shall study regular three-dimensional meshes to use the equations of Section 4.6. Figure 4.11 shows the index conventions. The solution box has rectangular sides with dimensions $(x_{max} - x_{min})$, $(y_{max} - y_{min})$, and $(z_{max} - z_{min})$. The volume is subdivided into elements with dimensions given by mesh spacings along each axis, Δx_i $(i = 0, I)$, Δy_j $(j = 1, J)$, and Δz_k $(k = 1,J)$. The spacings may be uniform or varied to give enhanced resolution. They can also be weighted to provide a better fit to critical surfaces. For example, Figure 4.10 shows the x-y projection of a weighted mesh optimized for a cylindrical object.

There are $(I+1)(J+1)(K+1)$ vertices in the solution box and $I \times J \times K$ elements. A set of indices (i,j,k) refers to a mesh point and the element directly below (in the negative x, y, and z directions). Following Section 4.5 we can increase the versatility of the code by defining a global index n and storing vertex and element quantities in one-dimensional arrays. We can also write the subroutine for the index conversion of Equation 4.74 so that it returns n = -1 for vertices or elements outside the solution volume. If we define the dummy values $\phi(-1) = 0.0$ and $\epsilon(-1) = 0.0$, then the program will automatically implement special Neumann conditions on any undefined boundary (Section 2.8). It is straightforward to determine input quantities for the coupling constants (Equations 4.73) at a vertex point. The mesh dimensions near the point (i,j,k) are given by

$$h_{xu} = \Delta x_{i+1}, \quad h_{xd} = \Delta x_i$$

$$h_{yu} = \Delta y_{j+1}, \quad h_{yd} = \Delta y_j \tag{5.1}$$

$$h_{zu} = \Delta z_{k+1}, \quad h_{zd} = \Delta z_k$$

The indices of elements surrounding the point are

$$(u, u, u) \rightarrow (i, j, k),$$

$$(u, u, d) \rightarrow (i, j, k-1),$$

$$(u, d, u) \rightarrow (i, j-1, k),$$

$$(u, d, d) \rightarrow (i, j-1, k-1),$$

$$(d, u, u) \rightarrow (i-1, j, k),$$

$$(d, u, d) \rightarrow (i-1, j, k-1),$$

$$(d, d, u) \rightarrow (i-1, j-1, k),$$

$$(d, d, d) \rightarrow (i-1, j-1, k-1).$$

(5.2)

The next step is to set the vertex and element properties to represent the physical system. We shall begin with vertices. There are two types: (1) variable points with potential determined by the solution process, and (2) fixed points where the potential retains its initial value.

Fixed points correspond to positions inside and on electrodes. Suppose we represent the fixed condition by one bit of a status array variable. To replicate the geometry of a set of electrodes we loop through the mesh, setting the fixed bit and the potential value for appropriate points. The points could be set by hand, but the work becomes prohibitive for large meshes with complex shapes.

We can automate vertex setting with the method of elementary volumes. Here, the user supplies information on the size, shape, and orientation of a sequence of elementary solid objects like cubes, spheres, boxes, and wedges. The program then analyzes the vertices to find which ones are inside. With the convention that marked points over-write any previous definitions, it is possible to weld objects or to cut holes. In this way, complex shapes can be assembled from a library of simple objects. Figure 5.1 shows an example for a three-dimensional electrostatic simulation of a field distortion spark gap with an offset trigger electrode.

There are three steps in the creation of an elementary object in the solution volume.

- Fabricate the object. As an example, fabrication of a cylinder involves giving a height, a radius, and the characteristic of enclosed vertices (fixed or variable). By convention, we create the object in a reference coordinate system (x', y', z') with Cartesian axes that lie

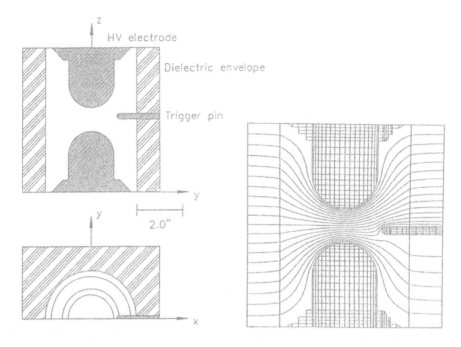

FIGURE 5.1
Three-dimensional calculation of electrostatic fields in a field-distortion spark gap. (left) Mechanical
drawing of the device. (right) Equipotential lines in a longitudinal section passing through the
trigger pin.

along those of the solution volume. Suppose the base of the cylinder lies in the x′–y′ plane so that the height extends in z.

- Orient the object. Orientation consists of rotation about one or more
  of the reference coordinate axis. By convention, we shall take rotations in order about the x′, y′, and z′ axes.

- Move the object to position. The final operation is translation along
  one or more axes of the solution coordinates.

The definition of a cylinder requires the following parameter sets:

- Physical properties, R (radius), H (height), FFlag (TRUE if fixed
  point), V_o (fixed potential)
- Rotation properties, θ_x, θ_y, θ_z
- Translation properties, L_x, L_y, L_z

Given the parameters, we can tell if a point (x,y,z) in the solution space is
inside the cylinder by first transforming to the object definition space (x′,y′,z′)
and then checking whether the following three conditions are satisfied,

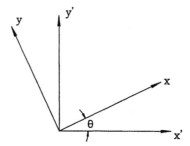

FIGURE 5.2
Transformation of coordinates between a Cartesian coordinate system (x′,y′,z′) and a system (x,y,z) rotated by an angle θ_z.

$$z' \geq 0$$

$$z' \leq H \tag{5.3}$$

$$\sqrt{x'^2 + y'^2} \leq R$$

The coordinate transformation is performed by subtracting L_x, L_y, and L_z from the position coordinates and then applying rotation matrices in the order $M(-\theta_z)$, $M(-\theta_y)$, and then $M(-\theta_x)$.

Regarding rotation matrices, Figure 5.2 shows the relationship between a Cartesian coordinate system (x′,y′,z′) and a system (x,y,z) rotated by an angle θ_z. A set of coordinate values transforms according to

$$\begin{bmatrix} x \\ y \\ z \end{bmatrix} = M(\theta_z) \begin{bmatrix} x' \\ y' \\ z' \end{bmatrix}, \tag{5.4}$$

where the rotation matrix is

$$M(\theta_z) = \begin{bmatrix} \cos\theta_z & \sin\theta_z & 1 \\ -\sin\theta_z & \cos\theta_z & 1 \\ 0 & 0 & 1 \end{bmatrix}. \tag{5.5}$$

The matrices for rotations about the x and y axes are

$$M(\theta_x) = \begin{bmatrix} 1 & 0 & 0 \\ 0 & \cos\theta_x & \sin\theta_x \\ 0 & -\sin\theta_x & \cos\theta_x \end{bmatrix},$$

$$M(\theta_y) = \begin{bmatrix} \cos\theta_y & 0 & -\sin\theta_y \\ 0 & 1 & 0 \\ \sin\theta_y & 0 & \cos\theta_y \end{bmatrix}. \tag{5.6}$$

5.2   Two-Dimensional Conformal Triangular Meshes

We now turn to the more challenging problem of creating conformal triangular meshes in two dimensions. To begin, we need to clarify what constitutes a good conformal mesh. There are two measurements of quality.

- Field solutions are most accurate when the mesh triangles are all approximately equilateral. In this case, the coupling constants have about the same magnitude. Acute and distorted triangles give a wide disparity in coupling constant values, degrading accuracy and sometimes leading to numerical instabilities (Section 12.5).

- For the best accuracy and minimum run time, the scale size of triangles should vary in space to give good resolution of small objects with a moderate number of elements.

We shall discuss the technique of *logical mesh* deformation (see, for instance, Warren et al., 1987). The method has the advantage that it processes most geometries with little user intervention. Further, the regular mesh logic speeds calculations of field quantities. The disadvantages are that the fixed mesh logic may use memory inefficiently and the fitting procedure may yield distorted triangles for convoluted geometries.

The process starts with a collection of logically connected triangles that fill a solution region. The term *logically connected* means that the indices of elements and points surrounding a vertex are given by simple formulas. The critical step is to shift certain vertices so that the sides of elements lie along the boundaries of electrodes or dielectrics. Logical connections are preserved if the shifts are not too large; otherwise, some triangles may be turned inside-out. We apply the term *region* to a section of the mesh that corresponds to a particular geometric object. After the shifts are completed, vertices and elements are marked with an integer number to designate their region. The final step is to smooth variations between points that are not fixed on boundaries.

In this section, we will concentrate on the creation of logical meshes with variable resolution and review a convention for indexing. The following section covers procedures to shift vertices and to smooth the mesh. Initially, we let the logical mesh fill a rectangular region (Figure 5.3) with dimensions x_{min}, x_{max}, y_{min}, and y_{max}. The axes represent x–y for rectangular systems and z–r for cylindrical problems. The solution region can be any shape that fits inside and may include some or all of the available triangles. Note that the triangles in Figure 5.3 have a regular order — every internal vertex is surrounded by six elements.

The vertex indices range from $i = 1$ to I along the x (or z) axis and $j = 1$ to J along y (or r). As we saw in Section 4.6, storing mesh quantities in two-dimensional arrays uses memory inefficiently. It is better to use one-dimensional arrays with a global index, n. We shall use the convention that the mesh index increases most rapidly along the x (or z) direction.

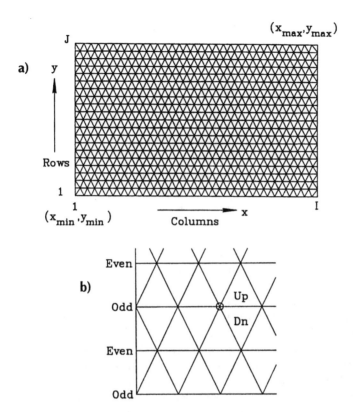

FIGURE 5.3

Mesh of triangular elements with constrained connection logic — six elements surround each internal vertex. (a) Indices of a logical mesh filling a rectangular region (x_{min}, y_{min}) to (x_{max}, y_{max}). (b) Detail of mesh showing vertex connections in odd and even rows.

$$n = jI + i. \tag{5.7}$$

The global index has the range $1 \leq n \leq N$, where

$$N = IJ. \tag{5.8}$$

The inverse transformations are

$$j = \frac{n-1}{I} + 1,$$
$$i = (n+I) \bmod I. \tag{5.9}$$

A good practice is to organize properties of the mesh and electrostatic solution in a data structure. For example, the one-dimensional array *Record/Mesh-Point/Mesh(0:NStoreMax)* has the following organization:

```
Type      MeshPoint
Real      x
Real      y
Real      Phi
Real      W5
Real      W6
Real      W1
Integer*1     RegNo
Integer*1     RegUp
Integer*1     RegDn
Real      RhoUp
Real      RhoDn
End Type
```

Note that the index n of the array starts at zero. All components of the zero element equal zero. We construct the index conversion routines so that they return n = 0 when vertices are outside the solution volume. These precautions ensure that (1) there are no index or floating point errors that could lead to a program crash and (2) unspecified boundaries automatically assume the specialized Neumann condition. The first two quantities in the structure are the x and y (or z and r) coordinates of the vertex and the third is the electrostatic potential. The following three quantities are coupling constants (Equation 2.68) to points n_5, n_6, and n_1. Because of a redundancy of values, it is necessary to store only three coupling constants for each point. The sixth quantity is the region number associated with the vertex point, a one-byte integer. Invalid vertex points (such as those outside the solution volume) are marked with region number 0 leaving 127 numbers to label solution regions. Later, we shall apply the sign bit of the integer to mark materials with field-dependent properties (Chapter 9). An inspection of Figure 5.3 shows that there are roughly two elements for each vertex point. By convention we shall associate the two triangles to the right with a vertex point. They are above and below the line that connects point (i,j) with (i+1,j). The quantities *RegUp* and *RegDn* are the region numbers of the elements. For problems with space-charge distributions we can include *RhoUp* and *RhoDn*, the charge densities in the upper and lower triangles.

To compute the coupling coefficients to neighboring vertices we need to identify (1) the six neighboring points to find the element cotangents and (2) the six surrounding elements to find the values of dielectric constant. Regarding vertices, inspection of Figure 5.3 gives the indices for surrounding points shown in Table 5.1. Turning to elements, all values of *Mesh(n).RegUp* and *Mesh(n).RegDn* are set to zero when the logical mesh is created. We then use the procedures described in the next section to mark elements inside the solution volume with appropriate region numbers. The numbers may correspond to constant potential electrodes or to a value of the relative dielectric constant. When calculating the coupling coefficients of Equations 2.67, we use a function

REAL FUNCTION Epsi(n,UpFlag)

to find the dielectric constant of the upper and lower elements at vertex n. If the element region number corresponds to a dielectric volume, the function returns the corresponding value of ϵ_r. The function returns 0.0 if the region number equals zero or corresponds to a volume inside an electrode. This feature ensures implementation of automatic Neumann boundaries (Section 2.8). It is easy to create rectangular logical meshes starting from sets of vertex coordinates along each axis, *XMesh(i)* and *YMesh(j)*. The values, ranging from x_{min} to x_{max} and y_{min} to y_{max}, need not be uniformly spaced. The first step is to fill the rectangular region by making a loop through all values of i and j and setting

$$n = NVal(i,j),$$

$$Mesh(n).x = XMesh(i), \qquad (5.10)$$

$$Mesh(n).y = YMesh(j).$$

If the index scheme of Table 5.1 is used in the function NVal, the process produces the set of right triangles shown in Figure 5.4a. With this triangle geometry it is sometimes impossible to fit curved or slanted boundaries. We can improve the performance of the fitting process by converting the elements to isosceles triangles. The subroutine of Table 5.2 produces the set of triangles shown in Figure 5.4b.

To conclude, we should note that the rectangular logical mesh of Figure 5.3 is a special case. There is a broad range of alternatives consistent with the fitting process of Section 5.3. For example, in electromagnetic problems, ordering vertices in planes parallel to the x-axis may produce numerical effects similar to interference in a crystalline medium. We can create glassy amorphous meshes by adding loops to the routine of Table 5.2 to introduce random vertex displacements in the y direction (Figure 5.5a). It is also possible to generate logical meshes that are optimized to an application. For example, Figure 5.5b shows a logical mesh to model a segment of a cylindrical system. Note that there is logarithmic weighting in the radial direction.

TABLE 5.1

Indices of Vertices Connected to Point (i,j)

	j odd	j even
Point 1	i+1, j+1	i,j+1
Point 2	i,j+1	i−1,j+1
Point 3	i−1,j	i−1,j
Point 4	i,j−1	i−1,j−1
Point 5	i+1,j−1	i,j−1
Point 6	i+1,j	i+1,j

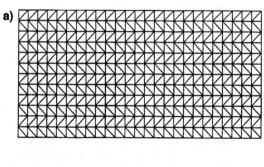

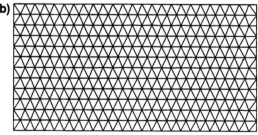

FIGURE 5.4
Options for element geometries in a logical mesh. (a) Right-angle elements; (b) relaxed mesh of isosceles triangles.

5.3 Fitting Triangular Elements to Physical Boundaries

Given a logical mesh extending over a solution area, the next step is to shift vertex positions so that the mesh conforms to the physical system. Another task is marking vertices and elements with appropriate region numbers. The numbers are associated with physical properties during the solution process.

Suppose region boundaries are defined by a set of geometric *segments* supplied by the user: points, lines, arcs, or other curves. Regions can be of two types: *open* or *filled*. Open regions represent boundary conditions or planar electrodes. They may consist of any combination of connected or unconnected segments. The vertices of open regions are marked with region numbers during processing, but no elements are marked. Filled regions represent volumes like dielectrics, space-charge clouds, and electrodes. The boundary segments must connect and form a closed path in space. All vertices and enclosed elements of a filled region are marked with the region number. This section discusses four steps of mesh generation.

- Shifting and marking logical mesh vertices so that they lie on boundary segments

- Testing elements and marking those that lie inside closed boundaries

TABLE 5.2

Routine to Convert Right Triangles to Isosceles Triangles
Aligned Along x

```
DO j = 1, JMax
   JOdd = MOD(j,2)
C--- i = 1 column ----
   n = NVal(1,1)
   DShift = 0.125*(Mesh(n+1).x-Mesh(n).x)
   IF (JOdd.EQ.1) THEN
      Mesh(n).x = Mesh(n).x + DShift
   ELSE
      Mesh(n).x = Mesh(n).x - DShift
   ENDIF
C--- i = IMax column ----
   n = NVal(IMax,1)
   DShift = 0.125*(Mesh(n).x-Mesh(n-1).x)
   IF (JOdd.EQ.1) THEN
      Mesh(n).x = Mesh(n).x + DShift
   ELSE
      Mesh(n).x = Mesh(n).x - DShift
   ENDIF
C--- General columns
   DO k = 2, (KMax-1)
      n = NVal(i,j)
      DShift = 0.25*AMIN1((Mesh(n+1).x-Mesh(n).x),
   x            Mesh(n).x-Mesh(n-1).x))
      IF (JOdd.EQ.1) THEN
         Mesh(n).x = Mesh(n).x + DShift
      ELSE
         Mesh(n).x = Mesh(n).x - DShift
      ENDIF
   END DO
END DO
```

- Relaxing vertex positions to achieve a smooth set of triangles
- Confirming the integrity of the mesh

To illustrate vertex shifting, consider first the operation for a point segment. Open regions may contain a set of unconnected points defined by coordinates (x_k, y_k) or (z_k, r_k). For example, this type of region may represent thin wires in a grid. To process a point (x_k, y_k) in a region with number *NReg*, the first step is to find the index n of the closest logical mesh vertex that is not already marked with the region number. Then, it is simply a matter of setting Mesh(n).x = x_k, Mesh(n).y = y_k, and Mesh(n).RegNo = NReg.

A line segment is specified by start and end coordinates: (x_{ks}, y_{ks}) and (x_{ke}, y_{ke}). Here, the first step is to set the start and end vertices using the point method. Then, it is necessary to shift a series of intervening vertices to make a logically connected path between the ends. The term *logically connected* means that the set of vertices that constitute a line are neighbors on the mesh. Starting at (x_{ks}, y_{ks}), we check mesh neighbors to find which unmarked vertex

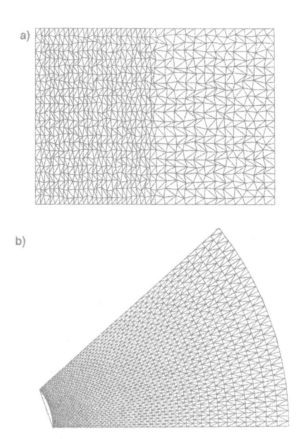

FIGURE 5.5
Options for element geometries in a logical mesh. (a) Amorphous (glass-like) mesh;
(b) cylindrical-section mesh with logarithmic weighting in the radial direction.

is closest to the line segment. The vertex is moved to the closest point on the
line and marked with the current region number. We continue by checking
neighbors of the new point and relocating the one closest to the line. The process ends when the set of neighbors includes the segment end point, (x_{ke}, y_{ke}).
This method is reliable except when the geometry of the logical mesh is inap-
propriate, such as an attempt to fit a shallow angle line to a set of tall triangles. The process is similar for arcs, except that a different method is used to
find the distance from the test point to the curve.

We can develop the mathematics to find the minimum distance between a
test point and a line segment succinctly in vector notation. Figure 5.6 shows
the geometry with three coordinate vectors: the test point $\mathbf{X}_k = (x_k, y_k)$, the
start point of the line $\mathbf{X}_s = (x_s, y_s)$, and the end point $\mathbf{X}_e = (x_e, y_e)$. The line segment is the vector $\mathbf{L} = \mathbf{X}_e - \mathbf{X}_s$. The goal is to find the point on the line closest
to the test point, $\mathbf{X}_c = (x_c, y_c)$. For the derivation we also define vectors that
point from the line ends to the test point, $\mathbf{P}_s = \mathbf{X}_k - \mathbf{X}_s$ and $\mathbf{P}_e = \mathbf{X}_k - \mathbf{X}_e$. If the
line has infinite length, the coordinates of the closest point are given by

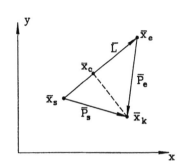

FIGURE 5.6
Minimum distance from a point to a line segment.

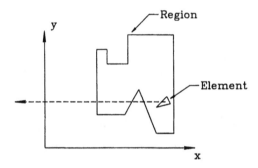

FIGURE 5.7
Method to check whether an element is inside a closed set of boundary segments.

$$X_c = X_s + (P_s \cdot L)L, \qquad (5.11)$$

where the quantity in parenthesis is a dot product. The distance to the line is $|X_c - X_k|$. For a finite length line, we must check whether the point X_c lies on the segment L. The condition is true if the dot products $P_e \cdot L$ and $P_s \cdot L$ have opposite signs. Otherwise, we set the spacing equal to the shortest distance to one of the end points.

For filled regions we must mark the enclosed elements with the current region number. The task appears challenging if the point lies inside a complex re-entrant figure, as shown in Figure 5.7. Fortunately, there is a simple procedure that works with any boundary. The first step to check an element is to calculate the triangle center-of-mass coordinates. Next, we draw a line between this point and an arbitrary point well outside the solution region (dashed line in Figure 5.7). Finally, we check the set of line and arc segments that comprise the closed boundary, counting the number of times that the test line intersects a segment. If the number of intersections is odd, the point is inside the region.

The vertices shifted to boundaries are marked with a flag; their positions should not be changed in any subsequent smoothing operations. All other points are fluid. We can shift them to produce a relaxed mesh with a reduced

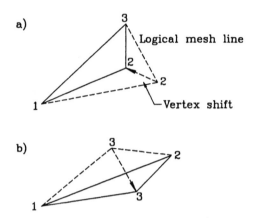

FIGURE 5.8
Checking mesh integrity. (a) Valid triangular element with vertex points ordered according to positive rotation. (b) Vertex points are out of order in an inverted triangle.

probability of acute triangles. A simple procedure is to adjust the positions of fluid points toward the average location of their nearest neighbors. If (x_{av}, y_{av}) represents the average of coordinates of surrounding vertices, then the displacement from the average is

$$x_d = x_{av} - \text{Mesh}(n).x,$$
$$y_d = y_{av} - \text{Mesh}(n).y. \tag{5.12}$$

The coordinates are then corrected according to

$$\text{Mesh}(n).x = \text{Mesh}(n).x + \omega\, x_d,$$
$$\text{Mesh}(n).y = \text{Mesh}(n).y + \omega\, y_d. \tag{5.13}$$

where ω is a smoothing parameter. Typically, the process requires about 10 iterations with $\omega = 0.95$.

The final task is to check the integrity of the mesh. Small vertex shifts (Figure 5.8a) preserve the mesh logic. The problem to avoid is shown in Figure 5.8b. Here, the logical mesh resolution was too coarse and a vertex shift produced an inverted triangle. It is easy to check for triangle inversion by applying the area formula of Equation 2.39. The equation should yield a positive number if the three vertices of the element are ordered with positive rotation. The distorted triangle of Figure 5.8b has an inverted order so that the area equation returns a negative number. Upon detecting a bad element the mesh generator can attempt to correct the distorted triangle or give the user the option to move vertices.

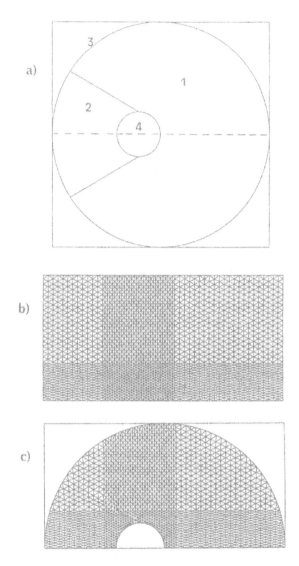

FIGURE 5.9
Offset cylindrical electrode with a dielectric support inside a grounded pipe. (a) Full cross-section of system with region numbers — dashed line shows possible Neumann boundary. (b) Logical mesh for one half of the system with enhanced resolution near the electrode. (c) Mesh with points clamped on boundary segments. (d) Smoothed mesh.

An example will illustrate the mesh generation process. Figure 5.9a shows a cross section of an offset cylindrical electrode attached to a dielectric support inside a grounded pipe. Because of field symmetry, we need simulate only half the system by assigning a specialized Neumann boundary along the dashed line. The first step is to create a logical mesh suitable to the problem (Figure 5.9b), marking all vertices and elements with region number 0.

d)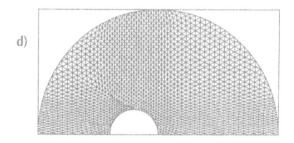

FIGURE 5.9 (continued)

Note the fine resolution near the center to give enhanced accuracy. The problem has four regions that must be entered in order.

> Region 1, the general solution space. This filled dielectric region with $\epsilon_r = 1$ covers the semicircle of Figure 5.9a.
>
> Region 2, the dielectric support. The filled region with $\epsilon_r = 2.78$ overwrites the region number of several of the elements of region 1.
>
> Region 3, the grounded boundary. The open region overwrites the region numbers of vertices on the curved outer surface of region 1.
>
> Region 4, the central electrode with a fixed potential of 50 kV. The filled region overwrites several elements of region 1 and also changes the region numbers of the vertices shared with region 2. In the solution, these vertices will be interpreted as fixed potential points.

It is not necessary to define a region for the lower boundary because it automatically assumes a specialized Neumann boundary condition. Figure 5.9c shows the state of the mesh with clamped points after processing the region boundary segments. The logical mesh vertices and elements outside the solution region are ignored for the field solution. Figure 5.9d illustrates the smoothed mesh that results from the application of Equations 5.12 and 5.13.

5.4   Neumann Boundaries in Resistive Media

We saw in Section 2.8 how unspecified boundaries of finite element solutions satisfy the specialized Neumann condition

$$\frac{\partial \phi}{\partial n} = 0, \tag{5.14}$$

if the solution volume is surrounded by a set of dummy elements with $\rho_i = 0$ and $\epsilon_i = 0$. We can achieve the same effect without wasting storage by adopting the following conventions.

- Elements of the logical mesh that are not part of the solution region have NReg = 0
- The subroutine to calculate the global index returns n = 0 and NReg = 0 for indices outside the logical mesh
- The functions to calculate relative dielectric constant and space-charge density return the value 0.0 for any element with NReg = 0

Equation 5.14 implies that the gradient of potential and the electric field are parallel to the boundary. In this section, we shall discuss electrostatic solutions for current flow in resistive media. Here, the ability to apply the condition along arbitrary curved or slanted boundaries offers a significant advantage.

Consider a region of contiguous solid or liquid conductors. In steady state the net flow of current into or out of a volume equals zero, or

$$\nabla \cdot \mathbf{j} = 0. \tag{5.15}$$

In Equation 5.15, $\mathbf{j}$ is the current density in A/m^2. The vector quantity is parallel to the local electric field and is proportional to the product of the field and the local electrical conductivity,

$$\mathbf{j} = \sigma \, \mathbf{E}. \tag{5.16}$$

The role of Neumann boundaries follows from Equation 5.16. The boundaries represent the surfaces of electrical insulators immersed in the resistive media. Because current cannot penetrate the volume of insulators, it must be parallel to the surface.

Again, we can write the electric field as a gradient of a scalar potential,

$$\mathbf{E} = -\nabla \phi. \tag{5.17}$$

Combining Equations 5.15, 5.16 and 5.17 leads to Laplace's equation

$$\nabla \cdot (\sigma \nabla \phi) = 0, \tag{5.18}$$

where σ may be a function of position. We can solve Equation 5.18 in two dimensions with the finite-element methods of Section 2.7 by replacing ϵ with σ in the expressions for the coupling coefficients (Equation 2.67).

In the resistive model, a good conductor has a high value of σ and an insulator has $\sigma = 0.0$. As an illustration, Figure 5.10 shows equipotential lines and electric field streamlines for current flow through a homogeneous resistive solution around an array of insulating rods. The top and the bottom boundaries are at fixed potential and the left and right boundaries automatically

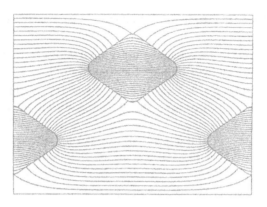

FIGURE 5.10
External and internal curved Neumann boundaries — modeling steady-state current flow around an array of insulating rods in a conducting medium.

assume the specialized Neumann condition. The solution also contains internal Neumann boundaries created by setting $\sigma = 0.0$ inside the circular regions.

5.5   Boundary Value Solutions by Successive Over-Relaxation

To this point, we have discussed several methods to generate difference equations that represent the Poisson equation or Gauss's law in two and three dimensions. They resolve to a coupled set of linear equations of the form

$$\phi_n = \frac{\sum_i W_{in}\phi_{in} + Q_n}{\sum_i W_{in}}. \tag{5.19}$$

Equation 5.19 applies to all vertices that do not have fixed potential. The sum depends on the mesh geometry and extends over the connected vertices according to the indexing scheme.

Methods to solve large sets of linear equations could occupy several texts. We shall concentrate on two approaches that handle most field problems. Chapter 6 discusses direct matrix solutions. In this section, we shall study an approach based on iterative correction called the *method of successive over-relaxation*. Although the procedure may require many cycles, it is usually faster than direct solutions. The entire process can be carried out in the random-access memory of a personal computer without intermediate storage on disk. Furthermore, relaxation routines are easy to understand and to program.

Suppose we set the potential at fixed vertices and start with an initial guess of ϕ at the variable points. The initial values can be far from the final result — one option is to set all unknown values to zero. At each variable point there is a present value of the potential (ϕ_{n0}). We could also calculate a value ϕ'_{n0} determined by the potentials of neighboring points. The relationship to determine ϕ'_{n0} depends on the difference scheme and system geometry. For example, Equation 2.69 applies in two-dimensional finite-element calculations. The difference between the two values is the *residual* (or error) at the point,

$$R_{n0} = \phi'_{n0} - \phi_{n0}. \qquad (5.20)$$

If $R_{n0} = 0$ at every mesh point, then the set of potential values is the correct electrostatic solution. Clearly, this condition does not hold initially.

The guessed solution would improve if we added a correction factor at each point proportional to the error,

$$\phi_{n1} = \phi_{n0} + \omega\, R_{n0}. \qquad (5.21)$$

The quantity ω is a constant on the order of unity. Applying Equations 5.20 and 5.21 sequentially over all vertices would continually improve the solution to a level of acceptable accuracy. For example, we may require that all residuals are less than 10^{-8} times the root-mean-squared potential variations in the solution.

One way to apply Equation 5.21 is to loop through the variable vertices, storing new potential values and then replacing the old ones. The process does not use memory efficiently — we must set aside two large arrays to store the old and new values during a cycle. A better approach is to correct values in place during the loop so that the process uses a mixture of old and new values. The problem is the relaxation process may be skewed depending on the order in which vertices are processed. To avoid this problem and to achieve fastest convergence, the points are corrected in odd-even order. For example, on a regular mesh we divide the mesh points like a checkerboard into odd (red) and even (black) vertices. The new values of odd vertices depend on the old values of the surrounding even vertices, and vice versa. The new values are independent of the loop order if we change all the odd points in a cycle and then correct the even points. The separation is not exact on an arbitrary mesh with more than four neighboring vertices. Nonetheless, it is still helpful to divide points into two uniformly mixed sets and to perform alternate relaxations.

The rate of convergence for the relaxation process depends on the choice of ω. The process is stable for values in the range

$$0 \le \omega \le 2. \qquad (5.22)$$

Small errors grow for values greater than 2, leading to numerical instability. Values of ω greater than unity give faster convergence. This choice, where the adjustments anticipate future corrections, is called *over-relaxation*.

For simple problems on a regular mesh, there is a prescription for the best choice of ω called *Chebyshev acceleration*. For a two-dimensional solution, suppose the solution region is a rectangle with a uniform mesh. The mesh has I divisions of width Δx and J divisions of width Δy. The parameter called the *spectral radius* of the solution is given by

$$\rho = \frac{\cos\left(\frac{\pi}{I}\right) + \left(\frac{\Delta x}{\Delta y}\right)^2 \cos\left(\frac{\pi}{J}\right)}{1 + \left(\frac{\Delta x}{\Delta y}\right)^2}. \tag{5.23}$$

The method is to apply an odd-even relaxation order, changing ω at each half step according to the following equations,

$$\omega_0 = 1,$$

$$\omega_{\frac{1}{2}} = \frac{1}{1 - \rho^2/2}, \tag{5.24}$$

$$\omega_{n+\frac{1}{2}} = \frac{1}{1 - \rho^2 \omega_{\frac{n}{2}}/4}.$$

The relaxation parameter approaches a optimum value after several cycles,

$$\omega_{opt} = \frac{2}{1 + \sqrt{1 - \rho^2}}. \tag{5.25}$$

The value of ω_{opt} approaches 2 for large values of I and J.

The prescription of Equations 5.23, 5.24, and 5.25 is helpful for simple boundary value problems such as electrostatics in a rectangular region with a few internal fixed points. On the other hand, analytic theory gives little guidance for realistic problems with complex boundaries, variable mesh resolution, conformal elements, and nonlinear material characteristics. Generally, relaxation solutions converge for almost all well-posed boundary value problems with a choice of ω between 1.5 and 2.0. In practice, the best way to pick a value of the relaxation parameter for a class of problems is to experiment with different values, noting the effect on stability and the number of cycles to produce a given residual error. Long Neumann boundaries generally increase the required number of cycles in electrostatic problems.

Chapter 5   Exercises

5.1. Derive Equation 5.5 for a coordinate transformation with rotation θ_z about the z-axis.

5.2. Use a spreadsheet to experiment with defining geometric parameters on a regular mesh. Set up a two-dimensional mesh with square elements to represent an area with dimensions 1 m by 1 m with 20 elements per side. The goal is to set element region numbers to 1 for vacuum and 2 for a cylindrical electrode with radius R at position (x_o, y_o). Assume an element is part of the electrode if its center of mass is within a distance R of (x_o, y_o). Make R, x_o, and y_o adjustable parameters and experiment with different values. Enter cell formulas so that the elements assume the correct region number.

5.3. Consider a conformal triangular mesh with *column logic*, starting from the right-angle set illustrated in Figure E5.1.

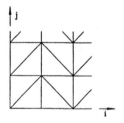

FIGURE E5.1

(a) Give the indices of vertex points connected to point (i,j).

(b) Sketch a procedure to convert the logical mesh to a set of isosceles triangles aligned along y.

5.4. Confirm that it is necessary to store only three coupling constants per vertex point in a conformal triangular mesh. Prove the following relationships applying the index conventions described in Section 5.2.

Even values of j: $W2(i,j) = W5(i-1,j+1)$, $W3(i,j) = W6(i-1,j)$, $W4(i,j) = W1(i-1,j-1)$

Odd values of j: $W2(i,j) = W5(i,j+1)$, $W3(i,j) = W6(i-1,j)$, $W4(i,j) = W1(i,j-1)$

5.5. Develop equations to set the vertex locations of the cylindrical mesh with logarithmic radial weighting shown in Figure 5.5b.

5.6. Find an expression for the minimum distance between a point (x,y) and an arc that extends from (x_1,y_2) to (x_2,y_2) with center at (x_c,y_c).

5.7. The space between coaxial electrodes with radii $R_i = 0.05$ m and $R_o = 0.15$ m is filled with two resistive materials. The material in one half ($\theta = 0$ to π) has volume resistivity $\rho = 500$ Ω-m and the material in the other ($\theta = \pi$ to 2π) has $\rho = 300$ Ω-m. What is the total resistance between cylinders?

5.8. The region between parallel plates separated by distance d is filled with a material with a spatially varying conductivity given by $\sigma = \sigma_o \exp(x/d)$. The potentials on the plates are $\phi(0) = 0$ and $\phi(d) = V_o$. Find the variation of potential between the plates.

5.9. Figure E5.2 shows a uniform medium with conductivity σ between co-axial cylinders. There is a metal bar of unknown potential immersed in the medium. Region numbers for an electrostatic solution are shown. What material properties should be associated with the regions?

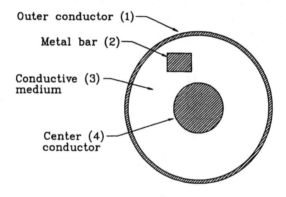

Outer conductor (1)
Metal bar (2)
Conductive (3) medium
Center (4) conductor

FIGURE E5.2

5.10. Twenty-one people have lined up at the bookstore to purchase field solution texts. The first person in line has decided to buy $150 worth while the last person is determined to spend only $50. The people in between are easily swayed by those around them. Although they originally planned to spend $100, they decide to base their decisions on the amount paid by those around them according to a sensible formula. The amount equals 20% of the amount paid by the second person ahead, 40% of the payment of the person immediately ahead, 30% for the person immediately behind, and 10% for the second person behind. The second and penultimate people in line are special cases. The second person gives 60% weighting to the first person, while the penultimate person gives 40% weighting to the last. Apply the method of successive relaxation on a spreadsheet to find out the amounts that people will pay after considerable discussion. Use a relaxation constant of $\omega = 1.00$ and assume that everybody changes their mind at the same time. (After 100 discussions, the eighteenth person in line has decided to pay \$133.55.)

5.11. Parallel plates with spacing $d = 0.05$ m have the applied potentials $\phi(0) = 0$ and $\phi(d) = 2000$ V. A uniform space-charge density $\rho_o = 2.5 \times 10^{-5}$ coulombs/m^3 extends between $x = 0.01$ and 0.04 m. Use a spreadsheet to solve for the potential by successive over-relaxation with $\omega = 1.9$. Confirm that the potential at $x = 0.025$ is 1741V.

5.12. Solve a simple two-dimensional electrostatic problem with a spreadsheet. Use a regular square mesh with $\Delta = 0.05$ m. The solution volume is a vacuum-filled square 1 m on a side with grounded walls. A centered internal electrode at potential 100 V is a rectangle with dimensions 0.3 m in x and 0.2 m in y. Program a spreadsheet to apply the method of successive over-relaxation. Investigate convergence with different values of ω and confirm instability for $\omega > 2$.

6

Matrix Methods for Field Solutions

I would advise you Sir, to study algebra, if you are not already an adept in it: your head would be less muddy, and you will leave off tormenting your neighbours about paper and packthread, while we all live together in a world that is bursting with sin and sorrow.

Samuel Johnson

We have seen that boundary-value field problems can be expressed as a large set of coupled linear equations. The previous chapter discussed the solution of such sets by successive over-relaxation, an iterative process. In this chapter, we shall review solutions using matrix algebra that proceed to an *exact* solution in a single step. The problem is that the first step is a big one. There is considerable work involved in matrix inversions and a large amount of intermediate data to store, even for two-dimensional problems. Nonetheless, matrix methods have advantages in certain circumstances.

- The number of iterations for relaxation solutions may be sensitive to the choice of the relaxation parameter ω. In all but the simplest geometries, the choice of ω is a guess. A direct matrix solution avoids trial-and-error in picking an acceptable relaxation parameter.

- In solutions that involve a search, we may need to determine the best value of a parameter by comparing the properties of field solutions. Sometimes, simultaneous iterations to find both the field solution and the parameter may not converge to the desired solution.

Matrix solutions entail considerably more effort to program. On the positive side, it is only necessary to make the effort once. A well-written solution unit can be applied to any boundary value problem. The goal in this chapter is to cover the necessary mathematics for matrix solutions of the finite-element equations for triangular meshes. In the process, we shall develop methods for linear equations with broad applicability. For example, we shall use matrix inversion in the least-squares fitting procedures of Section 7.2.

Section 6.1 reviews Gauss-Jordan elimination with pivoting, a standard method to invert matrices and to solve moderate sets of linear equations. With the speed and memory capacity of personal computers, it is practical to apply the method to about 500 equations. Many physical problems give rise to matrices where most of the elements equal zero. For example, the matrix that represents the difference equations for the one-dimensional Poisson equation has non-zero elements only on the matrix diagonal and adjacent rows. This matrix form is called *tridiagonal*. Section 6.2 covers special fast methods to solve for tridiagonal matrices. Section 6.3 illustrates direct matrix solutions for one-dimensional electrostatics with space-charge. The example illustrates some important points: setting up matrices, automatic implementation of Neumann boundary conditions, and handling constant potential points. We proceed to two-dimensional solutions in Section 6.4. The full coefficient matrix may be very large — a 100×100 mesh gives a matrix with about 10^8 elements. On the other hand, most of the elements are zero in electrostatic problems because of nearest-neighbor coupling in the difference equations. The elements in the coefficient matrix for the two-dimensional finite-element difference equations are clumped in blocks near the diagonal. For this case, inversion of the matrix is within the capability of a personal computer. Section 6.5 describes solution techniques for tridiagonal block matrices. Critical issues are optimizing the use of random-access memory and minimizing scratch data storage on disk.

6.1   Gauss-Jordan Elimination

Consider a set of linear equations

$$a_{11}\phi_1 + a_{12}\phi_2 + \ldots + a_{1n}\phi_n = q_1,$$

$$a_{21}\phi_1 + a_{22}\phi_{21} + \ldots + a_{2n}\phi_n = q_2, \qquad (6.1)$$

$$a_{n1}\phi_1 + a_{n2}\phi_2 + \ldots + a_{nn}\phi_n = q_n.$$

The quantities a_{ij} are a known set of coefficients and the quantities q_i are known source terms. The goal is to find the unknown quantities ϕ_i. We can rewrite the equations in matrix notation as

$$
\begin{bmatrix}
a_{11} & a_{12} & a_{13} & \cdots & a_{1n} \\
a_{21} & a_{22} & a_{23} & \cdots & a_{2n} \\
& & L & & \\
a_{n1} & a_{n2} & a_{n3} & \cdots & a_{nn}
\end{bmatrix}
\begin{bmatrix}
\phi_1 \\
\phi_2 \\
L \\
\phi_n
\end{bmatrix}
=
\begin{bmatrix}
q_1 \\
q_2 \\
L \\
q_n
\end{bmatrix}. \qquad (6.2)
$$

or symbolically

$$A \, \phi = Q .$$ (6.3)

Here, the quantity A is a square matrix of dimension $n \times n$ while ϕ and Q are vectors with dimension n. The goal of Gauss-Jordan elimination is to perform operations on Equation 6.2 that preserve the equality and transform the matrix A to the identity matrix

$$I = \begin{bmatrix} 1 & 0 & 0 & \cdots & 0 \\ 0 & 1 & 0 & \cdots & 0 \\ & & \cdots & & \\ 0 & 0 & 0 & \cdots & 1 \end{bmatrix}.$$ (6.4)

An inspection of Equations 6.1, 6.2, and 6.3 imply the following are valid operations.

1. Interchange two rows of A and the corresponding rows of Q.
2. Multiply a row of A and the corresponding row of Q by a constant.
3. Replace a row of A with a linear combination of itself and any other row with the same replacement performed on Q.
4. Interchange two columns of A with a corresponding interchange of rows of ϕ and Q.

Operation 1 simply changes the order in which the equations are written, while operation 4 changes the order in which the variables are written.

We start by multiplying the first row by $1/a_{11}$. This leaves the value 1 at a_{11}, modified coefficients in the other columns of the first row of A, and the value q_1/a_{11} as the first element in Q. Next we replace the second row by the combination of itself minus the first row multiplied by a_{21}. This step leaves zero at a_{21} and modified values in the other columns and q_2. If we proceed downward this way, the first column entries of all rows except a_{11} are reduced to zero. Next, multiply the second row by $1/a_{22}$. The second row has 0 in column 1, 1 in column 2, and modified values for the other elements and q_2. We can now subtract appropriate multiples of row 2 from the other rows to reduce the element in column 2 to 0. The zeroes in column 1 are unaffected because the modified value of a_{21} equals 0. Clearly, continuing the process reduces all elements of A to 0 except for 1s on the diagonal. At the end, $A = I$. At this point, the modified values of q_i correspond to the desired values of ϕ_i.

The diagonal value in the currently processed row by which other values will be divided is called the *pivot element* because it plays a pivotal role. Clearly, the method fails if one of the pivot values equals zero. Even if the pivot values are all non-zero, the method is numerically inaccurate if the pivot value is much smaller than the other values in its row. We can avoid these problems by the process of *pivoting*. Here, we use operations 1 and 4 to move a desirable element to the pivot position. The best element has the highest magnitude. To avoid destroying the portion of the identity matrix already processed, we pick the element from rows below the current row and columns to the right of the current column. With no pivoting, the Gauss-Jordan elimination process is numerically unstable. The method is often sufficiently stable with *partial pivoting*. Here, only row interchanges are used (operation 1). *Full pivoting* uses both row and column interchanges. Column interchanges are somewhat more difficult to program because they change the order of the solution vector. These changes must be recorded and unscrambled after the transformation. Nonetheless, the extra effort is worthwhile. We shall see in Section 6.6 that the tridiagonal block reduction method used for two-dimensional electrostatic solutions is unstable with Gauss-Jordan elimination with partial pivoting.

An advantage of the Gauss-Jordan method is that several sets of linear equations can be solved simultaneously for a given set of coefficients. Suppose we have an $n \times n$ matrix of coefficients a_{ij}. Consider the special case where we want to find n sets of values for q_i that give n sets of solutions ϕ_i. We can write the sets of source and solution vectors as square matrices, where the columns of ϕ and Q correspond to different solutions.

$$
\begin{bmatrix}
a_{11} & a_{12} & a_{13} & \cdots & a_{1n} \\
a_{21} & a_{22} & a_{23} & \cdots & a_{2n} \\
& & \cdots & & \\
a_{n1} & a_{n2} & a_{n3} & \cdots & a_{nn}
\end{bmatrix}
\begin{bmatrix}
\phi_{11} & \phi_{12} & \phi_{13} & \cdots & \phi_{1n} \\
\phi_{21} & \phi_{22} & \phi_{23} & \cdots & \phi_{2n} \\
& & \cdots & & \\
\phi_{n1} & \phi_{n2} & \phi_{n3} & \cdots & \phi_{nn}
\end{bmatrix}
=
\begin{bmatrix}
q_{11} & q_{12} & q_{13} & \cdots & q_{1n} \\
q_{21} & q_{22} & q_{23} & \cdots & q_{2n} \\
& & \cdots & & \\
q_{n1} & q_{n2} & q_{n3} & \cdots & q_{nn}
\end{bmatrix}. \quad (6.5)
$$

Applying the rule of matrix multiplication, it is easy to see that Equation 6.5 is equivalent to n instances of Equation 6.2. We can write Equation 6.5 symbolically as

$$ A \, \phi = Q. \quad (6.6) $$

In this case, ϕ and Q are square matrices.

The extension of the Gauss-Jordan procedure to handle Equation 6.5 or 6.6 is to apply operations 1, 2, and 3 to all elements in a row of A and Q. The equation set is unchanged when two columns of A are interchanged as long as the corresponding rows of both ϕ and Q are interchanged. The procedure to reduce the matrix A to the identity matrix is equivalent to multiplying both sides of Equation 6.6 by A^{-1},

$$ A^{-1} A \, \phi = \phi = A^{-1} Q. \quad (6.7) $$

The transformed Q matrix equals the product of the inverse of A times the original Q matrix. For the special case where $Q = I$, the Q matrix contains A^{-1} after the Gauss-Jordan elimination.

The function listed in Table 6.1 performs a Gauss-Jordan elimination with full pivoting on a matrix A for a source matrix Q. Both matrices have dimension NMat × NMat. The procedure is performed in place with minimal storage. At the end of the procedure, the original matrix A has been replaced by A^{-1} and Q by the product $A^{-1}Q$.

6.2   Solving Tridiagonal Matrices

Tridiagonal matrices have non-zero components only on the diagonal and the adjacent columns. Because most of the elements are zero, tridiagonal matrices are called *sparse*. Solutions of sparse matrices require far fewer operations than the Gauss-Jordan elimination of a non-sparse matrix of the same dimensions. This section reviews the *LU decomposition* technique to invert tridiagonal matrices. The method is applied in Section 6.3 as an alternative to back-substitution to solve the one-dimensional Poisson equation. It is extended to large trigonal block matrices in Section 6.5 for application to general two-dimensional solutions of the Poisson equation.

The general tridiagonal matrix problem has the form

$$\begin{bmatrix} b_1 & c_1 & 0 & 0 & 0 & \cdots & 0 & 0 & 0 \\ a_2 & b_2 & c_2 & 0 & 0 & \cdots & 0 & 0 & 0 \\ 0 & a_3 & b_3 & c_3 & 0 & \cdots & 0 & 0 & 0 \\ 0 & 0 & a_4 & b_4 & c_4 & \cdots & 0 & 0 & 0 \\ & & & & & & & & \\ 0 & 0 & 0 & 0 & 0 & \cdots & 0 & a_N & b_N \end{bmatrix} \begin{bmatrix} \phi_1 \\ \phi_2 \\ \phi_3 \\ \phi_4 \\ \\ \phi_N \end{bmatrix} = \begin{bmatrix} q_1 \\ q_2 \\ q_3 \\ q_4 \\ \\ q_N \end{bmatrix} . \tag{6.8}$$

or

$$A \phi = Q . \tag{6.9}$$

We seek to express the matrix A as the product of two special matrices L and U. In this case Equation 6.9 becomes,

$$L U \phi = Q . \tag{6.10}$$

or

$$U \phi = \psi . \tag{6.11}$$

TABLE 6.1

FORTRAN Function for Gauss-Jordan Elimination

```
      LOGICAL FUNCTION GJInvert(A,Q,NMat)
C
C--- Gauss-Jordan inversion with full pivoting
C    A is a square matrix of dimensions NMat × NMat.
C    Q is a square matrix of NMat column vectors
C    of dimension NMat. On output, A is replaced by
C    Inv(A) and Q by Inv(A)*Q. The integer arrays
C    IndexCol,IndexRow, and IPivot are used to keep
C    track of pivot interchanges. Returns TRUE if
C    the elimination was successful. Routine adapted
C    from Numerical Recipes: The Art of Scientific
C    Computing published by Cambridge University Press,
C    used by permission (Copyright 1986, 1992, Numerical
C    Recipes Software)
C
      DOUBLE PRECISION A,Q
C
      INTEGER NMatMax
      PARAMETER (NMatMax = 320)
      DIMENSION A(1:NMatMax,1:NMatMax)
      DIMENSION Q(1:NMatMax,1:NMatMax)
      INTEGER I,ICol,IRow,J,K,L,LL,NMat
      INTEGER IndexCol,IndexRow,IPivot
      DIMENSION IndexCol(1:NMatMax)
      DIMENSION IndexRow(1:NMatMax)
      DIMENSION IPivot(1:NMatMax)
      DOUBLE PRECISION BigNum,Dummy,PivotInv
C
C--- Initialize
      GJInvert = .TRUE.
      DO J = 1,NMat
         IPivot(J) = 0
      END DO
C--- Main loop over matrix columns
      DO I = 1,NMat
         BigNum = 0.0
C--- Search rows and columns for the pivot element
      DO J = 1,NMat
         IF(IPivot(J).NE.1) THEN
            DO K = 1,NMat
               IF (IPivot(K).EQ.0) THEN
                  IF (DABS(A(J,K)).GE.BigNum) THEN
                     BigNum = DABS(A(J,K))
                     IRow = J
                     ICol = K
                  ENDIF
C--- Singular matrix
               ELSE IF (IPivot(K).GT.1.0) THEN
                  GOTO 1000
               ENDIF
            END DO
         ENDIF
      END DO
      IPivot(ICol) = IPivot(ICol)+1
C--- Interchange rows and/or relabel columns to bring the
```

TABLE 6.1 (continued)

FORTRAN Function for Gauss-Jordan Elimination

```fortran
C     element to the pivot position
        IF (IRow.NE.ICol) THEN
          DO L = 1,NMat
            Dummy = A(IRow,L)
            A(IRow,L) = A(ICol,L)
            A(ICol,L) = Dummy
          END DO
          DO L = 1,NMat
            Dummy = Q(IRow,L)
            Q(IRow,L) = Q(ICol,L)
            Q(ICol,L) = Dummy
          END DO
        ENDIF
C--- Divide the pivot row by the pivot element at IRow and ICol
        IndexRow(i) = IRow
        IndexCol(i) = ICol
C--- Singular matrix, return FALSE
        IF (A(ICol,ICol).EQ.0.0) GO TO 1000
        PivotInv = 1.0/A(ICol,ICol)
        A(ICol,ICol) = 1.0
        DO L = 1,NMat
          A(ICol,L) = A(ICol,L)*PivotInv
        END DO
        DO L = 1,NMat
          Q(ICol,L) = Q(ICol,L)*PivotInv
        END DO
C--- Reduce the other rows
        DO LL = 1,NMat
          IF(LL.NE.ICol) THEN
            Dummy = A(LL,ICol)
            A(LL,ICol) = 0.0
            DO L = 1,NMat
              A(LL,L) = A(LL,L)-A(ICol,L)*Dummy
            END DO
            DO L = 1,NMat
              Q(LL,L) = Q(LL,L)-Q(ICol,L)*Dummy
            END DO
          ENDIF
        END DO
C--- Reorder the solution to remove the effect of
C    column interchanges
        DO L = NMat,1,-1
          IF(IndexRow(L).NE.IndexCol(L)) THEN
            DO K = 1,NMat
              Dummy = A(K,IndexRow(L))
              A(K,IndexRow(L)) = A(K,IndexCol(L))
              A(K,IndexCol(L)) = Dummy
            END DO
          ENDIF
        END DO
        RETURN
1000  GJInvert = .FALSE.
        RETURN
        END
```

and

$$L \psi = Q . \tag{6.12}$$

The L (*lower*) matrix has components on and below the diagonal, while the U (*upper*) matrix has components on the diagonal and above. We assume forms for the matrices and write the product as

$$
\begin{bmatrix}
L_1 & 0 & 0 & 0 & 0 & \cdots & 0 & 0 & 0 \\
a_2 & L_2 & 0 & 0 & 0 & \cdots & 0 & 0 & 0 \\
0 & a_3 & L_3 & 0 & 0 & \cdots & 0 & 0 & 0 \\
0 & 0 & a_4 & L_4 & 0 & \cdots & 0 & 0 & 0 \\
& & & & & & & & \\
0 & 0 & 0 & 0 & 0 & \cdots & 0 & a_N & L_N
\end{bmatrix}
\begin{bmatrix}
1 & U_1 & 0 & 0 & 0 & \cdots & 0 & 0 & 0 \\
0 & 1 & U_2 & 0 & 0 & \cdots & 0 & 0 & 0 \\
0 & 0 & 1 & U_3 & 0 & \cdots & 0 & 0 & 0 \\
0 & 0 & 0 & 1 & U_4 & \cdots & 0 & 0 & 0 \\
& & & & & & & & \\
0 & 0 & 0 & 0 & 0 & \cdots & 0 & 0 & 1
\end{bmatrix} . \tag{6.13}
$$

Carrying out the matrix multiplication in Equation 6.13 and comparing terms to the original matrix, we can find the components U_i and L_i in terms of a_i, b_i, and c_i.

$$L_1 = b_1, \quad U_1 = c_1/L_1,$$

$$L_2 = b_2 - a_2 U_1, \quad U_2 = c_2/L_2, \tag{6.14}$$

$$\cdots$$

$$L_i = b_i - a_i U_{i-1}, \quad U_i = c_i/L_i.$$

Evaluating the equations in the order shown gives necessary quantities as they are needed.

We can see the motivation for the forms of the L and U matrices by writing Equations 6.11 and 6.12 in component form. First, consider the equation for the components of the ψ vector,

$$
\begin{bmatrix}
L_1 & 0 & 0 & 0 & 0 & \cdots & 0 & 0 & 0 \\
a_2 & L_2 & 0 & 0 & 0 & \cdots & 0 & 0 & 0 \\
0 & a_3 & L_3 & 0 & 0 & \cdots & 0 & 0 & 0 \\
0 & 0 & a_4 & L_4 & 0 & \cdots & 0 & 0 & 0 \\
& & & & & & & & \\
0 & 0 & 0 & 0 & 0 & \cdots & 0 & a_N & L_N
\end{bmatrix}
\begin{bmatrix}
\psi_1 \\
\psi_2 \\
\psi_3 \\
\psi_4 \\
\\
\psi_N
\end{bmatrix}
=
\begin{bmatrix}
q_1 \\
q_2 \\
q_3 \\
q_4 \\
\\
q_N
\end{bmatrix} . \tag{6.15}
$$

The values of ψ can easily be determined. Starting at the first row, we see that

$$\psi_1 = q_1/L_1,$$

$$\psi_2 = \left(q_2 - a_2\psi_1\right)/L_2, \qquad (6.16)$$

$$\psi_i = \left(q_i - a_i\psi_{i-1}\right)/L_i.$$

Ordering the equations from the top to bottom of the matrix gives quantities as they are needed. Given the values of ψ, we can find the values of the ϕ matrix from Equation 6.12,

$$
\begin{bmatrix}
1 & U_1 & 0 & 0 & 0 & \cdots & 0 & 0 & 0 \\
0 & 1 & U_2 & 0 & 0 & \cdots & 0 & 0 & 0 \\
0 & 0 & 1 & U_3 & 0 & \cdots & 0 & 0 & 0 \\
0 & 0 & 0 & 1 & U_4 & \cdots & 0 & 0 & 0 \\
& & & & & & & & \\
0 & 0 & 0 & 0 & 0 & \cdots & 0 & 0 & 1
\end{bmatrix}
\begin{bmatrix}
\phi_1 \\ \phi_2 \\ \phi_3 \\ \phi_4 \\ \\ \phi_N
\end{bmatrix}
=
\begin{bmatrix}
\Psi_1 \\ \Psi_2 \\ \Psi_3 \\ \Psi_4 \\ \\ \Psi_N
\end{bmatrix}. \qquad (6.17)
$$

Starting at the bottom, we can identify the components of ϕ,

$$\phi_N = \psi_N,$$

$$\phi_{N-1} = \psi_{N-1} - U_{N-1}\psi_N, \qquad (6.18)$$

$$\phi_i = \psi_i - U_i\psi_{i+1}.$$

The method, involving upward and downward scans, is reminiscent of the method of back-substitution (Section 4.4). It is fortuitous that the procedure is stable for practical cases without the need for pivoting. To summarize, the first step is to find N values of L_i and $(N-1)$ values of U_i using Equations 6.14. Then the relationships of Equation 6.16 are applied to find the intermediate values ψ_i. Finally, the desired solution values ϕ_i are obtained from Equations 6.18. Table 6.2 shows a routine to implement a tridiagonal matrix inversion.

6.3 Matrix Solutions for One-Dimensional Electrostatics

Section 4.2 showed that the finite-element difference representation of Gauss' law in one dimension was

$$W_{i-1}\phi_{i-1} - \left(W_{i-1} + W_{i+1}\right)\phi_i + W_{i+1}\,\phi_{i+1} = \frac{\rho_{i+1}\left(x_{i+1} - x_i\right) + \rho_i\left(x_i - x_{i-1}\right)}{2\epsilon_o}, \quad (6.19)$$

where

$$W_{i+1} = \frac{\epsilon_{i+1}}{x_{i+1} - x_i}, \quad W_{i-1} = \frac{\epsilon_i}{x_i - x_{i-1}}. \quad (6.20)$$

Equation 8 represents a one-dimensional electrostatic solution if we make the association

$$a_i = W_{i-1},$$

$$b_i = -\left(W_{i-1} + W_{i+1}\right),$$

$$c_i = W_{i+1}, \quad (6.21)$$

$$q_i = \frac{\rho_{i+1}\left(x_{i+1} - x_i\right) + \rho_i\left(x_i - x_{i-1}\right)}{2\epsilon_o}.$$

TABLE 6.2

FORTRAN Function for Tridiagonal Matrix Inversion

```
 LOGICAL FUNCTION TriDiag(A,B)
C--- Tridiagonal inversion where A is a tridiagonal matrix of dimension
C    NMat x NMat and B is a square matrix consisting of NMat column
C    vectors of length NMat. No pivoting
C    On output, A is replaced by Inv(A) and B by Inv(A)*B
C
     INCLUDE 'BLOCK.BLK'
     DOUBLE PRECISION A,B
     DIMENSION A(1:NMatMax,1:NMatMax)
     DIMENSION B(1:NMatMax,1:NMatMax)
     DOUBLE PRECISION Alpha,Beta,Gamma
     DIMENSION Alpha(1:NMatMax)
     DIMENSION Beta(1:NMatMax)
     DIMENSION Gamma(1:NMatMax)
     DOUBLE PRECISION InvFactor
     INTEGER*4 I,J
     TriDiag = .TRUE.
C--- Store coefficients and replace A by the unit matrix
     Beta(1) = A(1,1)
     Gamma(1) = A(1,2)
     DO I=2,NMat-1
        Alpha(I) = A(I,I-1)
        Beta(I) = A(I,I)
        Gamma(I) = A(I,I+1)
     END DO
```

TABLE 6.2 (continued)

FORTRAN Function for Tridiagonal Matrix Inversion

```
      Alpha(NMat) = A(NMat,NMat-1)
      Beta(NMat) = A(NMat,NMat)
      DO I=1,NMat
        DO J=1,NMat
          IF (I.EQ.J) THEN
            A(I,J) = 1.0
          ELSE
            A(I,J) = 0.0
          ENDIF
        END DO
      END DO
C--- Convert to U format
C--- Row 1
      IF (Beta(1).EQ.0.0) GO TO 1000
      InvFactor = 1.0/Beta(1)
      Beta(1) = 1.0
      Gamma(1) = InvFactor*Gamma(1)
      DO J=1,NMat
        A(1,J) = InvFactor*A(1,J)
        B(1,J) = InvFactor*B(1,J)
      END DO
C--- Rows 2 through N-1
      DO I=2,NMat-1
        Beta(I) = Beta(I) - Alpha(I)*Gamma(I-1)
        IF (Beta(I).EQ.0.0) GO TO 1000
        InvFactor = 1/Beta(I)
        Gamma(I) = InvFactor*Gamma(I)
        DO J=1,NMat
          A(I,J) = InvFactor*(A(I,J) - Alpha(I)*A(I-1,J))
          B(I,J) = InvFactor*(B(I,J) - Alpha(I)*B(I-1,J))
        END DO
      END DO
C--- Last row
      Beta(NMat) = Beta(NMat) - Alpha(NMat)*Gamma(I-1)
      IF (Beta(NMat).EQ.0.0) GO TO 1000
      InvFactor = 1/Beta(NMat)
      DO J=1,NMat
        A(NMat,J) = InvFactor*(A(NMat,J) - Alpha(NMat)*A(NMat-1,J))
        B(NMat,J) = InvFactor*(B(NMat,J) - Alpha(NMat)*B(NMat-1,J))
      END DO
C--- Backsubstitution, leave bottom row
      DO I=NMat-1,1,-1
        DO J=1,NMat
          A(I,J) = A(I,J) - Gamma(I)*A(I+1,J)
          B(I,J) = B(I,J) - Gamma(I)*B(I+1,J)
        END DO
      END DO
      RETURN
1000  TriDiag = .FALSE.
      RETURN
      END
```

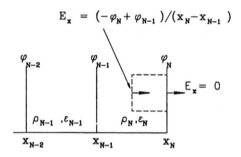

FIGURE 6.1
One-dimensional electrostatic calculation with Neumann condition on the right-hand boundary. Dashed line shows Gauss' law integration volume at vertex x_N.

Equation 21 applies to variable internal points. Points on the boundary require special consideration. We will discuss two possibilities: (1) the boundary has fixed potential value or (2) it represents a symmetry axis where $d\phi/dx = 0$ (Neumann boundary). The constant potential condition is relatively easy to represent. Suppose that the potential at an internal or boundary point I equals V_o, independent of the other potential values. We set row I of the matrix A equal to a corresponding row of the identity matrix ($a_I = 0$, $b_I = 1$, and $c_I = 0$) and replace q_I with V_o. This procedure holds for boundary points and constant potential internal points. Although internal constant potential points have little purpose in one-dimensional calculations, they play a significant role in the two-dimensional solutions of Section 6.4.

To understand the Neumann condition, consider Gauss' law applied at the boundary position x_N (Figure 6.1). In contrast to internal points, the integration extends only over the region shown to the left. Because of the Neumann condition, there is no contribution to the surface integral of electric field on the face at x_N. Carrying out the volume integral of charge and the surface integral of field gives an equation for the potential

$$\frac{\epsilon_N \left(-\phi_N + \phi_{N-1}\right)}{x_N - x_{N-1}} = \frac{\rho\left(x_N - x_{N-1}\right)}{2\epsilon_o}. \tag{6.22}$$

Using the definitions of Equation 6.21, we can write Equation 6.22 as

$$W_{N-1}\phi_{N-1} - W_{N-1}\phi_N = \frac{\rho_N\left(x_N - x_{N-1}\right)}{2\epsilon_o}. \tag{6.23}$$

Equation 6.23 has an important implication. If we simply ignore contributions of field and charge outside the boundary while setting up the matrix, the boundary automatically assumes the Neumann condition. This again illustrates the property of finite-element solutions discussed in Section 5.4.

To illustrate the implementation of boundary conditions, consider a one-dimensional solution with the following properties: (1) the mesh is uniform with spacing Δx, (2) the region has a uniform charge density ρ_o, (3) the right-hand boundary has the fixed potential V_o, and (4) the specialized Neumann condition holds on the left-hand boundary. The coefficient matrix and source vector have the following forms.

$$
\begin{bmatrix}
-W_1 & W_1 & 0 & 0 & \cdots & 0 & 0 \\
W_1 & -(W_1 + W_2) & W_2 & 0 & \cdots & 0 & 0 \\
0 & W_2 & -(W_2 + W_3) & W_3 & \cdots & 0 & 0 \\
 & & & & & & \\
0 & 0 & 0 & 0 & \cdots & 0 & 1
\end{bmatrix}
\begin{bmatrix}
\phi_1 \\ \phi_2 \\ \phi_3 \\ \\ \phi_N
\end{bmatrix}
=
\begin{bmatrix}
\rho_o \Delta x/2\epsilon_o \\ \rho_o \Delta x/\epsilon_o \\ \rho_o \Delta x/\epsilon_o \\ \\ V_o
\end{bmatrix}
. \quad (6.24)
$$

A benchmark test is useful to check the validity of the tridiagonal matrix method and the boundary representation. Consider a region that extends from $-x_o$ to $+x_o$ with uniform space-charge density ρ_o. If the boundaries are grounded, Poisson's equation has the solution

$$
\phi(x) = -\frac{\rho_o \, x_o^2}{2\epsilon_o}\left[1 - \left(\frac{x}{x_o}\right)^2\right]. \quad (6.25)
$$

For the choice $x_o = 0.05$ m and $\rho/\epsilon_o = 10^4$ the potential at $x = 0$ is 12.5 V. The first column of Table 6.3 shows numerical results of a matrix solution with a division of the region into 10 elements. The entries in rows 1 and 11 of the coefficient matrix were set to the identity values and $q_1 = q_{11} = 0$. The potential has a parabolic variation and reaches the expected value at $x = 0$. The second column corresponds to a Neumann boundary on the right-hand side implemented by changing row 11 of the coefficient matrix and q_{11} as shown in Equation 6.24. The symmetry condition is equivalent to doubling the size of the solution slab. Equation 6.25 implies that the maximum potential should rise to 50 V. Finally, to check the implementation of internal constant potential points, the third column of Table 6.3 shows a solution with ground points at $x = -0.05$, $x = 0.0$, and $x = 0.05$, effectively halving the width of the solution region. The predicted potential of 3.125 V occurs at $x = \pm 0.025$ m. Within the limit of the mesh resolution, the numerical procedure gives the correct result.

6.4 Matrices for Two-Dimensional Finite-Element Solutions

Two-dimensional difference representations involve many more equations than one-dimensional solutions and lead to correspondingly larger coefficient

TABLE 6.3

One-Dimensional Electrostatic Solution by Matrix Inversion Vacuum with
Uniform Space-Charge Density

Electrostatic Potential (V)		
Grounded Boundaries	**Neumann Boundary — Right**	**Grounded Midpoint**
0.00000	0.00000	0.00000
4.50000	9.50000	2.00000
8.00000	18.00000	3.00000
10.50000	25.50000	3.00000
12.00000	31.99999	2.00000
12.50000	37.49999	0.00000
12.00000	41.99999	2.00000
10.50000	45.49999	3.00000
8.00000	47.99999	3.00000
4.50000	49.49999	2.00000
0.00000	49.99999	0.00000

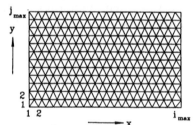

FIGURE 6.2
Indices for a conformal triangular mesh.

matrices. Fortunately, most of the elements are zero so we can use special
matrix inversion methods to make the procedure practical (Section 6.5). In
this section, we shall derive the matrix representation of Gauss' law on conformal triangular meshes. Figure 6.2 shows indices of a mesh in the x–y plane
following the conventions of Section 5.2. The logical mesh covers a rectangular area — elements are stretched and excluded to match arbitrary solution
boundaries. All internal vertices have six surrounding triangles and vertices.
The index i along the x direction varies from 1 to I and the j index along y varies from 1 to J. As before, we shall store values of the vertex coordinates and
electrostatic potential in a one-dimensional array with index n. We adopt the
convention that n increments most rapidly along rows as shown. The index
n is related to i and j by

$$n = j \ I + i \tag{6.26}$$

In this case, the solution vector Φ consists of $I \times J$ entries that we shall arrange
in order of increasing n. For convenience, the large vector Φ can be written as
a collection of J subvectors

$$\Phi = \begin{bmatrix} \Phi_1 \\ \Phi_2 \\ \cdots \\ \Phi_J \end{bmatrix}. \qquad (6.27)$$

Each of the subvectors represents a row of the mesh and has the entries

$$\Phi_j = \begin{bmatrix} \phi_{1j} \\ \phi_{2j} \\ \cdots \\ \phi_{Ij} \end{bmatrix}. \qquad (6.28)$$

The source quantity q_i equals one third of the sum of space-charge in the six triangles surrounding the point i divided by ϵ_o. The space-charge is taken to be zero in triangles outside the solution boundaries. As with Φ, we can represent the source vector Q in terms of a set of row subvectors,

$$Q = \begin{bmatrix} Q_1 \\ Q_2 \\ \cdots \\ Q_J \end{bmatrix}. \qquad (6.29)$$

Using the notation that $W_1(i,j)$ denotes the coupling coefficient to point 1 relative to the point with indices (i,j), the difference equation at the vertex for an odd value of j is

$$-\sum_{k=1}^{6} W_k \ \phi(i,j) + W_1(i,j)\phi(i+1,j+1) + W_2(i,j)\phi(i,j+1) + W_3(i,j)\phi(i-1,j)$$

$$+ W_4(i,j)\phi(i,j-1) + W_5(i,j)\phi(i+1,j-1) + W_6(i,j)\phi(i+1,j) = q(i,j). \qquad (6.30)$$

On an even row, Equation 6.30 becomes

$$-\sum_{k=1}^{6} W_k \ \phi(i,j) + W_1(i,j)\phi(i,j+1) + W_2(i,j)\phi(i-1,j+1) + W_3(i,j)\phi(i-1,j)$$

$$+ W_4(i,j)\phi(i-1,j-1) + W_5(i,j)\phi(i,j-1) + W_6(i,j)\phi(i+1,j) = q(i,j). \qquad (6.31)$$

Consider the complete coefficient matrix. Because there is only coupling between adjacent rows, the matrix equation has the following form.

$$
\begin{bmatrix}
\mathbf{B_1} & \mathbf{C_1} & 0 & 0 & \cdots & 0 & 0 \\
\mathbf{A_2} & \mathbf{B_2} & \mathbf{C_2} & 0 & \cdots & 0 & 0 \\
0 & \mathbf{A_3} & \mathbf{B_3} & \mathbf{C_3} & \cdots & 0 & 0 \\
& & & & \\
0 & 0 & 0 & 0 & \cdots & \mathbf{A_J} & \mathbf{B_J}
\end{bmatrix}
\begin{bmatrix}
\Phi_1 \\ \Phi_2 \\ \Phi_3 \\ \cdots \\ \Phi_J
\end{bmatrix}
=
\begin{bmatrix}
Q_1 \\ Q_2 \\ Q_3 \\ \cdots \\ Q_J
\end{bmatrix}.
\qquad (6.32)
$$

As before, the quantity Φ_1 represents the I values of potential along row 1. The quantity $\mathbf{B_1}$ is a matrix of dimensions I^2 that represents couplings in Equations 6.30 or 6.31 to potential values on the same row. The quantities $\mathbf{A_1}$ and $\mathbf{C_1}$ contain terms that relate the potential to values on rows below and above. The submatrices in Equation 6.32 are called *blocks*. The arrangement of the submatrices in the equation resulting from nearest row coupling is called *block tridiagonal form*. In Section 6.5, we shall see how to apply the methods developed in Section 6.2 to large block matrices.

We can use Equations 6.30 and 6.31 to fill in the values of the blocks in Equation 6.32 for an internal point of a triangular mesh. The **B** matrix for row j is

$$
\mathbf{B_1} =
\begin{vmatrix}
-W_S(1,j) & W_6(1,j) & 0 & 0 & \cdots & 0 & 0 \\
W_3(2,j) & -W_S(2,j) & W_6(2,j) & 0 & \cdots & 0 & 0 \\
0 & W_3(3,j) & -W_S(3,j) & W_6(3,j) & \cdots & 0 & 0 \\
& & & & & & \\
0 & 0 & 0 & 0 & \cdots & W_3(I,j) & W_6(I,j)
\end{vmatrix}. \qquad (6.33)
$$

Equation 6.33 uses the notation

$$
W_S(i,j) = \sum_{k=1}^{6} W_k(i,j). \qquad (6.34)
$$

The **A** and **C** matrices have two possible forms, depending on whether 1 is odd or even. The forms for coupling to rows below and above an odd row are

$$
\mathbf{A_1} =
\begin{vmatrix}
W_4(1,j) & W_5(1,j) & 0 & 0 & \cdots & 0 & 0 \\
0 & W_4(2,j) & W_5(2,j) & 0 & \cdots & 0 & 0 \\
0 & 0 & W_4(3,j) & W_5(3,j) & \cdots & 0 & 0 \\
& & & & & & \\
0 & 0 & 0 & 0 & \cdots & 0 & W_4(I,j)
\end{vmatrix}, \qquad (6.35)
$$

and

$$
C_1 = \begin{vmatrix}
W_2(1,j) & W_1(1,j) & 0 & 0 & \cdots & 0 & 0 \\
0 & W_2(2,j) & W_1(2,j) & 0 & \cdots & 0 & 0 \\
0 & 0 & W_2(3,j) & W_1(3,j) & \cdots & 0 & 0 \\
& & & & & & \\
0 & 0 & 0 & 0 & \cdots & 0 & W_2(I,j)
\end{vmatrix}. \quad (6.36)
$$

The forms for an even row are

$$
A_1 = \begin{vmatrix}
W_5(1,j) & 0 & 0 & 0 & \cdots & 0 & 0 \\
W_4(2,j) & W_5(2,j) & 0 & 0 & \cdots & 0 & 0 \\
0 & W_4(3,j) & W_5(3,j) & 0 & \cdots & 0 & 0 \\
& & & & & & \\
0 & 0 & 0 & 0 & \cdots & W_4(I,j) & W_5(I,j)
\end{vmatrix}, \quad (6.37)
$$

and

$$
C_1 = \begin{vmatrix}
W_1(1,j) & 0 & 0 & 0 & \cdots & 0 & 0 \\
W_2(2,j) & W_1(2,j) & 0 & 0 & \cdots & 0 & 0 \\
0 & W_2(3,j) & W_1(3,j) & 0 & \cdots & 0 & 0 \\
& & & & & & \\
0 & 0 & 0 & 0 & \cdots & W_2(I,j) & W_1(I,j)
\end{vmatrix}. \quad (6.38)
$$

To complete the setup, we must account for boundary and constant potential points. The specialized Neumann boundary condition is easy. The form of the matrices guarantees that there is no coupling to vertices outside the solution region, equivalent to zero coupling constants. This condition guarantees that the integral of normal electric field around the region is zero. The process for a constant potential point sounds overwhelming, but is straightforward to program. For example, suppose the potential at vertex $i = 10$ and $j = 30$ equals 25.0. The tenth row of the block matrix B_{30} is set with identity matrix components (all 0.0 except for 1.0 on the diagonal). The tenth rows of matrices A_{30} and C_{30} are set to 0.0 and the tenth entry of Q_{30} is set to 25.0.

6.5 Solving Tridiagonal Block Matrix Problems

The final step is to solve the large set of linear equations represented by the tridiagonal block matrix of Equation 6.32. The procedure is logically similar to that of Section 6.2. The difference is that operations are performed on block matrices rather than scalar elements. Again, we seek to transform the coefficient matrix to the block LU form

$$\begin{bmatrix} L_1 & 0 & 0 & 0 & 0 & \cdots & 0 & 0 & 0 \\ A_2 & L_2 & 0 & 0 & 0 & \cdots & 0 & 0 & 0 \\ 0 & A_3 & L_3 & 0 & 0 & \cdots & 0 & 0 & 0 \\ 0 & 0 & A_4 & L_4 & 0 & \cdots & 0 & 0 & 0 \\ & & & & & & & & \\ 0 & 0 & 0 & 0 & 0 & \cdots & 0 & A_N & L_N \end{bmatrix} \times$$

$$\begin{bmatrix} 1 & U_1 & 0 & 0 & 0 & \cdots & 0 & 0 & 0 \\ 0 & 1 & U_2 & 0 & 0 & \cdots & 0 & 0 & 0 \\ 0 & 0 & 1 & U_3 & 0 & \cdots & 0 & 0 & 0 \\ 0 & 0 & 0 & 1 & U_4 & \cdots & 0 & 0 & 0 \\ & & & & & & & & \\ 0 & 0 & 0 & 0 & 0 & \cdots & 0 & 0 & 1 \end{bmatrix}.$$

(6.39)

The block matrix components are given in terms of the blocks of the original matrix as

$$L_1 = B_1, \quad U_1 - L_1^{-1} C_1,$$

$$L_2 = B_2 - A_2 U_1, \quad U_2 = L_2^{-1} C_2, \tag{6.40}$$

$$L_j = B_j - A_j U_{j-1}, \quad U_j = L_j^{-1} C_j.$$

In comparison to Equation 6.14, the matrix inversion process replaces simple division. The quantities $A_j U_{j-1}$ represent matrix multiplication. Note that the inverse of a tridiagonal matrix is generally not a tridiagonal matrix. Therefore, the inversion and multiplication to derive the U matrices must be performed by the Gauss-Jordan procedure or an equivalent. The L matrices can be used to derive components of a subsidiary vector ψ that consists of J row vectors of length I,

$$\psi_1 = L_1^{-1} Q_1,$$

$$\psi_2 = L_2^{-1} (Q_2 - A_2 \psi_1), \tag{6.41}$$

$$\psi_j = L_j^{-1} (Q_j - A_j \psi_{j-1}).$$

Here, quantities $L_{j-1} Q_j$ represent multiplication of a square matrix times a column vector. Finally, the ψ vectors can be used to derive the Φ vectors, the desired result.

$$\Phi_{JMax} = \Psi_{JMax}$$

$$\Phi_{JMax-1} = \Psi_{JMax-1} - U_{JMax-1}\Psi_{JMax} \tag{6.42}$$

$$\Phi_j = \Psi_j - U_j\Psi_{j+1}$$

Equations 6.40, 6.41 and 6.42 deal with large data structures. For example, with a 100×100 mesh the Φ, Q, and ψ vectors each have 10,000 elements. There are 298 instances of the block matrices A_j, B_j, and C_j with 10,000 elements. There are 100 L_j matrices and 99 U_j matrices. If we attempted to store all the variables in random-access memory with 8 byte double precision numbers, the required storage is 40 MB. A 300×300 mesh would require about 1.1 GB. To perform the operation on the average personal computer, we need intermediate storage on a hard disk. Memory use should be efficient and hard disk operations should be kept to a minimum.

We can organize the procedure so that only four matrices need be stored in memory at a time. Furthermore, only the components U_j are transferred to and restored from disk. Suppose we want to solve problems that fit within a mesh of dimensions $N \times N$. We shall set aside three $N \times N$ block matrices in memory: M_1, M_2, and M_3. We also need space for another matrix V that consists of N vectors of length N, one for each row of the mesh. Tables 6.4 and 6.5 show how the tridiagonal block matrix inversion is carried out within this storage limit. Using double precision numbers on a 100×100 mesh, the required memory is only 320 kB, with 7.9 MB of information transferred to disk. For a 300×300 mesh, the numbers are 2.88 and 215 MB, respectively.

For a logical mesh with dimensions I and J, the number of operations for a tridiagonal block solution with row organization is proportional to I^3J. The process is efficient for square or tall solution volumes where $I \le J$, but wastes time for short solution volumes. One resolution of the problem is to organize data in column format when $I > J$. In this case, the global index is given by

$$n = i\ J + j \tag{6.43}$$

The only changes necessary to implement column organization are to extend the subroutines that set matrix coefficients (Equation 6.34 through 6.38) and the source terms (Equation 6.29).

TABLE 6.4

Tridiagonal Block Matrix Inversion: Computation of ψ_j and U_j

M1	M2	M3	V
	Fill $B_1 \rightarrow L_1$	Fill C_1	Fill $Q_1 \dots Q_N$
		$U_1 = L_1^{-1} C_1$	$L_1^{-1} Q_1 \psi_1$
Fill A_2	Fill B_2		
	$B_2 - A_2 U_1$		$\psi_2 = L_2^{-1} (Q_2 - A_2 \psi_1)$
		Store U_1	
		Fill C_2	
		$U_2 = L_2^{-1} C_2$	
...	...	...	...
Fill A_j	Fill B_j		
	$B_j - A_j U_{j-1}$		$\Psi_j = L_j^{-1} (Q_j - A_j \Psi_{j-1})$
		Store U_{j-1}	
		Fill C_j	
		$U_j = L_j^{-1} C_j$	

TABLE 6.5

Tridiagonal Block Matrix Inversion: Computation of Φ_j

M1	M2	M3	V
$\psi_1, \psi_2, \dots, \psi_N$			$\Phi_N = \psi_N$
		Recall U_{N-1}	$\phi_{N-1} = \psi_{N-1} - U_{N-1}\psi_N$
		Recall U_{N-2}	$\phi_{N-2} = \psi_{N-2} - U_{N-2}\psi_{N-1}$
...	...	...	...
		Recall U_1	$\phi_1 = \psi_1 - U_1\psi_2$

Chapter 6 Exercises

6.1. Use a spreadsheet to understand the operations in Gauss-Jordan elimination with partial pivoting. Find the inverse of the matrix

$$A = \begin{vmatrix} 3 & 2 & 1 \\ 5 & -1 & 7 \\ 6 & -2 & 3 \end{vmatrix}$$

Set up the matrix and a unity matrix ($B = I$) as described in Section 6.1. Perform simultaneous operations on the two matrices that reduce A to the unity matrix. Proceed one step at a time, making copies of the modified matrices, At the end, confirm that the modified matrix B contains A^{-1}.

6.2. Most spreadsheets have the capability to invert moderate-sized matrices. Use this feature to solve the following set of coupled linear equations.

$$13.40x_1 + 2.70x_2 - 2.40x_3 + 4.30x_4 = 78.49,$$

$$2.40x_1 - 12.40x_2 + 8.90x_3 + 6.80x_4 = -15.00,$$

$$3.60x_1 - 6.90x_2 + 4.80x_3 + 0.50x_4 = -24.65,$$

$$11.00x_1 + 2.10x_2 + 0.75x_3 - 5.30x_4 = -2.57.$$

6.3. Find the L and U matrices for the tridiagonal matrix

$$A = \begin{bmatrix} 5.3 & 4.8 & 0.0 & 0.0 \\ 1.5 & 3.4 & 2.6 & 0.0 \\ 0.0 & 2.7 & -3.5 & 6.2 \\ 0.0 & 0.0 & 9.2 & -2.7 \end{bmatrix}.$$

Confirm the validity of the factorization by computing the product LU.

6.4. For the matrix of Exercise 6.4, use the L and U matrices to find the solution to the set of coupled linear equations

$$A \begin{bmatrix} \phi_1 \\ \phi_2 \\ \phi_3 \\ \phi_4 \end{bmatrix} = \begin{bmatrix} 36.2 \\ 20.94 \\ 15.87 \\ 15.93 \end{bmatrix}.$$

Use Equation 6.16 to find $\psi_1, \ldots \psi_4$ and then apply Equation 6.18 to find $\phi_1, \ldots \phi_4$.

6.5. Give expressions for the components of block matrices for a two-dimensional electrostatic solution on a regular mesh following the method in Section 6.4. Assume difference equations of the form of Equation 4.67. The mesh has dimensions I and J in the x and y directions with vertex coordinates x_i and y_j. Use J blocks with dimensions $I \times I$.

6.6. Use the matrix inversion capability of a spreadsheet to solve a simple one-dimensional electrostatic problem. Consider a region of width $d = 0.06$ m between grounded plates. A uniform space-charge density $\rho_o = 3.5 \times 10^{-8}$ coulombs/m³ fills the space from $x = 0.00$ to 0.03 m.

(a) Divide the region into six elements and find a numerical solution from Equation 6.21.

(b) Derive an expression for the potential from the Poisson equation. Compare the numerical and analytic values at $x = 0.03$.

7

Analyzing Numerical Solutions

Mathematics may be compared to a mill of exquisite workmanship, which grinds your stuff to any degree of fineness; but, nevertheless, what you get out depends on what you put in; and as the grandest mill in the world will not extract wheat flour from peascods, so pages of formulae will not get a definite result out of loose data.

Thomas Henry Huxley

In preceding chapters we studied the mechanics of numerical electrostatic solutions: setting up a mesh, generating difference equations, and solving them. The next issue is how to use the information that comes out. This chapter covers some techniques for analyzing numerical data, with emphasis on finite-element solutions on triangular meshes. The first three sections deal with interpolations and spatial derivatives of quantities. In electrostatic solutions, we want to find accurate values of potential between vertices and derivatives of the potential to determine the electric fields. The methods discussed extend to the full spectrum of physical systems covered in this book. Section 7.1 describes the first step for an interpolation, location of the element of an arbitrary mesh that contains a test point. The next step is to collect information on the potential at neighboring vertices for the interpolation. On an arbitrary mesh we cannot be sure how many points will be available. For this situation least-squares interpolation methods are the best approach. They give reasonably accurate answers with flexibility on the number of input points. Section 7.2 reviews the theory of least-squares fits and Section 7.3 covers the application to electrostatic fields.

The final two sections discuss techniques to display field information graphically. Section 7.4 covers the analysis of meshes to create plots of elements and the boundaries of solution regions. Section 7.5 explains techniques to generate plots of field information. In addition to basic plots of potential and field amplitude contours, advanced techniques are introduced. These techniques include color-coded element plots and three-dimensional representations.

7.1 Locating Elements

To find the field at a target point we must first identify the element where it resides. Element location is easy on a regular mesh. We simply find the indices of the vertex just below the target point. For example, consider a one-dimensional solution where the mesh points are the set x_i with $1 \leq i \leq N$. If the target point x is in the range

$$x_i \leq x \leq x_{i+1}, \tag{7.1}$$

then the potential at the point is approximately

$$\phi(x) \cong \phi_i \frac{x_{i+1} - x}{x_{i+1} - x_i} + \phi_{i+1} \frac{x - x_i}{x_{i+1} - x_i}. \tag{7.2}$$

Finding the index value that satisfies Equation 7.1 involves a search through the array x_i. For small arrays, the search can be carried by testing all values x_i from $i = 1$ to N. There are much faster techniques for large arrays. The *method of bisection* is a good all-purpose choice. We assume that the entries x_i are arranged in increasing order. The first step is to divide the set in half with the dividing value $x_{mid} = x(N/2)$. If the target value is less than x_{mid}, it is in the lower set; otherwise, it is in the upper set. The next step is to divide the set containing the target value in half and then to test which subset contains the value. This process continues until the subset contains only one entry. The total number of tests to locate the value of x_i just below x is approximately equal to the power of 2 that gives a number equal to or greater than N. For example, it takes about eight bisections to find the position of a value in an array with 200 entries. In contrast, it is necessary to make an average of 100 tests with a simple sequential search. Table 7.1 shows a succinct routine to carry out a quick bisectional search.

Locating an element in three-dimensional meshes with variable resolution (Section 4.6) is straightforward. The vertex positions are given by three arrays: x_i, y_j, and z_k. Three calls to the search routine give indices for the element that contains the target point.

Element location on a conformal triangular mesh where the vertex positions do not follow a regular pattern is more challenging. The only procedure that guarantees complete success is to check each triangle sequentially. The first question is how to tell if a point is inside a given triangle. The following procedure is elegant and fast — it requires only simple multiplications. It uses the function of Table 7.2 to return the area of a triangle determined by any three points in space.

Suppose we have the coordinates of a test point and the three vertices of an element arranged in counterclockwise order (positive rotation): (x_1, y_1),

TABLE 7.1

Bisection Search Routine

```
      LOGICAL FUNCTION Search(XArray,N,x,i)
      INTEGER I,N
      REAL XArray(1:N),x
      INTEGER IDn,IMid,IUp

C Inputs:  Monotonically increasing one-dimensional array XArray
C          of dimension (1:N) and target value x
C Outputs: Function value TRUE if search successful - returns
C          index i such that x(i) < x < x(i+1). Returns FALSE
C          if x out of range

C--- Initialize upper and lower indices
      IDn = 0
      IUp = N+1
C--- Continue to bisect interval until reduced to unity
  100 IF ((IUp-IDn).GT.1) THEN
         IMid = (IUp+IDn)/2
C--- Pick the subdivision that contains the target value
         IF (x.GT.XArray(IMid)) THEN
            IDn = IMid
         ELSE
            IUp = IMid
         ENDIF
         GO TO 100
      ENDIF
C--- Set up return values
      IF ((IDn.EQ.0).OR.(IDn.EQ.N)) THEN
         Search = .FALSE.
         I = 0
      ELSE
         Search = .TRUE.
         I = IDn
      ENDIF
      RETURN
      END
```

TABLE 7.2

Triangle Area Function

```
      REAL FUNCTION TriArea(xa,ya,xb,yb,xc,yc)
      REAL xa,ya,xb,yb,xc,yc
C--- Returns positive value for points arranged in counterclockwise
C    orientation
      TriArea = 0.5*((ya*(xc-xb) + yc*(xb-xa) + yb*(xa-xc)))
      RETURN
      END
```

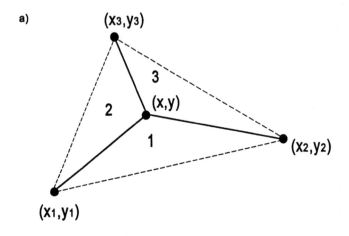

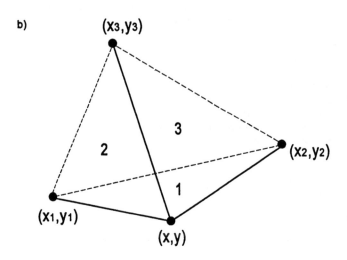

FIGURE 7.1
Test to determine if a point is inside an element. (a) Point inside — the area formula yields positive values for triangles 1, 2 and 3. (b) Point outside — the formula gives a negative area for triangle 1.

(x_2, y_2), and (x_3, y_3). If the test point is inside the triangle, as in Figure 7.1a, it defines three subtriangles. We make three calls to the function TriArea,

$$\text{Area1} = \text{TriArea}(x_1, y_1, x_2, y_2, x, y)$$

$$\text{Area2} = \text{TriArea}(x_2, y_2, x_3, y_3, x, y) \qquad (7.3)$$

$$\text{Area3} = \text{TriArea}(x_3, y_3, x_1, y_1, x, y)$$

Note that the points have positive rotational order if the test point is inside the element. Therefore, the three calls return positive numbers. On the other hand, suppose the test point is outside the element as in Figure 7.1b. The values for Area1 and Area2 are positive, but the ordering of points for the triangle defined by (x_3,y_3), (x_1,y_1), and (x,y) is negative. Therefore, the routine returns a negative number. In summary, if the three quantities of Equation 7.3 are greater than or equal to zero then the point is either inside the triangle or on a boundary.

Usually, we can avoid checking every triangle of a mesh to locate an element. The mesh generation procedure of Sections 5.2 and 5.3 starts with a logical mesh and shifts certain vertices to fit object boundaries. If the shifting is not too severe, the x-axis position of a vertex, $x(i,j)$, will be close to $x(i,1)$. Similarly, $y(i,j) \approx y(1.j)$. In this case, we can make a preliminary search of coordinates along the x and y boundaries using the bisectional method of Table 7.1 to estimate the indices i and j. We can then narrow in on the correct element by making a local search in the vicinity. It is usually necessary to check only about 25 triangles compared to thousands for a full mesh search. Local search techniques are particularly valuable in applications such as charged-particle orbit tracking that demand an extended set of searches.

7.2 Generalized Least-Squares Fits

Field interpolation is relatively easy on a regular mesh far from boundaries. Each vertex has a known number of neighbors. We can fit polynomial functions (Section 8.3) of different orders, depending on how many points are included. The problem is more difficult with arbitrary boundaries and meshes. Here, we do not know in advance how many points will be available. Consider field calculations on the mesh of Figure 7.2. Point A is a point internal to a dielectric. We can use a full complement of surrounding points, the number determined by the order of the interpolation and the mesh logic. Fewer points are available near point B inside an electrode indentation. We cannot use points internal to the electrode for an interpolation because the field is discontinuous at the surface. A similar problem occurs at point C near a boundary between two dielectrics. Only points on one side of the boundary should be used for a continuous interpolation function. Finally, point D is near a symmetry boundary of the solution region. We can improve accuracy by adding extra points projected to the external region according to the boundary symmetry condition.

Clearly, we need an interpolation method that tolerates variations in the amount of available data. The least-squares method is a good solution because it seeks a best fit to a data set rather than one that passes exactly through a given number of points. The application of least-squares fits to

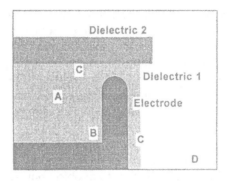

FIGURE 7.2
Collecting points for field interpolations in a complex geometry. A, General interior point; B, point in recess has fewer neighboring vertices; C, points on dielectric boundary; D, point near a Neumann symmetry boundary.

straight line data or simple curves is a familiar process. In this section, we extend the technique to include generalized functions in multiple dimensions.

Consider a data set consisting of N values of a function. To illustrate the problem of potential interpolation, we shall use the set $\Phi_i(x_i,y_i,z_i)$ consisting of potential values at N positions in space, (x_i,y_i,z_i). We want to fit a continuous function $\phi(x,y,z,a_m)$ to the data. The function has M free parameters a_m that are chosen to achieve the best fit. The working definition of best fit is that the parameters give the *maximum likelihood* for observing the data set. Suppose we had an exact solution to the physical problem, $\phi(x,y,z,a_m)$, and sought the probability of observing the data set Φ_i. If observations of potential differences followed a normal probability distribution with a *uniform standard deviation* σ, then the probability of a certain data set would be

$$P = A\prod_{i=1}^{N}\left[\exp\left[-\frac{1}{2}\left(\Phi_i - \phi\left(x_i,y_i,z_i,a_m\right)\right)^2\right]\right] \tag{7.4}$$

In Equation 7.4, the quantity A is a normalizing parameter and the symbol $\prod$ denotes a product of N terms. We seek values for a_m that maximize the probability. Taking the logarithm of Equation 7.4, the peak probability corresponds to the minimum value of the sum,

$$\chi^2 = \sum_{i=1}^{N}\left[\Phi_i - \phi\left(x_i,y_i,z_i,a_m\right)\right]^2 \tag{7.5}$$

The extreme value of χ^2 is determined by the M equations

$$\frac{\partial\left(\chi^2\right)}{\partial a_m} = 0 \tag{7.6}$$

Suppose that the potential variation over a spatial region is approximated by the expression

$$\phi(x,y,z) = a_1 f_1(x,y,z) + a_2 f_2(x,y,z) + a_3 f_3(x,y,z) + \ldots + a_M f_M(x,y,z). \quad (7.7)$$

for a set of N values Φ_i. The quantities $f_m(x,y,z)$ can represent any choice: polynomials, trigonometric functions, logarithms, or combined functions. Because the number of free parameters cannot exceed the amount of data, the condition $M \leq N$ must hold. In this case, the expression for χ^2 is

$$\chi^2 = \sum_{i=1}^{N} \left[\Phi_i - a_1 f_1(x_i,y_i,z_i) - a_2 f_2(x_i,y_i,z_i) - \ldots - a_M f_M(x_i,y_i,z_i) \right]^2. \quad (7.8)$$

Following Equation 7.6, the partial derivatives give a set of M linear equations for the unknown coefficients,

$$a_1 \sum_{i=1}^{N} f_1(x_i,y_i,z_i)^2 + a_2 \sum_{i=1}^{N} f_1(x_i,y_i,z_i) f_2(x_i,y_i,z_i) + \ldots$$

$$+ a_M \sum_{i=1}^{N} f_1(x_i,y_i,z_i) f_M(x_i,y_i,z_i) = \sum_{i=1}^{N} f_1(x_i,y_i,z_i) \Phi_i,$$

$$a_1 \sum_{i=1}^{N} f_2(x_i,y_i,z_i) f_1(x_i,y_i,z_i) + a_2 \sum_{i=1}^{N} f_2(x_i,y_i,z_i)^2 + \ldots$$

$$+ a_M \sum_{i=1}^{N} f_2(x_i,y_i,z_i) f_M(x_i,y_i,z_i) = \sum_{i=1}^{N} f_2(x_i,y_i,z_i) \Phi_i, \quad (7.9)$$

$$\ldots$$

$$a_1 \sum_{i=1}^{N} f_M(x_i,y_i,z_i) f_1(x_i,y_i,z_i) + a_2 \sum_{i=1}^{N} f_M(x_i,y_i,z_i) f_2(x_i,y_i,z_i) + \ldots$$

$$+ a_M \sum_{i=1}^{N} f_M(x_i,y_i,z_i)^2 = \sum_{i=1}^{N} f_M(x_i,y_i,z_i) \Phi_i.$$

We can write Equation 7.9 more succinctly in matrix notation as

$$C \cdot A = D, \quad (7.10)$$

where

$$A_m = a_m,$$

$$D_m = \sum_{i=1}^{N} f_m(x_i, y_i, z_i)\, \Phi_i, \qquad (7.11)$$

$$C_{mn} = \sum_{i=1}^{N} f_m(x_i, y_i z_i)\, f_n(x_i, y_i, z_i).$$

The coefficients C_{nm} and the constants D_m are determined from the known data values and coordinate locations. The equations can be solved by the Gauss-Jordan reduction method described in Section 6.1 to find the best values of a_m.

As an example, suppose we want to make a second-order polynomial fit to a two-dimensional data set $\Phi_i(x_i, y_i)$ for interpolation at the point (x, y). The second-order fit has the form

$$\phi = a_1\, f_1 + a_2\, f_2 + a_3\, f_3 + a_4\, f_4 + a_5\, f_5 + a_6\, f_6. \qquad (7.12)$$

Defining the quantities

$$X_i = x_i - y, \quad Y_i - y_i - y, \qquad (7.13)$$

we can write the six functions in Equation 7.12 as

$$f_1 = 1,\ f_2 = X,\ f_3 = Y,\ f_4 = X^2,\ f_5 = XY,\ f_6 = Y^2. \qquad (7.14)$$

Therefore, a least-squares fit requires six or more data points. The entries in Equation 7.11 are

$$D_1 = \sum_{i=1}^{N} \Phi_i, \ D_2 = \sum_{i=1}^{N} \Phi_i X_i, \ D_3 = \sum_{i=1}^{N} \Phi_i Y_i, \ D_4$$

$$= \sum_{i=1}^{N} \Phi_i X_i X_i, \ D_5 = \sum_{i=1}^{N} \Phi_i X_i Y_i, \ D_6 = \sum_{i=1}^{N} \Phi_i Y_i Y_i. \qquad (7.15)$$

Because of the form of the functions, some of the coefficients C_{mn} have the same values. We can speed up the interpolation by avoiding redundant calculations.

$$C_{12} = \sum_{i=1}^{N} X_i, \ C_{13} = \sum_{i=1}^{N} Y_i, \ C_{14} = \sum_{i=1}^{N} X_i X_i,$$

$$C_{15} = \sum_{i=1}^{N} X_i Y_i, \ C_{16} = \sum_{i=1}^{N} Y_i Y_i, \ C_{24} = \sum_{i=1}^{N} X_i X_i X_i,$$

$$C_{25} = \sum_{i=1}^{N} X_i X_i Y_i, \ C_{26} = \sum_{i=1}^{N} X_i Y_i Y_i, \ C_{36} = \sum_{i=1}^{N} Y_i Y_i Y_i,$$

$$C_{44} = \sum_{i=1}^{N} X_i X_i X_i X_i, \ C_{45} = \sum_{i=1}^{N} X_i X_i X_i Y_i,$$

$$C_{46} = \sum_{i=1}^{N} X_i X_i Y_i Y_i, \ C_{56} = \sum_{i=1}^{N} X_i Y_i Y_i Y_i, \ C_{66} = \sum_{i=1}^{N} Y_i Y_i Y_i Y_i, \quad (7.16)$$

$$C_{11} = N, \ C_{21} = C_{12}, \ C_{22} = C_{14}, \ C_{23} = C_{15}, \ C_{31} = C_{13}, \ C_{32} = C_{15},$$

$$C_{33} = C_{16}, \ C_{34} = C_{25}, \ C_{35} = C_{26}, \ C_{41} = C_{14}, \ C_{42} = C_{24}, \ C_{43} = C_{25},$$

$$C_{51} = C_{15}, \ C_{52} = C_{25}, \ C_{53} = C_{26}, \ C_{54} = C_{45}, \ C_{55} = C_{46},$$

$$C_{61} = C_{16}, \ C_{62} = C_{26}, \ C_{63} = C_{36}, \ C_{64} = C_{46}, \ C_{65} = C_{56}.$$

The expressions in Equation 7.16 are straightforward but laborious to derive. The coefficient matrix for a second-order polynomial fit in three dimensions has 81 coefficients. An analysis reveals that only 41 coefficients are independent.

7.3 Field Calculations on a Two-Dimensional Triangular Mesh

This section covers electric field calculations on triangular meshes. The three-point formula of Equations 2.53, 2.57, and 2.58 (equivalent to linear interpolation) gives a quick estimate of the field. Assuming that potential values are stored in a one-dimensional array with global index n (Section 5.2), the routine of Table 7.3 returns values for the interpolated potential and field components (E_x, E_y) or (E_z, E_r).

Because of the condition of constant electric fields in elements, the values from the routine ETri are coarse. As an illustration, Figure 7.3 shows results for a benchmark test of the field between two concentric spheres of radii 1.0 and 3.0.

TABLE 7.3

Linear Interpolation on a Triangular Mesh

```
SUBROUTINE ETri (N1,N2,N3,Xin,Yin,ExOut,EyOut,PhiOut)
INTEGER N1,N2,N3
REAL  Ex,Ey
REAL  X1,X2,X3,Y1,Y2,Y3
REAL  Q1,Q2,Q3
REAL  A,B
X1 = Mesh(N1).x
X2 = Mesh(N2).x
X3 = Mesh(N3).x
Y1 = Mesh(N1).y
Y2 = Mesh(N2).y
Y3 = Mesh(N3).y
Q1 = Mesh(N1).Phi
Q2 = Mesh(N2).Phi
Q3 = Mesh(N3).Phi
C --- Find slopes
A = ((Q2-Q1)*(Y3-Y1)-(Q3-Q1)*(Y2-Y1))/
&     ((X2-X1)*(Y3-Y1)-(X3-X1)*(Y2-Y1))
B = ((Q2-Q1)*(X3-X1)-(Q3-Q1)*(X2-X1))/
&     ((Y2-Y1)*(X3-X1)-(Y3-Y1)*(X2-X1))
PhiOut = A*(XIn-X1) + B*(YIn-Y1) + Q1
ExOut = -A
EyOut = -B
RETURN
END
```

The open circles show the relative field error for linear interpolations. Fields were calculated at 50 positions uncorrelated with the element boundaries.

Higher-order interpolations applied to several vertex points near the test point give better accuracy. For example, consider the second-order polynomial described in Section 7.2. Following Figure 7.4, the first step is to find the element that contains the target point and note the region number that gives the material type. The next step is to locate the nearest vertex and collect points in the vicinity until there are at least six potential values. The order of preference is the nearest point, the six adjacent points, and then the next 12 surrounding points. A vertex is rejected if it is outside the solution region or if it is not connected to at least one triangle that has the same region number as the target element. This condition ensures that the interpolation function is unaffected by field discontinuities at dielectric boundaries. At Neumann boundaries an external point is added for each valid point inside the boundary. The new point has the same potential and the mirror position relative to the boundary. After data collection, a least-squares fit to a function of the form of Equation 7.12 gives the polynomial coefficients and the output values $\phi = a_1$, $E_x = -a_2$, $E_y = -a_3$. The filled squares in Figure 7.3 show improved results applying a second-order routine to the cylindrical benchmark test.

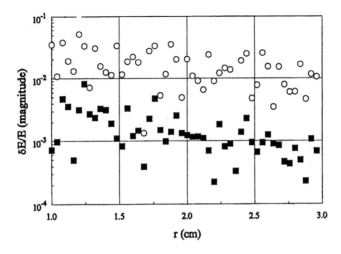

FIGURE 7.3
Relative error for field interpolations between concentric spheres of radii 1.0 cm and 3.0 cm. Approximately 50 elements along the solution dimension. Open circles, linear interpolation; filled squares; second-order, least-squares fit.

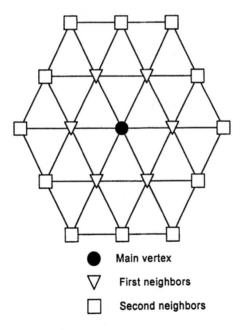

●   Main vertex

▽   First neighbors

☐   Second neighbors

FIGURE 7.4
Order of preference to collect vertex values for a second-order least-squares interpolation on a conformal triangular mesh. Filled circle; main vertex, closest to target point; open triangles, 6 neighbors to main vertex; open squares, 12 second neighbors to main vertex.

Although the second-order polynomial of Equations 7.12 and 7.14 is good for general-purpose interpolations, sometimes we can improve the efficiency of calculations by picking functions matched to the physical system. For electrostatic problems in a homogeneous medium with no space-charge, the function should be consistent with the Laplace equation,

$$\nabla^2 \phi = 0. \tag{7.17}$$

Consider a polynomial fitting function for the potential of a planar system near the target point (x_o, y_o) written in terms of the coordinates $X = x - x_o$ and $Y = y - y_o$,

$$F(X, Y) = a_1 + \sum_{n=2}^{\infty} a_n \, f_n(X, Y). \tag{7.18}$$

Instead of picking arbitrary polynomials for the terms of Equation 7.18, we shall use combinations of powers of x and y that are consistent with the equation

$$\frac{\partial^2 F}{\partial X^2} + \frac{\partial^2 F}{\partial Y^2} = 0. \tag{7.19}$$

It is easy to show that the following terms independently satisfy Equation 7.19,

$$a_1, \; a_2 X, \; a_3 Y, \; a_4 XY. \tag{7.20}$$

The entries are the same as those in Equation 7.14. On the other hand, the individual terms X^2 and Y^2 should be replaced with the combination

$$a_5 \left(X^2 - Y^2 \right). \tag{7.21}$$

The general third-order expansion has the terms

$$A \; X^3 + B \; X^2 Y + C \; XY^2 + D \; Y^3. \tag{7.22}$$

Substitution in Equation 7.18 shows that there are two independent third-order polynomials consistent with the Laplace equation,

$$a_6 \left(X^3 - 3XY^2 \right), \; a_7 \left(Y^3 - 3X^2 Y \right). \tag{7.23}$$

TABLE 7.4

Harmonic Polynomials for Electrostatic Potential

Rectangular Geometry	Cylindrical Geometry
1	1
x	z
y	$z^2 - r^2/2$
xy	$z^3 - 3zr^2/2$
$x^2 - y^2$	$z^4 - 3z^2r^2 + 3r^4/8$
$x^3 - 3xy^2$	$z^5 - 4z^3r^2/3 + zr^4/2$
$y^3 - 3x^2y$	
$4x^3y - 4xy^3$	
$x^4 - 6x^2y^2 + y^4$	
$x^5 - 10x^3y^2 + 5xy^4$	
$y^5 - 10y^3x^2 + 5yx^4$	

If there is no space-charge and we are careful to pick data points on one side of a boundary, we can achieve third-order accuracy using seven independent terms compared to ten terms for the general polynomial. We can continue the process for higher-order terms. The resulting expressions, the *harmonic polynomials*, are listed in Table 7.4 through the fifth order. The calculation for cylindrical coordinates uses the cylindrical form of the Laplace operator. Table 7.4 also lists harmonic polynomials for cylindrical problems through the fifth order. Because of the cylindrical symmetry only even powers of r appear.

7.4 Mesh and Boundary Plots

Plots can efficiently communicate information on two- and three-dimensional field solutions. We have seen several examples in previous chapters. This section and the next discuss plotting routines. We shall concentrate on two-dimensional finite-element solutions on conformal triangular meshes. The extension to regular meshes is straightforward. This section covers two types of plots to display the computational mesh and region boundaries. The following section reviews methods to display physical information about the field and materials.

To generalize the discussion, suppose we have two fundamental subroutines that handle the specific graphics device:

```
SetBound(XMin,YMin,XMax,YMax)
```

```
MakeLine(XStart,YStart,XEnd,YEnd)
```

The parameters passed to *SetBound* are the maximum and minimum values along the horizontal and vertical directions anticipated for the plot. The routine determines how to map the area on a device and sets up scaling factors. For example, to plot the full logical mesh discussed in Section 5.2 we associate x_{min}, x_{max}, y_{min}, and y_{max} with the edges of the mesh. Alternatively, we could zoom in on a region by setting a smaller view rectangle. The function *MakeLine* plots a line from (*XStart*,*YStart*) to (*XEnd*,*YEnd*) scaled to the plot boundaries with clipping if necessary. Beyond these essentials, we could define additional routines to set plot color and other parameters.

Mesh plots are important in finite-element solutions. They show the quality of the element fit to boundaries, the resolution for interpolations, and trouble spots where triangles may have unfavorable geometry (Section 12.5). A mesh plot requires the following steps.

- Make a loop through every possible line that connects two adjacent vertices of the logical mesh
- Decide whether each line is part of the solution space
- Pass the vertex coordinates to *MakeLine* if the line is valid

The following procedure applies to meshes following the conventions described in Sections 5.2 and 5.3. Three lines are checked at each vertex of the logical mesh: to the right, up and to the right, and up and to the left. The outer loop extends over all rows j = 1 to J and the inner loop over columns i = 1 to I. The following lines are checked for each vertex:

```
(i,j) to (i+1,j)
(i,j) to (i,j+1)
(i,j) to (i,j+1) for j odd, or
(i,j) to (i,j-1) for j even
```

A line is rejected if the region number of either vertex equals zero of if the region numbers of both elements adjacent to the line are zero. Depending on the type of solution, we can add other criteria to improve the plot. For example, in electrostatic solutions the field inside an electrode is meaningless so there is no reason to plot the mesh. These lines are suppressed by rejecting vectors when both adjacent elements have region numbers corresponding to a fixed potential region. Figure 7.5a shows an example of such a plot.

A boundary plot outlines the physical regions of the solution. The method checks lines in the same order as the mesh plot. The difference is that it rejects vectors if adjacent elements do not have different region numbers. Figure 7.5b shows the result. An alternate method that produces dramatic results is an element plot with color set by region number. This type of plot is easy to implement with a graphics library that includes a routine to fill a bounded

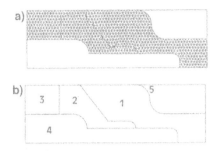

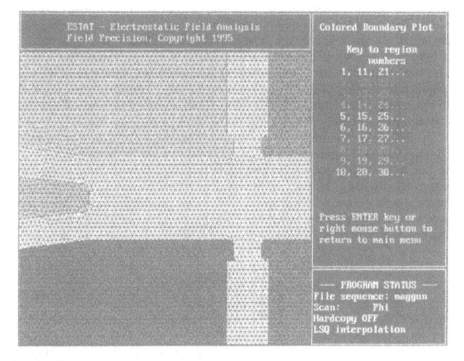

FIGURE 7.5
Plots to display the computational mesh and boundaries. Test problem of a high voltage bushing. Region 1, full solution volume, vacuum; region 2, alumina; region 3, transformer oil; region 4, inner electrode at 50 kV; region 5, outer electrode, 0 kV. (a) Mesh plot excluding elements inside electrodes. (b) Boundary plot. (c) Colored boundary plot with mesh lines.

area. The parameters passed to the routine usually include a boundary color, a fill color, and the coordinates of a point inside the boundary. The procedure is as follows.

- Loop through all elements of the logical mesh, rejecting those with zero region number
- Draw lines between the element vertices in the boundary color

- Average the vertex coordinates to find the internal point
- Set up the information and call the fill routine

Figure 7.5c shows the results.

7.5   Contour, Element, Elevation, and Field Line Plots

In this section we shall discuss graphical displays of field information. In electrostatic problems the quantities of interest may include the electrostatic potential, electric field vectors, or electric field amplitude. In magnetostatic solutions, it may be useful to plot variations of material properties. The contour plot is a common method to display two-dimensional variations of a scalar quantity. In electrostatics, the plot consists of a series of curves corresponding to constant values of potential. The electric field is normal to the contours with an amplitude inversely proportional to the distance between lines.

The first step to create an electrostatic equipotential plot is to select a set of contour values. We can set the values manually or let the program make the choice. In the latter case, the first step is to find minimum and maximum values of potential, ϕ_{min} and ϕ_{max}. The procedure starts by setting $\phi_{min} = BigNum$ and $\phi_{max} = -BigNum$, where $BigNum$ is the largest floating point number that can be represented by the computer. Then the program loops through all values of potential in the solution applying the tests

$$\text{if} \left(\phi_{ij} < \phi_{min}\right) \text{ then } \phi_{min} = \phi_{ij}$$

$$\text{if} \left(\phi_{ij} > \phi_{max}\right) \text{ then } \phi_{max} = \phi_{ij} \tag{7.24}$$

Finally, the program fills an array of contour values. For *NInt* equal potential intervals, the index n is in the range 0 to NInt and the values are

$$\phi_{cn} = \phi_{min} + n\,\frac{\phi_{max} - \phi_{min}}{NInt} \tag{7.25}$$

The following procedure generates a set of vectors that lie along the contour value ϕ_{cn}. The vectors are stored in an array of starting and ending coordinates: $\mathbf{v} = (x_s, y_s, x_e, y_e)$. The program makes a loop through all elements in the solution area, rejecting any with a region number that corresponds to a constant potential. The program checks the three sides of each triangle to see if ϕ_{cn} lies between the potential values at the ends. If so, the program stores the coordinates of the position given by a linear interpolation of potential

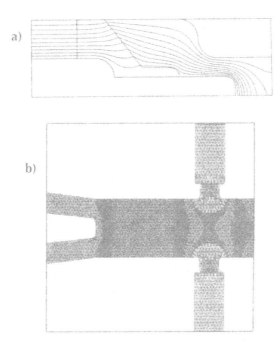

FIGURE 7.6
Solution plots for a high-voltage bushing example. (a) Equipotential plot. (b) Color-coded element plot of electric field amplitude.

along the line. When the program finds two coordinates for an element, it stores the corresponding vector. Three vectors are added to the list if there are three intersecting coordinates. The program takes no action if there is one intersection — the corresponding contour line will be represented in an adjacent element. The result is an array of vectors that can be passed to the *MLine* routine of the previous section for plotting. As an example, Figure 7.6a shows an equipotential plot for a high-voltage bushing.

The procedure described stores contour vectors in the order they are found. This depends on the mesh search order and does not necessarily produce a connected set of vectors. The vector order is not important for raster-type graphics devices (i.e.,video screens, laser printers) but may be critical for a pen plotter. To make a versatile routine, it is a good idea to sort the vectors before plotting. A contour may consist of several closed or open curves. The best we can do is to connect as many vectors as possible. The process requires the following functions and subroutines. Note that the pass parameters are four-component vector records.

```
Subroutine Exchange (v1,v2)

Subroutine Reverse (v1)

Integer Function Connect (v1,v2)
```

The first subroutine exchanges two vectors in the storage array, while the second reverses the starting and ending points of a vector. The third function compares the starting and ending points of two vectors and returns the following values:

```
0:  No connection

1:  Vector 1 Start connected to Vector 2 Start

2:  Vector 1 Start connected to Vector 2 End

3:  Vector 1 End connected to Vector 2 Start

4:  Vector 1 End connected to Vector 2 End
```

As with any comparison of floating point numbers, we must be careful in the definition of the term *connected*. It is unlikely that the coordinate values for connected vectors will be exactly equal to within the floating point precision of the computer. Instead, we must define a function *ApproxEqual* that checks for equality within a length tolerance.

The first step in sorting a contour set is to find the endpoint if it exists. Starting at the top of the vector list, we count connections with the remaining vectors. If there are no connections, the vector is isolated. We leave it in place and move down the list. If there is a single connection, the vector at the top of the list is an endpoint. Otherwise, we exchange the vector with the next one on the list and repeat the search. The procedure places an endpoint at the current top of the list which we shall call position i. If no endpoint is located the contour must be closed and it is sufficient to start with any vector. Once the endpoint is set, we process through the remaining vectors in the list (denoted with index j), applying the function Connect(v(i),v(j)) and taking the following actions.

Connect = 0 » Continue the search, j = j + 1

Connect = 1 » Reverse(v(i)), Exchange(v(i+1),v(j)), i = i+1

Connect = 2 » Reverse(v(i)), Reverse(v(j)), Exchange(v(i+1),v(j)), i = i+1

Connect = 3 » Exchange(v(i+1),v(j)), i = i+1

Connect = 4 » Reverse(v(j)), Exchange(v(i+1),v(j)), i = i+1

The process continues until there are no connections to succeeding vectors. If the index i is at the end of the vector list, the sort is complete. Otherwise, the remaining vectors belong to one or more disconnected lines. In this case, the search for an endpoint and connecting vectors is repeated.

Quantities like the field amplitude can be displayed effectively in a color-coded element plot. This plot is similar to the boundary plot discussed in the previous section. Here, we loop through all elements in the solution region,

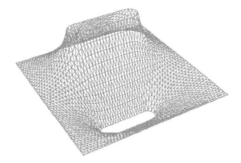

FIGURE 7.7
Three-dimensional wire frame projection plot of potential variation over a two-dimensional region, bipolar high-voltage electrodes with dielectric supports inside a grounded box.

assigning a plot color according to the value of the quantity of interest. Figure 7.6b shows an example — a plot of electric field amplitude in a high-voltage bushing. Although the plot does not convey as much information as an equipotential plot, it gives a dramatic indication of areas of concern.

A third way to display information about a scalar function over a two-dimensional space is with an elevation plot. Figure 7.7 shows an example for a planar electrostatic solution. The display shows potential as an elevation (in the z direction) over a region in the x–y plane. For a conformal mesh the potential surface consists of triangular facets. The *wire frame* plot of Figure 7.7 shows the edges of all facets. In a *hidden surface* plot, the facets are opaque. The surfaces closer to the viewer obscure those behind.

To understand the elevation plot, consider the general problem of representing a three-dimensional object on a two-dimensional surface. The object surface is approximated by a discrete set of coordinates (X_i', Y_i', Z_i') that define facets. The object coordinates are projected into a drawing plane using the method of perspective shown in Figure 7.8. This projection involves connecting the object coordinates to an observation point and finding the intersection coordinates in the drawing plane. For simplicity we take the drawing plane normal to the Z' axis at $Z' = 0$. The observation point is on the axis at position $Z' = D_2$. Inspection of Figure 7.8 gives the drawing plane coordinates as

$$X_i = \frac{X_i'}{1 - \dfrac{Z_i'}{D_2}},$$

$$Y_i = \frac{Y_i'}{1 - \dfrac{Z_i'}{D_2}}.$$

(7.26)

Now, suppose an electrostatic solution on a triangular mesh is plotted as a three-dimensional object with the vertex coordinates along axes X'' and Y''

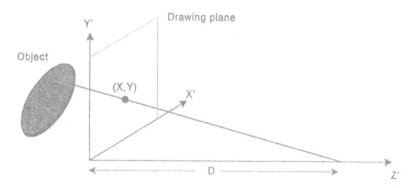

FIGURE 7.8
Geometry for projection of a three-dimensional object on to a two-dimensional plane with perspective.

and the relative potential plotted in Z″. To make the plot look interesting we include the option to transform these coordinates to the (X',Y',Z') coordinates of Figure 7.8 through rotations. A general rotation can be written in terms of the rotational matrices of Section 5.1,

$$X_i' = M(\theta_z)\, M(\theta_y)\, M(\theta_x)\, X_i'' , \qquad (7.27)$$

where $X_i' = (X_i',Y_i',Z_i')$. Suppose that the electrostatic solution region fits inside a rectangle with dimensions x_{min}, y_{min}, x_{max}, and y_{max} and the minimum and maximum potential values are ϕ_{min} and ϕ_{max}. For convenience, we normalize quantities so that all points lie in a three-dimensional space in the range $-0.5 \le X_i'' \le 0.5$, $-0.5 \le Y_i'' \le 0.5$, $-0.5 \le Z_i'' \le 0.5$. If $S_{xy} = \text{Max}(x_{max} - x_{min}, y_{max} - y_{min})$ and $S\phi = \phi_{max} - \phi_{min}$, then the normalizing equations are

$$X'' = -0.5 + \frac{x_i - x_{min}}{S_{xy}} ,$$

$$Y'' = -0.5 + \frac{y_i - y_{min}}{S_{xy}} , \qquad (7.28)$$

$$Z'' = -0.5 + \frac{\phi_i - \phi_{min}}{S_{\phi}} .$$

We now have the tools necessary to make elevation plots. To start, consider the wire frame plot. We identify facet lines by the same method used to create the mesh plot of Section 7.4. The coordinates and potential values at the vertices give the coordinates (X_i'',Y_i'',Z_i'') for the ends of each vector. Rotations are applied to transform these to (X_i',Y_i',Z_i') and then to the drawing plane following Equations 7.25. The final step is to plot the vector with the *MLine* routine.

For a surface plot with hidden facets, it is more convenient to use elements as the basic plotting units. The basic tool is a subroutine that determines the three vertex points of an element facet, plots the outline in the drawing plane, and colors the enclosed space. Again the procedure is straightforward with a graphics library with a flood fill routine that overwrites previous information. Before making the plot the elements are arranged in order of their average distance D from the viewer,

$$X'_c = \frac{X'_1 + X'_2 + X'_3}{3},$$

$$Y'_c = \frac{Y'_1 + Y'_2 + Y'_3}{3},$$

$$Z'_c = \frac{Z'_1 + Z'_2 + Z'_3}{3},$$

$$D = \sqrt{X'^2_c + Y'^2_c + (D - Z'_c)^2}.$$

(7.29)

Plotting the elements in this order gives a hidden facet plot.

The final topic we shall discuss is the creation of electric field line plots. The ideal plot has two characteristics.

- Lines follow the electric field vector
- The spacing between adjacent lines is proportional to the field amplitude

In contrast to equipotential plots there is no general method to create field line plots. One compromise is the arrow plot consisting of an array of arrows drawn over the solution region. The arrow direction follows the electric field and the length is proportional to the field amplitude. Although these plots are sometimes dramatic, they are often confusing.

We can create electric field line plots for solutions without space-charge that have easily recognized symmetry axes. Consider a planar solution with field components E_x and E_y and assume that there is a function $U(x,y)$ such that the electric field equals the curl of $U\mathbf{z}$,

$$E_x = \frac{\partial U}{\partial y}, \quad E_y = -\frac{\partial U}{\partial x}.$$

(7.30)

It is easy to show that lines of constant U are parallel to the electric field and that the difference in the value of U between two points equals the total flux of electric field lines per meter through a surface that connects the points,

$$U_2 - U_1 = \int_1^2 dl \ \mathbf{E} \cdot \hat{n} \ . \tag{7.31}$$

These two properties imply that a contour plot of U is equivalent to an ideal field line plot.

The task is to find the function U(x,y) that corresponds to an electrostatic solution $\phi(x,y)$. We can rewrite Equations 7.29 as

$$\frac{\partial U}{\partial y} = -\frac{\partial \phi}{\partial x}, \quad \frac{\partial U}{\partial x} = \frac{\partial \phi}{\partial y} \ . \tag{7.32}$$

If ϕ satisfies the Laplace equation, Equations 7.32 imply that U does also,

$$\nabla^2 U = 0 \ . \tag{7.33}$$

We can use the same mesh and solution method used to find $\phi(x,y)$ to determine the function U(x,y). The trick is to invert the boundaries for the U solution, exchanging Neumann and Dirichlet conditions. The procedure is best illustrated with examples. Figure 7.9a shows equipotential lines for biased rods inside a grounded box. The left and right rods have constant potential Dirichlet boundary conditions of ±1 V, respectively, and the outer boundary has zero potential. To generate an electric field line plot, we recognize that the plane at x = 0.0 is a Neumann boundary. The electric field line plot of Figure 7.9b results from setting constant potentials on the midplane, 1.0 V along line A and 0.0 V along lines B and C. The outer boundary and rod surfaces satisfy the Neumann condition. The condition was implemented for the internal regions by taking $-\epsilon = 1.0 \times 10^{-7}$ in the enclosed elements. Figure 7.9c shows another example, electric field lines on a laser discharge control electrode. The electrode and lines A and B were set to the Neumann condition, while lines C and D had constant potentials of 1.0 and 0.0 V.

Chapter 7   Exercises

7.1. An isosceles triangle in the x–y plane has the vertex coordinates (4.66,3.44), (10.16,3.44), and (7.41,10.94).

   (a) Find the area of the triangle by applying the formula ½(base) (height).

   (b) Compare the result to the prediction of the routine of Table 7.2.

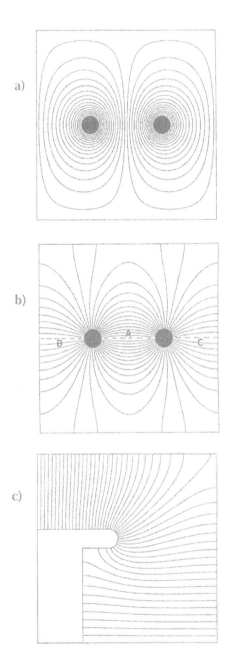

FIGURE 7.9
Electrostatic solution plots in planar geometry. (a) Equipotential plot for rods at ±1 V inside a grounded box. Dirichlet boundary conditions on the surfaces of the rods and box. (b) Electric field line plot for the same system. Neumann boundary condition of the surfaces of the rods and box. Constant potentials of 1 V along line A and 0 V along lines B and C. (c) Electric field line plot of a laser discharge control electrode. Neumann boundary condition on the electrode surface and lines A and B. Constant potentials of 1 V along line C and 0 V along D.

7.2. The cross-section of a toroidal element in a cylindrical system is a right triangle with the following coordinates in the r–z plane: (3.0,3.0), (4.5,3.0), and (3.0,4.5).

    (a) Find the exact volume of the element from an integration of radial slices weighted by $2\pi r$.

    (b) Estimate the volume by taking the product of the triangle area times $2\pi R$, where R is the center of mass radius of the element.

7.3. To understand the search algorithm of Table 7.1, work through steps to find the index of the target value x = 0.62 in the x(i) given below.

0.000	0.080	0.295	0.352	0.381	0.445	0.547	0.683	0.890	0.868	1.000

7.4. Write equations to find the area of an arbitrary quadrilateral element. One approach is to divide the area into two triangles and then to use cross-products to find the corresponding areas. Check the result by calculating the area in the x–y plane bounded by the following coordinates (in order of positive rotation): (1.15,2.03), (12.45,4.32), (13.59,9.44), (4.57, 9.80). Outline a procedure to determine if a test point is inside the element.

7.5. To determine the relative activities of two known radioactive isotopes in a mixture, consider fitting the time-decay of X-ray flux with a function of the form

$$F(t) = \alpha_1 \exp(-\lambda_1 t) + \alpha_2 \exp(-\lambda_2 t).$$

The decay constants are $\lambda_1 = 0.23$ s^{-1} and $\lambda_2 = 0.078$ s^{-1}. Find values for α_1 and α_2 from the following data using the least-squares procedure.

t	F(t)	t	F(t)	t	F(t)
0.0000	1.0089	9.0000	0.3544	18.0000	0.1568
1.0000	0.8766	10.0000	0.3424	19.0000	0.1563
2.0000	0.7878	11.0000	0.3013	20.0000	0.1327
3.0000	0.7000	12.0000	0.2813	21.0000	0.1368
4.0000	0.6121	13.0000	0.2525	22.0000	0.1089
5.0000	0.5424	14.0000	0.2325	23.0000	0.1040
6.0000	0.4975	15.0000	0.2146	24.0000	0.0891
7.0000	0.4566	16.0000	0.1933	25.0000	0.0867
8.0000	0.3986	17.0000	0.1853		

7.6. When an electrostatic solution has a symmetry boundary, we can achieve improved accuracy of nearby interpolations by adding extra points that reflect the symmetry condition. Suppose a planar solution has a boundary along the line y = 0.0 and that there is a nearby point with potential ϕ_i and coordinates (x_i, y_i). Give potential values and co-ordinates of mirror points for the following cases.

(a) Neumann boundary.

(b) Dirichlet boundary with $\phi = \phi_o$.

7.7. Use a spreadsheet with matrix inversion capability to perform a least-squares fit with three parameters. Fit a function of form $f(x) = a_1 + a_2 x + a_3 x^2$ to the following data. Calculate the quantity χ^2 from Equation 7.5 for the best values of a_1, a_2, and a_3. Make small changes in the fitting parameters and confirm that χ^2 has a minimum at the predicted values.

x	F(x)	x	F(x)	x	F(x)
0.000	4.539	2.000	16.309	4.000	16.137
0.500	8.584	2.500	17.379	4.500	14.381
1.000	11.892	3.000	17.900	5.000	11.592
1.500	14.448	3.500	17.276		

7.8. Verify that the harmonic potentials in the second column of Table 7.4 are solutions of the cylindrical Laplace equation.

7.9. We can use harmonic polynomials to fit three-dimensional electrostatic solutions in the absence of space-charge. Derive polynomials of x, y, and z through second-order that are solutions to the three-dimensional Laplace equation in Cartesian coordinates.

7.10. Verify the expressions in the linear interpolation routine of Table 7.3 by comparing them with the equations in Section 2.7.

7.11. Set up a spreadsheet program to create an interactive two-dimensional projection of a cube using Equations 7.26 and 7.27. Represent a cube one unit on a side centered at the origin, $(x'', y'', z'') = (0,0,0)$. The task in-volves transforming the eight vertex coordinates following the mathe-matics in Section 7.5 and plotting lines between them. Set up D_2, θ_x, and θ_z as adjustable parameters.

7.12. Verify that Equation 7.30 implies that lines of constant U are parallel to electric field lines. (Show that the condition that the total derivative of U equals 0 implies that $dy/dx = E_y/E_x$.)

7.13. An electrostatic quadrupole field centered at $(0.0, 0.0)$ is defined by the equations,

$$E_x(x, y) = E_o \frac{x}{a},$$

$$E_y(x, y) = -E_o \frac{y}{a}.$$

(a) Verify that the equations represent a valid electrostatic solution by showing that $\nabla \cdot E = 0$ and $\nabla \times E = 0$.

(b) Find an equation for the electrostatic potential $\phi(x,y)$ assuming $\phi(0,0) = 0$.

(c) Find the function U by applying Equation 7.30.

(d) Use a spreadsheet or plotting program to plot lines of constant ϕ and U (electric field lines).

7.14. A useful technique based on complex numbers generates families of free space electrostatic field solutions in Cartesian coordinates. Consider the complex variable $u = y + jx$, where $j = (-1)^{¼}$. Let f(u) represent any smoothly varying function of u.

(a) Apply the chain rule of partial derivatives to show that f(u) automatically satisfies the Laplace equation.

$$\frac{\partial^2 f}{\partial x^2} + \frac{\partial^2 f}{\partial y^2} = 0$$

(b) If we designate $\phi = \text{Re}(f)$ and $U = \text{Im}(f)$, show that the functions satisfy Equation 7.32.

(c) Show that the choice $f(u) = u^2$ gives the functions ϕ and U for the quadrupole field of Exercise 7.11.

8

Nonlinear and Anisotropic Materials

The road to wisdom?—Well, it's plain
and simple to express:
Err
and err
and err again
but less
and less
and less.

Piet Hein

The dielectric materials we studied in previous chapters had relatively simple properties. They were homogeneous and isotropic — the dielectric constant was independent of position and direction. In this chapter we tackle boundary-value problems that involve more complex materials. The properties of nonlinear materials depend on the field solution. For example, the value of ϵ_r in a dielectric may vary with the electric field magnitude. In this case the Poisson equation becomes nonlinear. Numerical solutions are more difficult with field-dependent materials and may sometimes fail. Iterative techniques are essential because the process is circular. The material properties depend on the field solution and the field solution depends on the materials. Anisotropic materials have characteristics that vary in direction. An example is a birefringent crystal with distinct values of the dielectric constant along two axes. Anisotropic materials may also be nonlinear. The most familiar example, permanent magnets, is discussed in the next chapter.

Section 8.1 describes iterative techniques for numerical solutions with nonlinear materials. Calculations for field-dependent dielectrics illustrate some of the stability problems that may arise. Section 8.2 explains how to represent material properties in numerical tables. Good interpolation techniques are essential to achieve solution convergence. The section concentrates on the cubic spline method which ensures continuity of the interpolated values and first derivatives of the dependent variable. Section 8.3 shows how to represent anisotropic materials in finite-element electrostatic calculations on a conformal triangular mesh. The only change is a modification of the coupling constants. To illustrate an alternate application of the methods, Section 8.4 describes finite-element solutions for steady-state flow of compressible gases. The variation of gas density introduces a nonlinearity.

8.1 Iterative Solutions to Boundary Value Problems

The solutions of previous chapters involved *linear* equations. With known variations of dielectric constant over the solution volume the terms in Poisson's equation depend on the first power of ϕ. The equation becomes *nonlinear* when the dielectric constant is a function of the potential. Often the dielectric constant depends on the amplitude of the electric field,

$$\epsilon = \epsilon\left(|\nabla\phi|\right). \tag{8.1}$$

The challenge of nonlinear boundary value problems is that we do not know the material properties until we have the correct field solution. It is impossible to find *self-consistent solutions* analytically except for the simplest cases. The standard approach in numerical solutions is to apply gradual corrections over many cycles. We make an initial guess of the material properties and then determine a field solution. The initial solution gives the potential gradient and other quantities that enable a better estimate of the properties in each element. Additional field solutions and material corrections follow. Usually the solution approaches the correct self-consistent values after several cycles. We can check convergence by ensuring that spatially averaged values of the adjustment factors approach zero. For example, a useful indicator of relative errors in the dielectric constant in nonlinear electrostatic problems is

$$\delta\epsilon = \left[\frac{\sum_j \left(\epsilon_j^{n+1} - \epsilon_j^n\right)^2}{\sum_j \left(\epsilon_j^n\right)^2} \right]^{1/2}. \tag{8.2}$$

The sums in Equation 8.2 extend over all elements of variable materials. The quantity ϵ_j^{n+1} is the relative dielectric constant in element j after n+1 corrections.

Iterative solutions may fail if the material properties have discontinuities. In this case, the interpolation in elements may bounce back and forth on alternate cycles between values on either side of the discontinuity. The problem can usually be avoided by smoothing discontinuities in the material tables and by averaging corrected values. The following averaging algorithm is effective in electrostatic problems:

$$\epsilon_j^{n+1} = \alpha_1 \ \epsilon_j^n + \alpha_2 \ \epsilon \left(E_j^n \right). \tag{8.3}$$

Here, the quantity ϵ_j^{n+1} is the new estimate of dielectric constant of element j for iteration cycle n+1, while ϵ_j^n and E_j^n are the dielectric constant and electric field from the previous cycle. The quantity $\epsilon(E_j^n)$ is the interpolated field-dependent dielectric constant based on a table or model. Finally, α_1 and α_2 are adjustment constants where $\alpha_1 + \alpha_2 = 1$. A value $\alpha_2 = 1$ gives full correction on each cycle, while $\alpha_2 \ll 1$ implies gradual correction. Low values of α_2 are often necessary for numerical stability.

We can use either matrix (Chapter 6) or relaxation methods (Section 5.5) for iterative field solutions. Relaxation is usually fast and efficient because the field and material corrections are carried out simultaneously. Material adjustment can be applied on each field relaxation cycle or at specified intervals. Solution convergence can be obtained for most problems by experimenting with values for the relaxation parameter ω and material adjustment parameters (Equation 8.2). Section 9.5 discusses procedures to minimize the number of operations involved in material correction.

To illustrate the rich undergrowth of phenomena that may occur in nonlinear solutions, consider the geometry of Figure 8.1a, a dielectric block of height 0.05 m between a ground plane and an electrode at 2000 V. The solution for constant ϵ is a uniform variation of potential as shown. The behavior is quite different for a saturable dielectric where ϵ decreases with increasing field amplitude. Figure 8.1b shows the consequences of such a variation at high field. Numerical modeling is difficult because the system is inherently unstable. The electric field concentrates in regions of low ϵ. Therefore, a slight field imbalance can grow. Figure 8.1c shows equipotential lines for the nonlinear solution. The relaxation solution required about 3500 cycles with $\omega = 1.7$ and $\alpha_2 = 0.05$. The vertical electric field is no longer uniform. There is a high field region with $\epsilon \approx 1$ at the top and a low field region with $\epsilon \approx 20$ at the bottom. The problem is indeterminant — the solution is one of a continuum of valid equilibria with the high-field band at different positions in the gap. The particular solution of Figure 8.1c occurs consistently because of slight asymmetry resulting from the order of relaxation. We can define a unique solution by introducing geometric variations. For example, the top electrode in Figure 8.1d is shaped so that the high-field region is forced to the bottom. Lowering the electrode potential leads to interesting solutions. Figure 8.1e shows the field distribution for an applied voltage of 1000 V. The high-field region compresses to a narrowed band with reduced potential, ultimately reverting to the uniform field of Figure 8.1a. The phenomenon is related to self-focusing of intense laser light in a saturable medium. The example illustrates

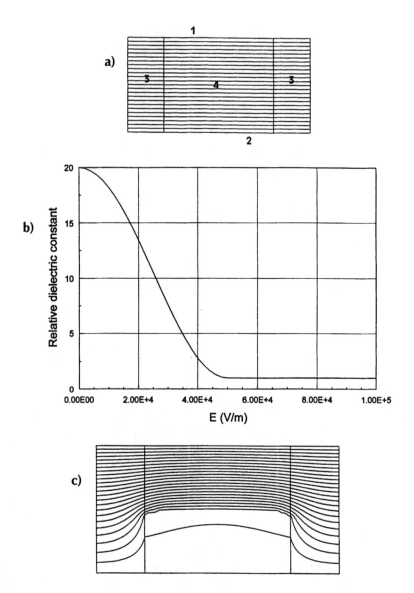

FIGURE 8.1
Nonlinear field solutions in a dielectric block. (a) Geometry and equipotential lines for uniform
relative dielectric constant $\epsilon_r = 20.0$. 1, Top electrode, 2000.0 V; 2, bottom electrode, 0.0 V; 3, air
spaces; 4, dielectric block. Neumann boundaries at left and right. (b) Nonlinear dielectric,
variation of ϵ_r with field amplitude. (c) Equipotential lines for the nonlinear solution.
(d) Equipotential lines for a nonlinear solution with a shaped top electrode. (e) Solution with
top electrode voltage lowered to 1000.0 V.

that even the simplest nonlinear solutions can get out of hand. Section 8.4
shows another example related to gas flow. Considerable vigilance is
required to ensure that results are physically meaningful.

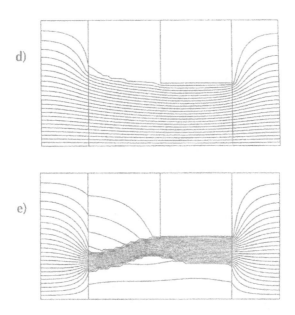

d)

e)

FIGURE 8.1 (continued)

8.2 Numerical Data for Material Properties

This section covers representations of materials with variable properties. One option is to hardwire subroutines that contain parametric models for different materials into the solution program. This approach has several disadvantages.

- The program is locked to one type of physical problem
- The programmer must anticipate all future user needs; new materials require program modification, recompilation, and revised documentation
- Parameters must be determined for new experimental or numerical material data

A better approach is to enter all material data in a standard numerical format. This usually takes the form of a one-dimensional table consisting of an array of independent and dependent variables. As an example, Table 8.1 shows the dielectric data table used for the example of Figure 8.1. The solution program includes a unit of routines to organize and to interpolate tables. There are several advantages to this approach.

TABLE 8.1

Field-Dependent Dielectric Constant

E (V/m)	ϵ_r	E (V/m)	ϵ_r
0.00	20.00000	28000.00	8.71988
2000.00	19.92509	30000.00	7.56434
4000.00	19.70154	32000.00	6.45510
6000.00	19.33288	34000.00	5.40965
8000.00	18.82491	36000.00	4.44447
10000.00	18.18566	38000.00	3.57480
12000.00	17.42520	40000.00	2.81434
14000.00	16.55553	42000.00	2.17509
16000.00	15.59036	44000.00	1.66712
18000.00	14.54490	46000.00	1.29846
20000.00	13.43566	48000.00	1.07491
22000.00	12.28012	50000.00	1.00000
24000.00	11.09651	52000.00	1.00000
26000.00	9.90349	54000.00	1.00000
		100000.00	1.00000

- The program user takes responsibility for developing and documenting physical models
- A single program can often cover a wide variety of physical problems
- New materials can be included with no program modification
- Data entry from experiments, spreadsheets, or user programs is straightforward
- The interpolation unit can also be applied to analyze arbitrary time-dependent functions in initial value problems

We shall cover two common techniques to determine values from one-dimensional tables: polynomial and cubic spline interpolation. For well-behaved data, the cubic spline technique provides smoother values and ensures continuity of the first derivative. These features may improve convergence of iterative calculations. The method is also more efficient in applications involving large numbers of interpolations because coefficients are computed once at the beginning of a calculation for the full range of entries. On the negative side, the cubic spline method can give nonphysical values for tabular data with noise or discontinuities of slope. In this case, the more tolerant method of polynomial interpolation does a better job.

We begin with the polynomial method. Suppose a one-dimensional table consists of an array of the independent variable x_i and dependent variable y_i. By convention we take the index from $i = 1$ to N. The entries are arranged in order of increasing values of x but need to be spaced at a uniform interval. The procedure is to find a polynomial function of the form

$$y = f(x) = a_o + a_1 x + \; + a_2 x^2 + a_3 x^3 + \ldots . \tag{8.4}$$

such that it matches values of y at several data points near the target value of x. We evaluate the function at x to estimate y. The polynomial is a straight line if we pick two data points adjacent to x (*linear interpolation*). Three nearby points define a unique quadratic function, and so forth. Suppose we pick M data points near x: x_1, x_2, ... x_j, ... x_M. Lagrange's formula gives the following expression for the polynomial that passes through the points y_1, y_2, ... y_j, ... y_M,

$$f(x) = \frac{(x-x_2)(x-x_3)...(x-x_M)}{(x_1-x_2)(x_1-x_3)...(x_1-x_M)} y_1$$
$$+ \frac{(x-x_1)(x-x_3)...(x-x_M)}{(x_2-x_2)(x_2-x_3)...(x_2-x_M)} y_2 \quad\quad (8.5)$$
$$...$$
$$\frac{(x-x_2)(x-x_2)...(x-x_{M-1})}{(x_M-x_2)(x_M-x_3)...(x_M-x_{M-1})} y_M.$$

Equation 8.5 can be implemented in a subroutine that fits M points in an ordered array near the target value x. Alternative methods for polynomial interpolation and practical subroutines are covered in *Numerical Recipes in FORTRAN*, (Press, 1992).

We must decide how to pick the value of M. For instance, why not pick M = N and define a polynomial that passes through all points of the data array? The answer is that although such a choice would have advantages similar to a cubic spline interpolation, there may be severe drawbacks. The advantage is that the N coefficients of Equation 8.3 need be computed once for each table. Problems occur when the data are imperfect. The resulting polynomial passes through all the data points and maintains smoothness through derivatives of order (N–1). If the data are not smooth the resulting curve can be highly distorted. Figure 8.2a illustrates the problem. The data set consists of 11 points that follow a quarter cycle of a sine curve with a jog at about x = 0.55. The solid line is a linear interpolation between sequential points, while the dashed line curve is a 10th-order polynomial fit. If the data followed a smooth sinusoid, the 10th-order fit would be much closer to the theoretical curve. With the jog it is clear that although the high-order polynomial is mathematically correct it is useless as an interpolation. Linear interpolations are always safe and second- or third-order polynomials generally give good fits. The sensitivity to small errors increases with the order of the polynomial.

The goal of cubic spline interpolations is more modest. The coefficients are chosen so that the interpolation curve passes through all data points while maintaining smoothness of the first derivative and continuity of the second derivative between intervals. As a result the cubic spline is less sensitive to

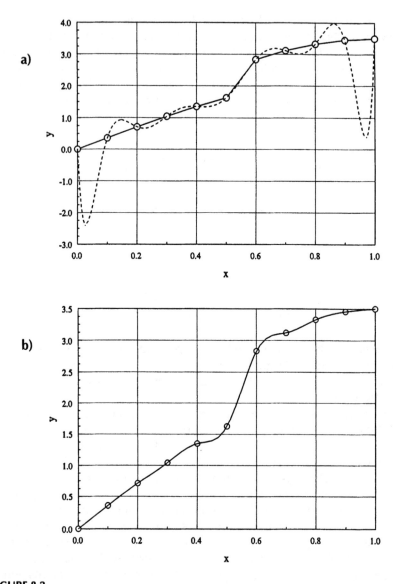

FIGURE 8.2
Global interpolations for a data set with a discontinuity (open circles). (a) Solid line, linear interpolation; dashed line, tenth-order polynomial fit. (b) Natural cubic spline fit.

data imperfections than a complete polynomial fit. Consider a tabular function with N data sets (x_i, y_i) arranged in order of increasing x. A linear interpolation over the interval between x_i and x_{i+1} can be written as

$$y = A\, y_i + B\, y_{i+1}, \tag{8.6}$$

where

$$A = \frac{x_{i+1} - x}{x_{i+1} - x_i}, \quad B = 1 - A = \frac{x - x_i}{x_{i+1} - x_i}. \tag{8.7}$$

Suppose we knew a set of second derivatives at x_i for an interpolating function,

$$y_i'' = \frac{d^2 y}{dx^2}\bigg|_{x_i} \tag{8.8}$$

To begin we consider the values in Equation 8.8 to be arbitrary coefficients in a cubic interpolation function expressed in terms of $\Delta x = x - x_i$. One possible form of the function is

$$y = Ay_i + By_{i+1} + Cy_i'' + Dy_{i+1}'', \tag{8.9}$$

where A and B are given by Equation 8.7 and

$$C = \frac{1}{6}(A^3 - A)(x_{i+1} - x_j)^2,$$
$$D = \frac{1}{6}(B^3 - B)(x_{i+1} - x_j)^2. \tag{8.10}$$

Note that the quantity C equals zero when A equals 0 or 1 and D equals zero when B equals 0 or 1. Therefore, the function in Equation 8.8 passes through the values y_i at x_i. The motivation for the factor 1/6 will soon be apparent.

The first derivative of Equation 8.9 is

$$\frac{dy}{dx} = y_i \frac{dA}{dx} + y_{i+1} \frac{dB}{dx} + y_i'' \frac{dC}{dA}\frac{dA}{dx} + y_{i+1}'' \frac{dD}{dB}\frac{dB}{dx}. \tag{8.11}$$

Substituting from Equations 8.7 and 8.10, we can rewrite Equation 8.11 as

$$\frac{dy}{dx} = \frac{y_{i+1} - y_i}{x_{i+1} - x_i} - \frac{3A^2 - 1}{6}(x_{i+1} - x_i)y_i'' + \frac{3B^2 - 1}{6}(x_{i+1} - x_i)y_{i+1}''. \tag{8.12}$$

Taking the derivative of Equation 8.12 gives the second derivative as

$$\frac{d^2 y}{dx^2} = Ay_i'' + By_{i+1}''. \tag{8.13}$$

The motivation for the choice of coefficients is clear in Equation 8.13. The cubic expansion ensures continuity of the second derivative of the interpolating function. The quantity varies linearly from y_i'' to y_{i+1}'' across the interval x_i to x_{i+1}.

For any choice of the set of y_i'' the curve will pass through the data points and maintain a continuous second derivative. It remains to decide how to make the best choice to fit the data. We shall seek a curve that guarantees continuity of the first derivative across the boundaries between intervals. This defines a set of equations to determine y_i''. Consider, for example, equality of the derivatives on both sides of the point x_i. Using Equations 8.7 and 8.12 we find

$$\frac{y_i - y_{i-1}}{x_i - x_{i-1}} + \frac{1}{3}\left(x_i - x_{i-1}\right) y_i'' - \frac{1}{6}\left(x_i - x_{i-1}\right) y_{i-1}'' =$$

$$\frac{y_{i+1} - y_i}{x_{i+1} - x_i} - \frac{1}{3}\left(x_{i+1} - x_i\right) y_i'' - \frac{1}{6}\left(x_{i+1} - x_i\right) y_{i+1}'' .$$

(8.14)

We can write Equation 8.14 as

$$\left[\frac{x_i - x_{i-1}}{6}\right] y_{i-1}'' + \left[\frac{x_{i+1} - x_{i-1}}{3}\right] y_i'' + \left[\frac{x_{i+1} - x_i}{6}\right] y_{i+1}'' =$$

$$\left[\frac{y_{i+1} - y_i}{x_{i+1} - x_i} - \frac{y_i - y_{i-1}}{x_i - x_{i-1}}\right] .$$

(8.15)

The above relationship represents a set of N-2 equations. We need two more conditions to determine the N unknown values of y_i''. A common choice, called the *natural cubic spline*, is to take zero second derivatives at the ends of the interpolation range, $y_1'' = 0$ and $y_N'' = 0$. Note that the set of equations represented by Equation 8.14 is in tridiagonal form. Given the values of y_i and x_i, we can solve for the spline coefficients y_i'' using the method of back-substitution (Section 4.4) or the tridiagonal matrix inversion of Section 6.2.

In summary, the following procedure yields interpolated values from a numerical table. The first step is to enter the data arrays of N values of x_i and y_i. It is good practice to include a sorting routine in case the user has supplied values out of order. The next step is to calculate a third array of splines, y_i''. This calculation is accomplished by computing coefficients in Equation 8.15 from the data set and solving the set of linear equations. To obtain interpolated quantities at x, a fast search (Table 7.1) is performed to find the data entries x_i and x_{i+1} below and above the target value. The quantities A, B, C, and D are then determined from Equations 8.7 and 8.10. Substitution in Equation 8.9 then gives the interpolated value of y. Equation 8.12 can be used in cases where the first derivative, dy/dx, is required. For example, the cubic spline can be used to find the value of μ for ferromagnetic materials from a table of B vs. H.

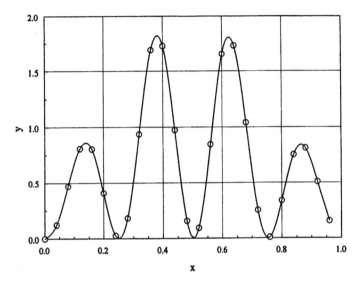

FIGURE 8.3
Natural cubic spline fit to a data set of 25 points given by the equation illustrates the quality of a cubic spline interpolation when the data set is consistent with smooth variations of the function and first derivative. The coarsely spaced set of 25 data values were generated from the function y(x) = exp [- (0.40x – 0.125)²] [1 – cos (25x)].

Figure 8.3 illustrates the quality of a cubic-spline interpolation when the data set is consistent with smooth variations of the function and first derivative. The coarsely spaced set of 25 data values were generated from the function

$$y_x = \exp\left[-(0.40x - 0.125)^2\right]\left[1 - \cos(25x)\right]. \qquad (8.16)$$

The interpolation (solid line) follows the theoretical variation closely. Finally, Figure 8.2b illustrates the cubic spline interpolation for the data set with a discontinuity. The nonphysical variations are not as severe as those for the complete polynomial fit. Nonetheless, the fit is poor near the discontinuity. It is a good practice to check cubic spline interpolations for new or modified material tables and presmooth the data if necessary.

8.3 Finite-Element Equations for Anisotropic Materials

Anisotropic materials have different properties along different directions. For example, shifts in polarization charge in dielectrics may differ along crystalline axes. This section covers two-dimensional finite-element models for simple anisotropic dielectrics. The crystalline axes are in the plane of solution

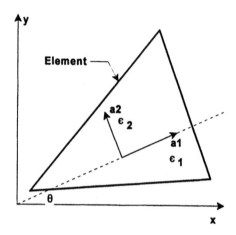

FIGURE 8.4
Global coordinate axes and crystalline axes a_1 and a_2 of a nonlinear dielectric element.

at a right angle to each other along unit vectors $\hat{a}_1$ and $\hat{a}_2$ (Figure 8.4). Constant values of relative dielectric constant ϵ_1 and ϵ_2 apply along the two directions. An applied field E_{o1} along $\hat{a}_1$ gives a total field E_1

$$E_{o1} = \epsilon_1 \, E_1. \tag{8.17}$$

A similar result holds along a_2. For arbitrary field orientation the applied field is related to the total field components projected along the axes by

$$E_o = \epsilon_1 \, E_1 \, \hat{a}_1 + \epsilon_2 \, E_2 \, \hat{a}_2 = E_{o1} \, \hat{a}_1 + E_{o2} \, \hat{a}_2. \tag{8.18}$$

To treat problems with several dielectrics at different orientations, we need field expressions in a global coordinate system. Figure 8.4 shows the geometry for a planar solution with a_1 inclined at an angle θ with respect to the x-axis. The total and applied fields are related by the matrix relationship

$$\begin{bmatrix} E_{ox} \\ E_{oy} \end{bmatrix} = \begin{bmatrix} \epsilon_{xx} & \epsilon_{xy} \\ \epsilon_{yx} & \epsilon_{yy} \end{bmatrix} \begin{bmatrix} E_x \\ E_y \end{bmatrix}, \tag{8.19}$$

which we can write symbolically as

$$E_o = \epsilon \cdot E. \tag{8.20}$$

The goal is to find expressions for the components of the dielectric tensor in Equation 8.19.

The total field can be expressed in terms of components along either the global or crystal coordinate systems

$$E = E_x \ \hat{x} + E_y + \hat{y} = E_1 \ a_1 + E_2 \ a_2. \tag{8.21}$$

Unit vectors in the two systems are related by

$$x = \cos\theta \ \hat{a}_1 - \sin\theta \ \hat{a}_2,$$
$$y = \sin\theta \ \hat{a}_1 + \cos\theta \ \hat{a}_2. \tag{8.22}$$

Substitution in Equation 8.21 gives

$$E = \left(E_x \ \cos\theta + E_y \ \sin\theta\right) \hat{a}_1 + \left(-E_x \ \sin\theta + E_y \ \cos\theta\right) \hat{a}_2. \tag{8.23}$$

Using Equation 8.18 we can write the applied field components along the crystalline axes as

$$E_{01} = \left(E_x \ \cos\theta + E_y \ \sin\theta\right) \epsilon_1,$$
$$E_{02} = \left(-E_x \ \sin\theta + E_y \ \cos\theta\right) \epsilon_2. \tag{8.24}$$

We can relate the applied field components of Equation 8.24 to those along the coordinate axes using the transformation of Equation 8.22,

$$E_{ox} = \cos\theta \ E_{o1} - \sin \ E_{o2},$$
$$E_{oy} = \sin\theta \ E_{o1} + \cos\theta \ E_{o2}. \tag{8.25}$$

Combining Equations 8.24 and 8.25 gives a relationship between the applied and total field components in the reference coordinate system,

$$E_{ox} = E_x \left[\epsilon_1 \ \cos^2\theta + \epsilon_2 \ \sin^2\theta\right] + E_y \left[\epsilon_1 \ \cos\theta \ \sin\theta - \epsilon_2 \ \sin\theta \ \cos\theta\right],$$
$$E_{oy} = E_x \left[\epsilon_1 \ \cos\theta \ \sin\theta - \epsilon_2 \ \sin\theta \ \cos\theta\right] + E_y \left[\epsilon_1 \ \sin^2\theta - \epsilon_2 \ \cos^2\theta\right]. \tag{8.26}$$

Comparing Equations 8.19 and 8.26, we can identify the components of the dielectric tensor,

$$\epsilon_{xx} = \epsilon_1 \ \cos^2\theta + \epsilon_2 \ \sin^2\theta,$$
$$\epsilon_{xy} = \epsilon_{yx} = \left(\epsilon_1 - \epsilon_2\right) \cos\theta \ \sin\theta,$$
$$\epsilon_{yy} = \epsilon_1 \ \sin^2\theta + \epsilon_2 \ \cos^2\theta. \tag{8.27}$$

The next step is to modify expressions for Gauss' law at a vertex (Section 2.7) to include the tensor dielectric constants. Figure 2.12 shows the geometry and notation. For simplicity, the coordinate values are given relative to the reference vertex. Gauss' law has the form

$$\iint_S (\epsilon \cdot \mathbf{E}) \cdot \mathbf{n} \, dS = \frac{\iiint_V dV \, \rho_o}{\epsilon_o}. \tag{8.28}$$

Again, we shall consider triangle 2 where the total electric field is given by

$$\mathbf{E}_2 = -u_2 \, \hat{\mathbf{x}} - v_2 \, \hat{\mathbf{y}} \,. \tag{8.29}$$

The applied field in the element is given by

$$\mathbf{E}_{o2} = -\left(u_2 \epsilon_{xx2} + v_2 \epsilon_{xy2}\right) \hat{\mathbf{x}} - \left(u_2 \epsilon_{yz2} + v_2 \epsilon_{yy2}\right) \hat{\mathbf{y}} \,. \tag{8.30}$$

The algebra is straightforward but disheartening. Again, we write out the surface integral and collect terms multiplying the potential at the test vertex and its neighbors to define a difference relationship. Equation 2.67 holds with modified expressions for the coupling constants,

$$W_i = w_{ia} + w_{i+1b} \,, \tag{8.31}$$

where

$$w_{ia} =$$

$$\frac{\epsilon_{xxi} y_i \left(y_{i+1} - y_i\right) - \epsilon_{xyi} x_i \left(y_{i+1} - y_i\right) - \epsilon_{yxi} y_i \left(x_{i+1} - x_i\right) + \epsilon_{yyi} x_i \left(x_{i+1} - x_i\right)}{4a_i}, \tag{8.32}$$

and

$$w_{ib} =$$

$$\frac{-\epsilon_{xxi} y_{i+1} \left(y_{i+1} - y_i\right) + \epsilon_{xyi} x_{i+1} \left(y_{i+1} - y_i\right) + \epsilon_{yxi} y_{i+1} \left(x_{i+1} - x_i\right) - \epsilon_{yyi} x_{i+1} \left(x_{i+1} - x_i\right)}{4a_i}. \tag{8.33}$$

The indices in Equations 8.32 and 8.33 should be interpreted as cyclical in the number of triangles surrounding the vertex. To illustrate the process, Figure 8.5 shows equipotential plots for an anisotropic dielectric rod between biased parallel plates. The dielectric constants along the primary axes are $\epsilon_1 = 3.0$ and $\epsilon_2 = 1.0$.

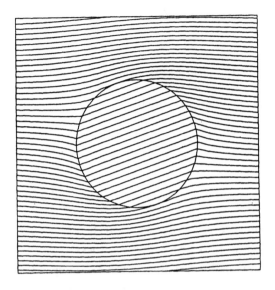

FIGURE 8.5
Equipotential plots for an anisotropic dielectric rod immersed in a uniform vertical field: $\epsilon_1 = 3.0$, $\epsilon_2 = 1.0$, and $\theta = 45.0°$.

8.4 Steady-State Gas Flow

To illustrate nonlinear solutions to the Poisson equation we review steady-state nonturbulent gas flow. The methods are useful for aerodynamics and pneumatic systems. We shall treat compressible fluid flow in the ideal gas limit where the kinetic energy of molecules accounts for most of the internal energy and the forces between molecules are small. An inherent assumption is that the gas is in local thermodynamic equilibrium with isotropic pressure. Ideal gases obey the law of Boyle and Gay-Lussac,

$$p = R\rho T. \tag{8.34}$$

The pressure p is in pascals, the density ρ in kg/m³, and the temperature T in degrees K. The quantity R equals

$$R = \frac{R_o}{M_o}, \tag{8.35}$$

where $R_o = 8.31451$ (Pa-m³-mol/K) and M_o equals the molar mass of the gas. For example, 1 mol of oxygen has $M_o = 32$ kg. *Internal energy* is the energy content per unit mass. For ideal gases, the internal energy is given by

$$u = \frac{R}{\gamma - 1} T, \qquad (8.36)$$

where the constant γ (the *adiabatic exponent*) depends on the gas. The value for air at moderate temperatures is $\gamma = 1.4$. The adiabatic exponent is related to the specific heat at constant pressure by

$$\gamma = \frac{C_P}{C_P - R}. \qquad (8.37)$$

We shall seek steady-state solutions for gas flow under the following conditions:

- Wall friction and boundary layer effects are negligible
- The flow is nonturbulent without vortices
- Velocities are small compared to the sound speed
- Changes in internal energy arise from work performed by and on the gas by compression and expansion

The third condition excludes shock phenomena. For reference, the sound speed in an ideal gas is

$$c_s = \sqrt{\gamma R T}. \qquad (8.38)$$

The last condition ensures that processes are reversible so there is no change in entropy.

For *isentropic* processes, the change in the total energy U of a volume V of gas at pressure p is

$$dU = -p \, dV, \qquad (8.39)$$

where dV is a change in volume. Dividing both sides of Equation 8.39 by the mass of the gas in the volume gives the change in internal energy

$$du = -p \, d\left(\frac{1}{\rho}\right) = \frac{p \, d\rho}{\rho^2} = RT \frac{d\rho}{\rho}. \qquad (8.40)$$

Equation 8.34 was used for the final form in Equation 8.40. We can also find the change in internal energy from Equation 8.36,

$$du = \frac{R}{\gamma - 1} dT. \qquad (8.41)$$

Equating the expressions of Equations 8.40 and 8.41 gives

$$\frac{d\rho}{\rho} = \left[\frac{1}{\gamma-1}\right]\frac{dT}{T},\qquad(8.42)$$

or

$$\left[\frac{\rho}{\rho_o}\right] = \left[\frac{T}{T_o}\right]^{\frac{1}{\gamma-1}}\qquad(8.43)$$

Consider a gas in isentropic flow with local velocity v and with no external energy source. Conservation of energy dictates that the sum of kinetic and internal energies for a unit mass of gas is constant, or

$$\frac{v^2}{2} + \left[\frac{R}{\gamma-1}\right]T = \left[\frac{R}{\gamma-1}\right]T_o.\qquad(8.44)$$

In Equation 8.44 the reference temperature T_o is the value for a stationary gas. The equation implies that there is a maximum gas speed that can be attained in any process

$$v_o = \sqrt{\frac{2RT_o}{\gamma-1}}.\qquad(8.45)$$

Finally, we can combine Equations 8.43 and 8.44 into a relationship that gives the gas density in terms of the local value of the gas speed

$$\frac{\rho}{\rho_o} = \left[1-\left(\frac{v}{v_o}\right)^2\right]^{\frac{1}{\gamma-1}}\qquad(8.46)$$

The assumption of no vorticity (circulating flow around a point) implies that

$$\nabla \times v = 0.\qquad(8.47)$$

By analogy with the theory of electrostatics (Section 2.3), Equation 8.47 implies that the velocity vector can be written as the gradient of a scalar quantity,

$$v = \nabla\Phi.\qquad(8.48)$$

In steady state the total mass flow into any volume of the solution must be zero, or

$$\nabla \cdot \left(\frac{\rho}{\rho_o} \mathbf{v} \right) = 0 . \tag{8.49}$$

Combining Equations 8.48 and 8.49 gives

$$\nabla \cdot \left(\frac{\rho}{\rho_o} \nabla \Phi \right) = 0 . \tag{8.50}$$

Equation 8.50 is Laplace's equation. In comparison with electrostatic solutions, the velocity potential Φ is the analogy of the electrostatic potential ϕ and the relative density ρ/ρ_o plays the same role as the relative dielectric constant. The problem is nonlinear because the density is a function of the velocity potential through Equation 8.46, $\rho = \rho(|\nabla \Phi|^2)$.

The Dirichlet condition Φ = constant implies that the gas velocity is normal to the surface. The condition is useful for symmetry boundaries or to define ideal entrance and exit planes. Fluid flow is parallel to a Neumann boundary. The Neumann condition represents impenetrable surfaces like the skin of an airfoil. As an example Figure 8.6 shows hydrodynamic flow of an incompressible fluid ($\rho/\rho_o = 1$) around one vane of a lattice. The solution involves four regions: a constant Φ condition on the entrance boundary at the left (region 1) and exit boundary on the right (region 2), the fluid volume (region 3), and the vane (region 4). The fluid velocity is normal to the two constant potential surfaces. If $\Phi = 0$ at the entrance then the exit value is $\Phi = -v_oL$, where v_o is the flow velocity at infinity and L is the distance between the planes. An internal Neumann boundary represents the surface of the vane. We can achieve the condition by setting the relative density of internal elements to a low value ($\rho/\rho_o = 10^{-9}$) and ignoring potential variations inside the structure. To determine the force on the vane, we must correctly represent the transverse motion of fluid. For this, the top and bottom boundaries have periodic conditions. This means that the coupling coefficients to and potential values of vertices above mesh row J are taken from corresponding vertices in row 2. Similarly, vertex properties below row 1 are taken from row (J −1). To implement this condition with the triangular mesh discussed in Chapter 5, we must ensure that vertex positions on the top and bottom rows are the same and that J is an odd number. The resulting solution of Figure 8.6a shows the velocity potential for fluid flow at an average velocity of 1 m/s. Note the fluid acceleration and displacement near the vane. Figure 8.6b is a plot of velocity streamlines created with the technique described in Section 7.5.

To illustrate problems that may occur with nonlinear materials, consider solutions for a rapidly flowing compressible gas around an obstruction in a duct (Figure 8.7). Neumann boundaries represent rigid walls at the top and

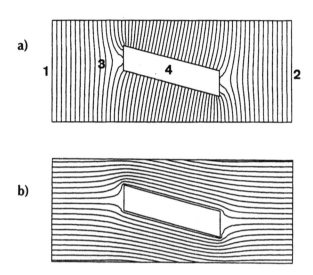

FIGURE 8.6
Flow of incompressible fluid past an array of vanes. Average velocity, 1 m/s. Region 1, entrance boundary at $\Phi = 0$; region 2, exit boundary at $\Phi = -v_oL$; region 3, fluid volume; region 4, impenetrable vane. (a) Contours of I. (b) Velocity streamlines.

bottom. We shall use argon gas at 1000 K with mass number A = 39.95. The gas constant in Equation 8.34 is R = 206.7 m²/K-s. The specific heat value C_p = 517 J/K-kg implies an adiabatic exponent of $\gamma = 1.668$. We can check the value by calculating the sound speed at room temperature (293 K) from Equation 8.38. The prediction is 317.8 m/s, close to the handbook value of 319 m/s. The maximum velocity in Equation 8.45 is v_o = 1017 m/s. We can use this information in Equation 8.46 to set up a table of relative density vs. gas speed (Table 8.2). Figure 8.7a shows a solution for a difference in velocity potential of 100 m²/s, corresponding to an average velocity of 200 m/s over the 0.5 m system. The well-behaved solution shows acceleration and rarefaction of the gas flowing around the object. We can find the force on the object by calculating the pressure in external elements from Equation 8.34 and integrating around the object boundary. With a small increase in velocity potential drop to 110 m²-s it is impossible to find a physical solution. Figure 8.7b shows the corresponding solution with diverging velocity in a small region. An inspection of the third column of Table 8.2 shows the cause of the problem. This column plots relative flux as a function of gas speed. While the flux increases with speed to $v = v_o/2$, it drops off at higher values. An increase in the applied velocity potential increases the total gas flux across the system. There must be a corresponding increase in velocity in flux in regions near the obstruction. When the gas velocity in this area exceeds $v_o/2$, further increases in velocity lower the flux. Therefore, it becomes impossible to satisfy global flux conservation and the solution crashes. Physically, this means that the assumption of isotropic pressure is no longer valid and the system must be described by a supersonic flow solution.

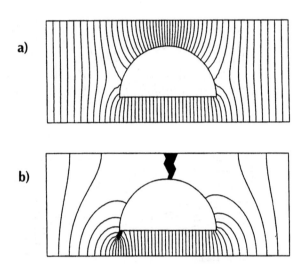

FIGURE 8.7
Flow of compressible fluid around an obstruction in a duct, contours of velocity potential. Argon gas at 1000 K. Horizontal dimension of solution region: 0.5 m. (a) Stable solution with difference in velocity potential between entrance and exit boundaries: $\Delta\Phi = 100$ m^2/s. (b) $\Delta\Phi = 110$ m^2/s.

TABLE 8.2

Argon Density (1000 K), $v_o = 1017$ m/s

v (m/s)	ρ/ρ_o	$(\rho/\rho_o)\,v$
0.0	1.000000	0.000
50.0	0.996384	49.819
100.0	0.985561	98.556
150.0	0.967611	145.142
200.0	0.942665	188.533
250.0	0.910911	227.728
300.0	0.872595	261.779
350.0	0.828023	289.808
400.0	0.777568	311.027
450.0	0.721673	324.753
500.0	0.660862	330.431
550.0	0.595752	327.664
600.0	0.527063	316.238
650.0	0.455641	296.166
700.0	0.382487	267.741
750.0	0.308802	231.601
800.0	0.236055	188.844
850.0	0.166107	141.191
900.0	0.101446	91.302
950.0	0.045765	43.477
1000.0	0.006098	6.098

Chapter 8 Exercises

8.1. Parallel plates that extend an infinite distance in y and z are separated by distance d in x. The plates have applied potentials $\phi(0) = 0$ and $\phi(d) = V_0$. The region $0 \le x \le d/2$ is filled with a material with a relative dielectric constant that depends on the electric field E_x,

$$\epsilon_r(E_x) = 1 - \alpha \frac{E_x d}{V_0}.$$

Find the maximum value of α consistent with a well-determined solution where the electric field has unique uniform values in the material and vacuum regions.

8.2. Set up a spreadsheet to model the system of Exercise 8.1 using successive over-relaxation. Use cell formulas that simultaneously relax the potential and adjust the dielectric constant. Experiment with different values of α, ω, and dielectric averaging parameters.

8.3. The two discrete data sets listed below model a continuous function. The first has a smooth variation and the second has superimposed random noise. Apply the Lagrange formula for polynomial interpolation to estimate a values for f at the point $x = 0.55$. (Use a spreadsheet to save work in applying Equation 8.5.) Compare values for linear, second-order, and fifth-order fits to the two data sets. The ideal value for the generating function is $f = 1.10714$.

x	f(x)	f(x)
0.00000	0.33000	0.34893
0.20000	0.60686	0.61953
0.40000	0.89754	0.87410
0.60000	1.17267	1.14996
0.80000	1.39519	1.39877
1.00000	1.51000	1.53165

8.4. Derive a general expression for second-order polynomial interpolation on a data set with uniform spacing Δx.

8.5. The discrete data set listed below approximates the function $f(x) = x^2/10 - \sin(x)$.

(a) Find the natural spline coefficients for the set.

(b) Estimate values for y, dy/dx, and d^2y/dx^2 at $x = 2.5$.

(c) Compare the results of Part (b) to the analytic values.

x	0.0000	1.0000	2.0000	3.0000	4.0000	5.0000
f(x)	0.0000	-0.7415	-0.5093	0.7589	2.3568	3.4589

8.6. An alternative to the natural cubic spline is to set values of y_1'' and y_N'' to give desired values of dy/dx at the ends of the interval. Use Equation 8.12 to find y_1'' and y_N'' in terms of the desired values y_1' and y_N'.

8.7. An anisotropic dielectric rod is immersed in a uniform applied electric field $\mathbf{E}_o = E_o\mathbf{y}$. Assume that the applied field $\mathbf{E}_o$ is uniform inside the rod. If $\epsilon_1 = 5$, $\epsilon_2 = 1$, and axis 1 of the dielectric inclines at an angle $\theta = 45°$ relative to the x-axis, show that the inclination of equipotential lines is 33.69°.

8.8. Describe how to adapt the theory of Section 8.3 to cover steady-state current flow in media with anisotropic conductivities σ_1 and σ_2.

8.9. The specific heat at constant pressure for neon at 25°C and 1 atm is $C_p = 1034$ J/kg-K.

(a) Find the adiabatic exponent γ for the gas.

(b) Find the predicted sound speed at 0°C and compare the result to the experimental value of 435 m/s.

(c) What is the internal energy of 1 m³ of neon at 1 atm and T = 100°C?

8.10. Investigate the stability of one-dimensional flow of air at 1 atm and T = 500 K. The specific heat is $C_p = 1030$ J/kg-K.

(a) Find the value of the adiabatic exponent γ and maximum gas speed v_o.

(b) For a region of width d = 1.0 m, find the limit on velocity potential for a valid solution with uniform velocity and density in x using the theory of Section 8.4.

9

Finite-Element
Magnetostatic Solutions

*The chief object of education is not to learn things
but to unlearn things.*

G.K. Chesterton

Having assembled a toolbox of mathematical methods, we can proceed to applications in electromagnetism and related areas of physics. We shall start with solutions for static magnetic fields. The main difference from electrostatic solutions is that magnetic materials usually have more complex properties than dielectrics. We shall make extensive use of the methods of the previous chapter. Sections 9.1 through 9.3 review the equations of magnetostatics and the properties of magnetic materials. Section 9.1 covers definitions of current, the magnetic force law, and Ampere's law. The latter is the basis of magnetostatic calculations with finite-difference and finite-element techniques. Accordingly, we shall discuss Ampere's law in differential and integral forms. Section 9.2 introduces the magnetic vector potential. Solving for this quantity greatly simplifies two-dimensional calculations. In planar and cylindrical geometries the vector potential has a single component that plays a role analogous to the scalar electrostatic potential.

The properties of magnetic materials may vary with direction and may depend on their history of exposure to magnetic fields. Section 9.3 concentrates on a subset of materials with relatively simple properties and broad applicability, namely isotropic media where the magnetic permeability is a single-valued function of the field amplitude. This class incudes soft iron, transformer steel, and many ferrites.

Section 9.4 derives finite-element equations for magnetostatics on a conformal mesh from the integral form of Ampere's law. The result is a set of linear equations that relate values of vector potential to those at neighboring vertices. The coupling coefficients are similar to those for electrostatics — the relative dielectric constant is replaced by the reciprocal of relative magnetic permeability. Section 9.5 covers iterative methods to solve magnetic equations with nonlinear materials.

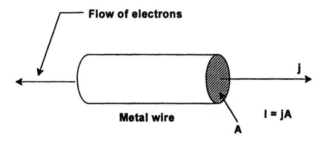

FIGURE 9.1
Relationship between current density j and current I in a wire.

The final two sections address systems with permanent magnets. Section 9.6 describes the properties of permanent magnets and representation of material data. We shall express demagnetization curves in terms of applied and material magnetic fields to make the physical meaning clear. Section 9.7 shows how to represent these highly anisotropic and nonlinear materials in two-dimensional finite-element simulations.

9.1 Differential and Integral Magnetostatic Equations

While electric forces act between charges, magnetic forces act between charges in motion. Moving charges constitute a *current*. A familiar example is the flow of electrons through a wire, illustrated in Figure 9.1. There is no net charge because the wire contains equal densities of ions and electrons. There is a flow of charge because the electrons move and the ions are stationary. The current in the wire equals the total charge that flows through a cross section per unit time,

$$I = \frac{dQ}{dt}. \tag{9.1}$$

The current density, j, is a vector quantity that points in the direction of the local charge flow. The magnitude of the current density equals the charge that crosses a unit area per unit time. Given the current density distribution in a wire, the total current is

$$I = \iint dA \; j \cdot \hat{n}. \tag{9.2}$$

In Equation 9.2 dA is a differential unit of cross section area, $\hat{n}$ is a unit vector normal to dA, and the integral extends over the wire cross section. In steady state, the net current that enters a volume equals zero, or

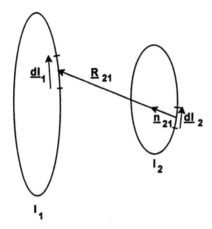

FIGURE 9.2
Ampere's force law between segments of two wire loops.

$$\nabla \cdot \mathbf{j} = 0 \tag{9.3}$$

Equation 9.3 implies that current must flow in a closed loop, or *circuit*.

Current-carrying wires exert forces on one another. Mathematical expressions for the force followed from experiments performed by Biot and Savart and Ampere. Consider the two wire loops of Figure 9.2 that carry currents I_1 and I_2. We divide each wire into a number of short segments. The direction and length of a segment of loop 1 is denoted by the vector $\mathbf{dl}_1$. Similarly, the quantity $\mathbf{dl}_2$ represents a segment of loop 2. The distance between the segments is R_{21} and n_{21} is a unit vector that points from segment 2 to 1. Ampere's law of force states that the force on segment 1 exerted by segment 2 is

$$d\mathbf{F}_{21} = I_1 \mathbf{dl}_1 \times \left[\frac{\mu_o}{4\pi} \frac{I_2 \mathbf{dl}_2 \times \hat{n}_{21}}{R_{21}^2} \right]. \tag{9.4}$$

In standard units, the force is in newtons, distances in meters, currents in amperes, and the constant μ_o equals $4\pi \times 10^{-7}$.

If we view segment 1 as a test segment, then we interpret the bracketed terms on the right-hand side of Equation 9.4 as the capacity for magnetic force at location 1 resulting from the presence of segment 2. In this case, we can write the force as

$$d\mathbf{F}_{21} = I_1 \mathbf{dl}_1 \times \mathbf{B}_{21}, \tag{9.5}$$

where

$$\mathbf{B}_{21} = \left(\frac{\mu_o}{4\pi} \right) \left(\frac{I_2 \mathbf{dl}_2 \times \hat{n}_{21}}{R_{21}^2} \right). \tag{9.6}$$

Equation 9.6 gives the *magnetic field* at location 1 created by segment 2. For current in amperes and distances in meters, the magnetic field has units of *tesla*.

Magnetic fields obey the principle of superposition; therefore, the total magnetic field created by loop 2 is the sum of Equation 9.6 over all segments dl_2. Noting that

$$I_2 \ dl_2 = j_2 \ dl_2 \ dA_2 = j_2 \ dV, \tag{9.7}$$

we can write the sum as a volume integral over the current density,

$$B(x) = \frac{\mu_o}{4\pi} \iiint dx'dy'dz' \ j(x') \times \frac{x - x'}{|x - x'|^3}. \tag{9.8}$$

Equation 9.8 gives the magnetic field at a spatial location $x = (x,y,z)$ in terms of a known current density distribution at positions $x' = (x',y',z')$.

Returning to Equation 9.5, the total force on loop 1 is the sum of contributions from all segments dl_1. An alternate form of Equation 9.5 gives the *volumetric force* (force/m³) on an object that carries current. Consider a segment of the object with current density j_1, cross section area A_1 and length along the direction of current dl_1. Recognizing that the total current is $I_1 = j_1A_1$, Equation 9.5 becomes $F_1 = j_1(A_1dl_1) \times B_1$, or

$$f = j \times B. \tag{9.9}$$

In Equation 9.9, f is the force per volume at a position where the local values of current density and magnetic field are j and B. Another useful expression is the force on a single charged particle in a magnetic field. Assume that the particle has charge q and velocity v. Consider the particle orbit for a small interval Δt. Here, the current is $I = q/\Delta t$ and the segment vector is $dl = v\Delta t$. The force on the particle is therefore

$$F = q \ v \times B. \tag{9.10}$$

We can apply Equation 9.8 to find the magnetic field around the long thin wire illustrated in Figure 9.3. To ensure continuity of current, assume that there is a return path at a large distance that makes negligible contributions to the integral. The contribution for the segment shown at position $(R,0,0)$ is

$$dB = \frac{\mu_o}{4\pi} \frac{Idz \ R}{\left(z^2 + R^2\right)^{3/2}} \ y. \tag{9.11}$$

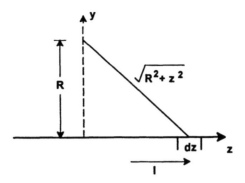

FIGURE 9.3
Calculation of the magnetic field generated by a thin wire.

Integrating Equation 9.11 from $z = -\infty$ to ∞ gives the total field. Recognizing that the results are independent of rotation about the z-axis, we can write the result in terms of cylindrical coordinates centered on the wire,

$$B = \frac{\mu_o I}{2\pi r}\, \theta,$$ (9.12)

where θ is a unit vector in the azimuthal direction.

Section 2.2 showed that Gauss's law for surface integrals in electrostatics was a consequence of the $1/r^2$ variation of electric field around a point charge. Similarly, *Ampere's law* for magnetostatic line integrals follows from the $1/r$ field variation of field generated by a thin wire. Consider a line integral of magnetic field around the circular path shown in Figure 9.4a. Applying Equation 9.12, the result is

$$\oint B \cdot dl = 2\pi r B_\theta = \mu_o I.$$ (9.13)

Next, consider the more complex path of Figure 9.4b. The radial sections are perpendicular to the field and do not contribute to the integral. While the azimuthal length of a section increases as r, the field magnitude decreases as $1/r$. Therefore, the contribution is independent of radius. We can extend the stairstep path of Figure 9.4b to approximate any path around the wire and also add displacements in z to create a tipped path. The following result,

$$\oint B \cdot dl = \mu_o I,$$ (9.14)

holds for any closed path.

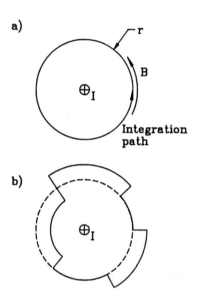

FIGURE 9.4
Line integrals of magnetic field around a wire. (a) Circular path; (b) irregular path with radial and azimuthal segments.

Because the left-hand side of Equation 9.14 is independent of the position of the wire, we can extend it to an arbitrary set of enclosed wires

$$\oint \mathbf{B} \cdot \mathbf{dl} = \mu_o \sum_i I_i \qquad (9.15)$$

The quantity $\sum I_i$ in Equation 9.15 is the total current passing through a surface bounded by the line integral. This quantity may be written as a surface integral of enclosed current density,

$$\oint \mathbf{B} \cdot \mathbf{dl} = \mu_o \iint dA \, \mathbf{j} \cdot \hat{\mathbf{n}}. \qquad (9.16)$$

The quantity $\hat{\mathbf{n}}$ is a unit vector normal to an element dA of the surface.

Equation 9.16 is Ampere's law in integral form. Our derivation was based on infinitely long wires. The proof that Equations 9.15 and 9.16 hold for any collection of wire loops requires several results from vector calculus and is covered in *Classical Electrodynamics* (Jackson, 1975). This reference also gives a useful alternative form of Equation 9.8,

$$\mathbf{B}(\mathbf{x}) = \nabla \times \left[\frac{\mu_o}{4\pi} \iiint dx' dy' dz' \frac{\mathbf{j}(\mathbf{x}')}{|\mathbf{x} - \mathbf{x}'|} \right]. \qquad (9.17)$$

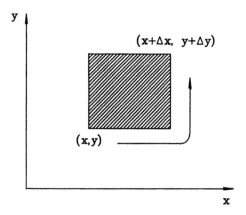

FIGURE 9.5
Application of Ampere's law around a small surface element in the x–y plane.

Application of Equation 9.16 around a small surface leads to the differential form of Ampere's law. Following Figure 9.5, we apply the equation to a surface element with projected area $\Delta x \Delta y$ in the x–y plane and a local component of the current density j_z. The result is

$$B_x(x)\Delta x + B_y(x + \Delta x)\Delta y - B_x(y + \Delta y)\Delta x - B_y(x + \Delta x)\Delta y = \mu_o \, j_z \, \Delta x \, \Delta y \, . \quad (9.18)$$

Again, we approximate field variations with a Taylor expansion,

$$B_x(y + \Delta y) \cong B_x(y) + \frac{\partial B_x(y)}{\partial y} \, \Delta y \, . \quad (9.19)$$

In the limit of a small element, Equation 9.18 becomes

$$-\frac{\partial B_x}{\partial y} + \frac{\partial B_y}{\partial x} = \mu_o \, j_z \, . \quad (9.20)$$

Similar expressions involving j_x and j_y hold for projections in the y–z and z–x planes. The result can be summarized in terms of the vector curl operator,

$$\nabla \times \mathbf{B} = \mu_o \, \mathbf{j} \, . \quad (9.21)$$

Equation 9.21 is the differential form of Ampere's law. For reference, the curl operators in Cartesian and cylindrical coordinates are given by the determinants of the following matrices.

$$\nabla \times \mathbf{B} = \begin{bmatrix} \hat{x} & \hat{y} & \hat{z} \\ \dfrac{\partial}{\partial x} & \dfrac{\partial}{\partial y} & \dfrac{\partial}{\partial z} \\ B_x & B_y & B_z \end{bmatrix}, \tag{9.22}$$

and

$$\nabla \times \mathbf{B} = \begin{bmatrix} \dfrac{\hat{r}}{r} & \hat{\theta} & \dfrac{\hat{z}}{r} \\ \dfrac{\partial}{\partial r} & \dfrac{\partial}{\partial \theta} & \dfrac{\partial}{\partial z} \\ B_r & rB_\theta & B_z \end{bmatrix} \tag{9.23}$$

The bold quantities in the first rows of Equations 9.22 and 9.23 represent unit vectors along the coordinate axes and the determinants are expanded along the top row.

Equation 9.17 shows that the magnetic field can be written as the curl of a vector function,

$$\mathbf{B} = \nabla \times \mathbf{A}, \tag{9.24}$$

where

$$\mathbf{A}(\mathbf{x}) = \frac{\mu_o}{4\pi} \iiint dx'dy'dz' \frac{\mathbf{j}(\mathbf{x}')}{|\mathbf{x} - \mathbf{x}'|}. \tag{9.25}$$

We will discuss the quantity $\mathbf{A}$, the vector potential, in the following section. It is easy to prove that

$$\nabla \cdot (\nabla \times \mathbf{A}) = 0 \tag{9.26}$$

holds for any vector function. Therefore, the magnetic field satisfies a second equation,

$$\nabla \cdot \mathbf{B} = 0. \tag{9.27}$$

We can understand the geometric implication of Equation 9.27 by inspecting Figure 9.6, which shows the field lines around current-carrying wires. In contrast to electric field lines, lines of $\mathbf{B}$ circulate around currents but do not terminate. The flux of magnetic field into any closed volume equals the outward flux, consistent with Equation 9.27.

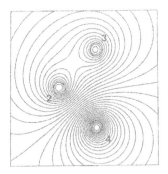

FIGURE 9.6
Magnetic field lines around three current-carrying wires with uniform current density. Region 2, +1000 A; region 3, +750 A; region 4, –1500 A.

9.2 Vector Potential and Field Equations in Two Dimensions

The magnetic vector potential defined in Equations 9.24 and 9.25 plays a role in magnetostatics similar to that of ϕ in electrostatics. We can combine the definition of **A** with Ampere's law to derive a differential equation that describes the vector potential,

$$\nabla \times (\nabla \times \mathbf{A}) = \mu_o \mathbf{j} \tag{9.28}$$

In Cortesian coordinates, Equation 9.28 can be written in the form

$$\nabla^2 \mathbf{A} = -\mu_o \mathbf{j} \tag{9.29}$$

when $\nabla \cdot \mathbf{A} = 0$.

One approach to numerical magnetostatic solutions is to find **A** and then to determine **B** by applying the curl operator of Equation 9.24. The motivation is clear for two-dimensional calculations. In planar or cylindrical systems there is only one non-zero component of vector potential and it satisfies simple boundary conditions. In this case, the solution resolves to the same scalar equation that we discussed for electrostatics.

We shall start with planar problems where derivatives in the z direction are zero. Inspection of the Cartesian form for Equation 9.24 shows that the vector potential components A_x and A_y give rise only to a field component B_z. In this case, the field magnitude is uniform in z and may vary in x and y. This field variation can be determined easily for a known current density. Nontrivial solutions involve the vector potential component A_z that arises from the current

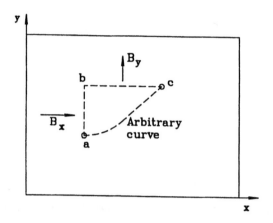

FIGURE 9.7
Relationship between differences of the vector potential A_z at two points and the magnetic flux passing between them.

density component j_z. The current creates field components B_x and B_y. In this case, the vector potential is given by

$$\frac{\partial^2 A_z}{\partial x^2} + \frac{\partial^2 A_z}{\partial y^2} = -\mu_0\, j_z. \qquad (9.30)$$

The form of Equation 9.30 is identical to the two-dimensional Poisson equation (2.24). Therefore, we can apply the finite-element techniques of electrostatics to magnetostatic problems. In preparation we shall first discuss the physical meaning of values of A_z.

Referring to Figure 9.7, consider calculating the magnetic flux per unit length in z across the surface defined by the dashed line between points *a* and *b*. The integral of **B·n** dA is given by

$$\Delta\Phi = \int_a^b B_x\, dy. \qquad (9.31)$$

Substituting from Equation 9.24, we can rewrite Equation 9.31 as

$$\Delta\Phi = \int_a^b \frac{\partial A_z}{\partial y}\, dy = A_z(b) - A_z(a). \qquad (9.32)$$

The flux equals the difference in vector potential between the two points. A similar result holds for the path *b–c* parallel to the x-axis. We can calculate the flux for arbitrary paths between any two points by constructing a stairstep

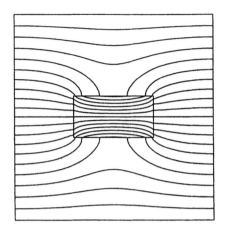

FIGURE 9.8
Magnetic fields near an iron bar of width 0.01 m immersed in a uniform field $B_{xo} = 0.10$ tesla. $x_{min} = 0.00$ m, $x_{max} = 0.05$ m, $y_{min} = 0.00$ m, $y_{max} = 0.05$ m. Neumann boundaries on the right and left, $A_z = 0.0$ on the bottom and $A_z = (0.05)(0.10)$ tesla-m on the top. Twenty field lines spaced at a flux interval of 5.0×10^{-4} tesla-m.

approximation of differential elements parallel to the axes. The general conclusion is the following: *the difference in vector potential between any two points in a planar solution equals the integral of magnetic flux per length between the points.*

We can now understand the meaning of vector potential boundary conditions in terms of the magnetic flux. Consider a solution inside a closed region of the x–y plane with the Dirichlet boundary condition A_z = constant. According to Equation 9.32, the flux integral is zero through any plane connecting two points on the boundary. Therefore, the magnetic fields are contained within the solution region, physically equivalent to perfectly conducting boundaries. The solution of Figure 9.6 illustrates this condition. A Neumann boundary has the normal derivative of vector potential equal to zero. The properties of the curl operator imply that the magnetic field is normal to such a boundary. Figure 9.8 illustrates an application of the Neumann condition to the calculation of magnetic fields near an iron bar immersed in a uniform field. The Neumann boundaries at the right and left approximate a uniform field B_{xo} at $x = \pm\infty$. The choice of Dirichlet conditions $A_z = 0$ on the bottom boundary and $A_z = LB_{xo}$ on the top gives a uniform field in the absence of the bar.

A field plot consists of a set of lines parallel to the local value of **B** separated by equal intervals of magnetic flux. The field magnitude is inversely proportional to the distance between lines. In planar geometry, a field line is defined by the equation

$$\frac{dy}{dx} = \frac{B_y}{B_x} \tag{9.33}$$

or,

$$B_x \, dy - B_y \, dx = 0 \, . \tag{9.34}$$

Substituting from Equation 9.24, Equation 9.34 can be written

$$\frac{\partial A_z}{\partial x} \, dx + \frac{\partial A_z}{\partial y} \, dy = dA_z = 0 \, . \tag{9.35}$$

Equation 9.35 implies that the value of the vector potential is constant along a magnetic field line. Therefore, a contour plot with equal intervals ΔA_z following the method of Section 7.5 is also a plot of field lines. The plot of Figure 9.8 has 20 intervals with spacing $\Delta A_z = 5.0 \times 10^{-4}$ tesla-m.

To conclude this section, we shall summarize relationships for magnetostatics in cylindrical geometry. Expanding Equation 9.24 with $\partial/\partial\theta = 0$ gives the field components,

$$B_r = -\frac{1}{r} \frac{\partial(rA_\theta)}{\partial z} = \frac{\partial A_\theta}{\partial z} \, ,$$

$$B_z = \frac{1}{r} \frac{\partial(rA_\theta)}{\partial r} \, , \tag{9.36}$$

$$B_\theta = \frac{\partial A_r}{\partial z} - \frac{\partial A_z}{\partial r} \, .$$

The third equation is equivalent to Equation 9.12. Here, we can find $B_\theta(R)$ simply by integrating the total axial current through the surface bounded by $r = R$. The top two equations usually require a numerical solution for A_θ. This component of vector potential arises from azimuthal currents such as those in solenoid coils. Equation 9.29 takes the form

$$\frac{1}{r} \frac{\partial}{\partial r} \, r \, \frac{\partial A_\theta}{\partial r} + \frac{\partial^2 A_\theta}{\partial z^2} = -\mu_o \, j_\theta \, . \tag{9.37}$$

The total magnetic flux passing through a plane normal to the z-axis between radii r_1 and r_2 is

$$\Delta\Phi = \int_{r_1}^{r_2} 2\pi r dr \, B_z \, . \tag{9.38}$$

Substituting from Equation 9.36, the enclosed flux is

$$\Delta\Phi - 2\pi \int_{r_1}^{r_2} \frac{\partial}{\partial r}(rA_\theta) = 2\pi \left([rA_\theta]_2 - [rA_\theta]_1 \right). \qquad (9.39)$$

The same result holds for the flux of magnetic field lines through a cylindrical surface between the points (r,z_1) and (r,z_2). Therefore, the magnetic flux through any surface is given by the difference in $[rA_\theta]$. The product of radius and azimuthal vector potential appears frequently in cylindrical magnetic field solutions and is designated as

$$\psi(r,z) = r\, A_\theta(r,z). \qquad (9.40)$$

The equation for a field line in cylindrical coordinates is

$$\frac{dr}{dz} = \frac{B_r}{B_z}. \qquad (9.41)$$

Substituting for the field components from Equation 9.36, we see that field lines lie on contours of constant ψ. For this reason, ψ is called the *stream function*. The magnetic flux between the lines is the same if the contours have equal separation $\Delta\psi$.

9.3 Isotropic Magnetic Materials

The dielectric materials discussed in Section 2.4 modify electric fields through redistribution of bound charge. Analogously, there are materials where a distribution of bound current modifies magnetostatic fields. The most interesting are ferromagnetic or ferrimagnetic materials where large currents result from relatively small applied fields. The response of magnetic materials is generally more complex than dielectrics. Therefore, it is useful to list some limitations of the treatment in this section.

- A rigorous description of magnetic materials requires quantum mechanics. Our discussion is limited to analogies from classical physics. This limited treatment will be sufficient to interpret experimental data and to perform numerical calculations.
- The mechanical and magnetic properties are often coupled so that an applied field can irreversibly change the material state. Time-dependent solutions may be difficult when the properties depend on the field history. We shall limit discussion to static fields in materials with an easily identified state.

- Magnetic materials often have a strong anisotropy in the currents generated in response to an applied field. In this section, we limit attention to *isotropic* materials and defer consideration of permanent magnets to Sections 9.6 and 9.7.

Most magnetically active materials are conductors that support real current in response to a time-varying magnetic field. Section 11.4 covers this effect. Here, we shall concentrate on a different type of current that results from the coordinated circulation of electrons constrained within atoms. We can imagine that the orbital electrons of certain atoms constitute small circuits. Figure 9.9a shows the field generated inside and outside an atomic current loop for positive electron rotation. The field from an atom exerts a force on a nearby atom that seeks to flip it to the opposite rotation polarity. The implication is that the atoms in a block of material have a mixture of orientations. In the absence of an applied field the material does not spontaneously generate a macroscopic field. An applied field exerts a force to align atomic currents to the same orientation as the current that produces the field (Figure 9.9b). This effect is called *paramagnetism* -the prefix *para* means *in the same direction*. Figure 9.9c shows that state of a block of the material when all atomic currents are aligned. The currents cancel internally but not on the surface. The surface current, in the same direction as the applied current, increases the magnetic field. In most materials, the magnetic energy to flip an atom is small compared to its thermal energy, so there is only a small fractional alignment. Contributions from paramagnetic materials typically make changes of only 10^{-6} to the field.

In magnetically active materials like iron or nickel, strong quantum exchange forces act to align nearby atoms, even at ambient temperature. Because it is energetically impossible to create large fields external to the material, the alignment does not extend to macroscopic scales. The resolution is that the currents are aligned in microscopic domains but the domains are randomly oriented (Figure 9.9d). An applied field can shift domains into alignment because (1) it supplies the energy for external magnetic fields and (2) it provides the forces to move domain boundaries.

The experiment in Figure 9.10a illustrates the response of magnetically active materials to an applied field. A magnetic coil encircles a torus of material with random domain orientation. The toroidal field that results from atomic alignment is contained entirely within the material; therefore, the function of the applied field is to exert force to shift domains. We determine the field contribution from the material as a function of the applied field. Initially the domains resist shifting so that there is little alignment with increasing current. Above a certain level of applied field, the domains begin to shift, producing a material field that may ultimately exceed the applied field by more than a factor of 1000. Increasing the applied field brings all domains into alignment so there is no further increase in atomic current — the material is in *saturation*. This behavior is illustrated in Figure 9.10b, which shows the total magnetic field in the material vs. applied field.

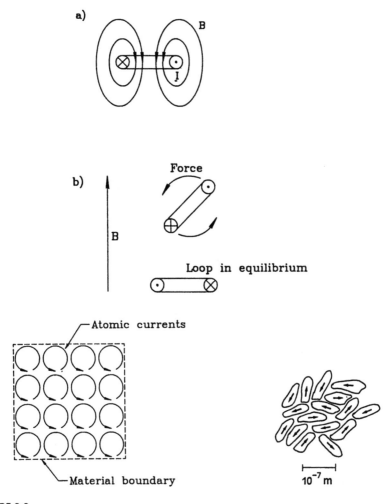

FIGURE 9.9
Properties of isotropic magnetic materials. (a) Magnetic field generated by an atomic current loop; (b) force on atomic current loops in an applied magnetic field; (c) generation of a surface current by aligned current loops in a magnetic materials; (d) random orientation of magnetic domains in unmagnetized material.

Next, suppose we reduce the applied current to zero. Because force is necessary to shift domains back to their original state, the material may retain magnetization. The remaining magnetic field generated by atomic currents is called the *remanence* field, B_r. A negative drive current demagnetizes the material. The nonreproducibility of material response in Figure 9.10 is called *hysteresis* (from the Greek word meaning *lagging*). The applied field necessary to demagnetize saturated material is called the *coercive* field, $-B_c$. Magnetic materials exhibit a wide range of values for the coercive field. In *hard* magnetic materials it takes strong applied fields to realign domains. Here the value of

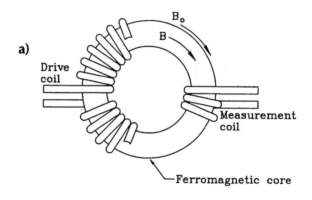

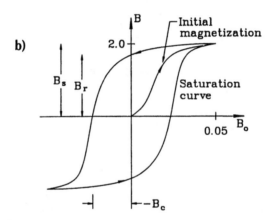

FIGURE 9.10
Measurement of the response of magnetic materials to an applied field. (a) Experimental set up. (b) Magnetization curve, total field in a magnetic material vs. applied field.

B_c may exceed a tesla. The value of the coercive field is low (~10^{-3} tesla) in *soft* magnetic materials.

In this section, we shall concentrate on soft materials. The hysteresis curve is narrow when $B_r \ll B_s$. In this limit, we can ignore the width of the curve so that the total and applied fields lie approximately on a single curve.

$$B_o = \frac{B}{\mu_r}. \tag{9.42}$$

Equation 9.42 applies to isotropic materials when the total field is in the same direction as an applied field. The quantity μ_r is the *relative magnetic permeability*. It equals the ratio of total to applied field and is generally much larger than unity. For the numerical calculations in the following sections, we view μ_r as a single-valued function of field magnitude $|B|$. It can be determined from experimental hysteresis curves. Figure 9.11 shows a plot of $\mu_r(|B|)$ for soft iron.

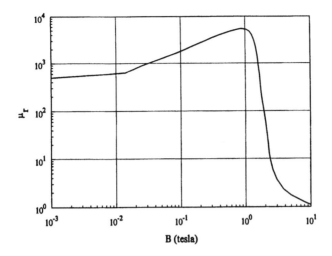

FIGURE 9.11
Plot of $\mu_r(B)$ for soft iron.

Equation 9.42 leads to a generalized form of Ampere's law for soft isotropic materials. We equate the circuital integral of applied field to the product of μ_o times the surface integral of applied current density,

$$\oint \frac{1}{\mu_r} \mathbf{B} \cdot d\mathbf{l} = \oint \frac{1}{\mu_r} (\nabla \times \mathbf{A}) \cdot d\mathbf{l} = \mu_o \iint_S \mathbf{j}_o \cdot \hat{\mathbf{n}} \ dA . \qquad (9.43)$$

Equation 9.43 forms the basis of the finite-element treatment in the following section. Note that the quantity μ_r is interpreted as an element quantity that may depend on the field magnitude in the element.

Conditions on fields adjacent to boundaries between magnetic materials are analogous to those for dielectrics (Section 2.4). Suppose the materials have relative magnetic permeabilities μ_{r1} and μ_{r2}. Applying Equation 9.27 to a narrow box at the interface implies that field component normal to the boundary is continuous,

$$B_{\perp 1} = B_{\perp 2} . \qquad (9.44)$$

Applying Ampere's law to a thin rectangle that includes the interface gives the following relationship for the parallel field component,

$$\frac{B_{\parallel 1}}{B_{\parallel 2}} = \frac{\mu_{r1}}{\mu_{r2}} . \qquad (9.45)$$

On a boundary between a material with $\mu_r \gg 1$ and vacuum with ($\mu_r = 1$), the parallel component of field is much smaller outside the material. Therefore,

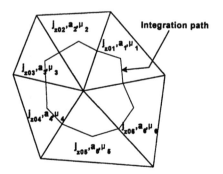

FIGURE 9.12
Path around a test vertex in a conformal triangular mesh for the application of Ampere's law.

magnetic field lines emerge almost perpendicular to the surface of a material with high μ_r.

9.4   Finite-Element Magnetostatic Equations

In this section, we shall consider the finite-element equations for magneto-statics in two dimensions. Before beginning, it is useful to review the reasons for the approach. An alternative is to integrate Equation 9.25 directly without setting up a computational mesh. This approach would be practical if we knew the complete current density distributions in advance. This is not the case when the solution region includes magnetic materials. We have no fore-knowledge of the atomic current distribution because it depends on the field solution. Furthermore, a mesh is essential for many applications. For exam-ple, charged-particle orbit tracking can be carried out quickly interpolating fields from potentials at mesh vertices. In contrast, evaluating Equation 9.25 at the particle position at each interval would be extremely time consuming.

We shall begin with planar geometry. The strategy is to apply the integral form of Ampere's law around the standard closed path encircling a mesh ver-tex (Figure 9.12),

$$\oint \frac{1}{\mu_r} \mathbf{B} \cdot d\mathbf{l} = \oint \frac{1}{\mu_r} \left(\nabla \cdot \mathbf{A}_z \right) \cdot d\mathbf{l} = \mu_o \iint_S j_{zo} \, dA . \qquad (9.46)$$

The result is a linear difference equation that relates the vector potential at the vertex to values at neighboring points. The path encircles one third of the area of the surrounding elements; therefore, we can immediately write an expres-sion for the right-hand side of Equation 9.46,

$$\mu_o \sum_i \frac{j_{zoi} \, a_i}{3} . \tag{9.47}$$

The sum extends over surrounding elements with cross section area a_i. We can extend the method of Section 2.7, again concentrating on the contribution from element 2. The magnetic field and relative magnetic permeability are assumed uniform over an element. The condition of constant field implies that the vector potential is a linear function of x and y,

$$A_z(x, y) = A_{zo} + ux + vy . \tag{9.48}$$

The constants u and v can be expressed in terms of the values of vector potential at the triangle vertices (A_{zo}, A_{z1}, and A_{z2}), analogous to Equations 2.57 and 2.58. Taking the curl, the element magnetic field is

$$\mathbf{B}_2 = v\hat{x} - u\hat{y} . \tag{9.49}$$

The contribution to the line integral of Equation 9.46 from triangle 2 is

$$\frac{1}{\mu_{r2}} \mathbf{B}_2 \cdot (\mathbf{L}_a + \mathbf{L}_b) = \frac{1}{\mu_{r2}} \mathbf{B}_2 \cdot \mathbf{L} . \tag{9.50}$$

Equation 2.62 gives an expression for L in terms of vertex coordinates centered at (x_o, y_o). Combining all terms, the contribution to the line integral is

$$\frac{1}{\mu_{rn}} \left[\frac{\left(A_{zo}(x_2 - x_1) - A_{z1}x_2 + A_{z2}x_1\right)(x_2 - x_1) + \left(A_{zo}(y_2 - y_1) - A_{z1}y_2 + A_{z2}y_1\right)(y_2 - y_1)}{4a_2} \right]. \tag{9.51}$$

Substituting the cotangent expressions of Section 2.6, Equation 9.51 becomes

$$\frac{1}{2\mu_{rn}} \left[A_{zo} \left(\cot\theta_{2b} + \cot\theta_{2a} \right) - A_{z1} \cot\theta_{2b} + A_{z2} \cot\theta_{2a} \right]. \tag{9.52}$$

Following the derivation of Section 2.7, the difference equation for the vector potential at point 0 is

$$A_{zo} \sum_{i=1}^{6} W_i - \sum_{i=1}^{6} A_{zi} W_i = \sum_{i=1}^{6} \frac{\mu_o j_{zi} a_i}{3} . \tag{9.53}$$

The coupling constants are

$$W_1 = \frac{\cot\theta_{2b}/\mu_{r2} + \cot\theta_{1a}/\mu_{r1}}{2},$$

$$W_2 = \frac{\cot\theta_{3b}/\mu_{r3} + \cot\theta_{2a}/\mu_{r2}}{2},$$

(9.54)

$$\ldots$$

$$W_6 = \frac{\cot\theta_{1b}/\mu_{r1} + \cot\theta_{6a}/\mu_{r6}}{2}.$$

Equations 9.53 and 9.54 are identical in form to those for electrostatics; therefore, we can apply techniques developed in previous chapters. The only changes are a reinterpretation of the element properties (j_z and μ_r) and boundary conditions.

Although Gauss' law and Ampere's law lead to similar expressions in planar geometry, we must be careful extending the results to cylindrical systems because of the asymmetric form of the curl operator in r and z. The field component expressions have the same form when written in terms of the stream function,

$$B_r = -\frac{1}{r}\frac{\partial\psi}{\partial z},$$

(9.55)

$$B_z = \frac{1}{r}\frac{\partial\psi}{\partial r}.$$

Instead of solving for A_θ, we take the ψ as a linear function of r and z in an element,

$$\psi(z,r) = \psi_o + uz + vr.$$

(9.56)

Modifying the above derivation, the contribution to the line integral for Ampere's law in element 2 is

$$\frac{1}{2\mu_{r2}R_2}\left[\psi_o\left(\cot\theta_{2b} + \cot\theta_{2a}\right) - \psi_1\cot\theta_{2b} + \psi_2\cot\theta_{2a}\right].$$

(9.57)

In Equation 9.57, the quantity R_2 is an average element radius. The right-hand side of Equation 9.46, an integral over a cross section in the r–z plane, is unchanged for cylindrical coordinates.

In summary, for cylindrical magnetostatics we solve for the stream function using the difference equations

$$\psi_o \sum_{i=1}^{6} W_i - \sum_{i=1}^{6} \psi_i W_i = \sum_{i=1}^{6} \frac{\mu_o j_{zi} a_i}{3} \qquad (9.58)$$

and determine the field components from Equations 9.55. The average element radii enter in modified coupling coefficients,

$$W_1 = \frac{\cot\theta_{2b}/\mu_{r2}R_2 + \cot\theta_{1a}/\mu_{r1}R_1}{2},$$

$$W_2 = \frac{\cot\theta_{3b}/\mu_{r3}R_3 + \cot\theta_{2a}/\mu_{r2}R_2}{2}, \qquad (9.59)$$

$$\cdots$$

$$W_6 = \frac{\cot\theta_{1b}/\mu_{r1}R_1 + \cot\theta_{6a}/\mu_{r6}R_6}{2}.$$

We must decide what average to use for R_2. Ideally, the quantity could be extracted from a weighted integral over the path in an element. In practice, it is sufficient to take an average of the radial positions of the element vertices. For a smooth field solution on the mesh described in Section 5.2, the average radius should be the same for adjacent elements on the same row, independent of whether the triangle base is at the top or bottom. The following expressions (referenced to the vertex and element definitions of Figure 9.12) give good results when used to find the coupling coefficients W_1, W_5, and W_6 at a vertex:

$$R_1 = \frac{r_o + r_1}{2}, \quad R_2 = \frac{r_o + r_1}{2},$$

$$R_5 = \frac{r_o + r_5}{2}, \quad R_6 = \frac{r_o + r_5}{2}. \qquad (9.60)$$

9.5 Magnetic Field Solutions

The example of Figure 9.13 illustrates techniques for magnetic solutions with field-dependent materials. The figure shows a cross section of half a C magnet. The goal is to produce a vertical magnetic field $B_o = -1.5$ tesla in the air gap between poles of the iron flux-return structure. The geometry is planar — the magnet extends out of the page. There are two multiturn coils in the upper and lower magnet sections with current I_o, each represented by two regions with current into and out of the page.

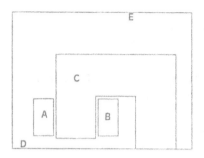 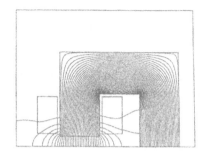

FIGURE 9.13
Calculation of the fields generated by a C-magnet that extends out of the page. (left) Geometry. A, Coil carrying +12 kA; B, coil carrying –12 kA; C, soft iron flux return structure; D, Neumann symmetry boundary; E, Dirichlet boundary ($A_z = 0$), $x_{min} = 0.0$ cm, $x_{max} = 24.0$ cm, $y_{min} = 0.0$ cm, $y_{max} = 13.0$ cm. (right) Calculated magnetic field lines, peak gap field of 1.23 tesla.

We can estimate the required coil current to generate the field from Ampere's law. The magnetic flux is continuous around the circuit that passes through the iron and crosses the air gap. The field magnitude is approximately uniform because the gap and flux return structure have uniform width. Application of Equation 9.43 gives

$$B_o \left(\frac{L_i}{\mu_{ri}} + L_g \right) = 2 \, \mu_o \, I_o . \tag{9.61}$$

In Equation 9.61, L_i is the length of the integration path through the pole (~0.42 m), L_g is the gap height (0.02 m), and μ_{ri} is the average relative magnetic permeability of the iron. If the iron is unsaturated with $\mu_{ri} \sim 1000$, the first term in parenthesis is negligible. The parameters imply a drive current of $I_o \approx 1.2 \times 10^4$ A-turns.

The lower boundary of the solution volume has a Neumann condition. In principle, the other boundaries should be at an infinite distance. In practice, because of field localization it is sufficient to locate a Dirichlet boundary at a distance equal to several gap widths. The main part of the solution volume is an air space ($\mu_r = 1$). The regions marked *A*, *B*, and *C* over-write portions of this space. Regions *A* and *B* are coils to the right and left with currents of $\pm I_o$ distributed uniformly over the cross sections. Region *C* is the flux-return structure with the field-dependent magnetic permeability given by Figure 9.11. Region D is a Neumann boundary and region E (comprising the top and side boundaries) is set to $A_z = 0$.

Solution methods are similar to those for electrostatics. In a relaxation solution (Section 5.5) the adjustment of μ_r takes place during the correction of field values. There is no cut-and-dried prescription for choosing the sequence of adjustments beyond experimentation with different combinations of parameters. As an example, the solution for the magnet of Figure 9.13 used a mesh with 4940 adjustable vertices. The solution reached a relative residual of 2.4×10^{-7} in 1900 cycles with an average relaxation parameter of $\omega = 1.875$. The relative magnetic permeability in elements of the flux return piece was initially set to the value for zero total field, $\mu_r = 480$. Adjustment of material properties was delayed for 150 relaxation cycles to allow some field smoothing and then was applied each 25 cycles. To ensure numerical stability, permeability values were corrected gradually. The following formula mixes predicted permeability values with the current element values,

$$\frac{1}{\mu_{new}} = \frac{1-\alpha_1}{\mu_{old}} + \frac{\alpha_1}{\mu_r(|B_{new}|)} . \tag{9.62}$$

The denominator in the right-most term is the value estimated from the data of Figure 9.11 using the current value of field amplitude. The solution of Figure 9.13 required slow correction because portions of the iron were close to saturation. The choice $\alpha_1 = 0.10$ provided good convergence.

Following Equations 9.54 and 9.59, coupling constants must be recalculated during adjustments of material properties. Corrections are necessary only for coupling constants along lines adjacent to variable elements. We can use the sign bit of the region number (Section 5.2) to mark elements with variable properties. The coupling constants associated with a vertex (W_1, W_5, and W_6) are updated if one or more of elements 1, 2, 5, or 6 (Figure 9.12) have negative region numbers. The region numbers of vertices that meet this condition are marked negative. The coupling constants of only these vertices are recalculated following adjustment of μ_r in the variable elements. In the mesh of Figure 9.13a there are 1418 vertices with negative region numbers. Figure 9.13b shows field lines of the completed solution. The peak gap field is only 1.23 tesla so a higher drive current is needed. The shortfall results partly from the expansion of the field lines at the gap (*fringing fields*). In addition, strong fields lower the permeability to $\mu_r \leq 100$ in sections of the flux return structure, modifying the prediction of Equation 9.61.

Given A_z in a planar solution, Equations 9.22 and 9.23 give the field components B_x and B_y. In cylindrical solutions we must be cautious applying Equations 9.55 that involve derivative of the stream function ψ. Because of the $1/r$ factor, the equations give invalid fields on the axis and inaccurate values nearby. For analysis of cylindrical solutions it is better to convert stream function values at vertices to A_θ. By convention, $A_\theta = 0$ at $r = 0$. Radial magnetic field values are easily evaluated from Equation 9.36. We can write the expression for the axial field as

$$B_z = \frac{\partial A_\theta}{\partial r} + \frac{A_\theta}{r} \cong 2\frac{\partial A_\theta}{\partial r} . \tag{9.63}$$

The first form on the right-hand side of Equation 9.63 applies at off-axis points. The second form, which follows from a Taylor's expansion of the vector potential using symmetry constraints, is used at points close to the axis.

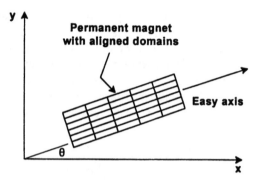

FIGURE 9.14
Aligned domains in a permanent magnet, definition of the easy axis.

Here, the term *close* means a small distance compared to the axial scale length for field variations. In practice, this usually means to a few element widths.

9.6   Properties of Permanent Magnet Materials

Permanent magnets are materials where the domains are aligned and locked in place during the manufacturing process (see, for instance, McCaig and Clegg, 1987). Figure 9.14 shows a schematic view of the configuration. Permanent magnets are inherently anisotropic — the direction of average alignment is called the *easy axis* of the material. Permanent magnets can produce strong magnetic fields in external volumes with no power input — the field energy was supplied during manufacturing. The quality of a permanent-magnet material depends on two factors: (1) how well the domains are aligned and (2) how strongly the domains are locked in alignment.

Application of a moderate applied field perpendicular to the easy axis has little effect on the domains. Therefore, the total magnetic field in this direction is approximately equal to the applied field. We can represent the material response with a relative magnetic permeability,

$$\mu_\perp = \frac{B_\perp}{B_{o\perp}} \cong 1. \tag{9.64}$$

The relationship between total and applied fields along the easy axis is more interesting. The response is usually represented with a *demagnetization curve,* a plot of applied field opposing the magnetization vs. total field in the material. The plot corresponds to the second quadrant of the hysteresis curve discussed in Section 9.3. The measurement is performed with the system of Figure 9.10a. A drive coil encloses a toroidal permanent magnet with the easy

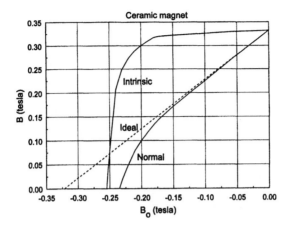

FIGURE 9.15
Demagnetization curves for a ceramic permanent magnet (Indiana General Indox 3).

axis in the azimuthal direction. With no drive current ($B_{o\parallel} = 0$), the magnetic field inside the torus has a value determined by the degree of alignment, a function of the material structure. The value is called the *remanence field* and is denoted B_r. For example, the ceramic magnet represented in Figure 9.15 has $B_r = 0.287$ tesla, a small value compared with the saturation field of iron (2 to 3 tesla).

To perform the measurement, we slowly apply a current to create an applied field in the direction opposite the remanence field until the total field in the material equals zero. The applied field value for this condition is called the *coercive field*, B_c. If the applied field had no effect on the configuration of the material, then it would simply subtract from the remanence field. In this case, the demagnetization curve is a straight line with slope –1 and $B_c = -B_r$. Modern magnetic materials like neodymium iron and samarium cobalt approach this ideal behavior. In contrast, there is considerable deviation for conventional materials like ceramics, Alnico, ferrites, and cobalt steel. The domains are not strongly locked and can shift with an applied field. The total field drops below the ideal value. The response of the material is apparent in a plot of the field contribution from the material,

$$B_{m\parallel} = B_\parallel - B_{o\parallel} , \qquad (9.65)$$

vs. the applied field. This plot, called an *intrinsic demagnetization curve*, is illustrated in Figure 9.15. Domains begin to shift at an applied field of about –0.17 tesla and the material is completely demagnetized at –0.26 tesla. In this state, the permanent magnet can produce no external field when the applied current is removed and must be remagnetized before use. The intrinsic demagnetization curves of modern materials like samarium cobalt are almost flat out to very large values of applied field. Commercial demagnetization curves are usually displayed in terms of the total $B_\parallel$ in kilogauss and the coercive force $H_\parallel$ in kilooersteds. Multiplying both axes by 0.1 converts the curves to $B_\parallel$ and $B_{o\parallel}$ in tesla. Finally, we should note that demagnetization curves apply only to quasi-static applications where the magnet is slowly introduced into its working configuration. The material properties can be quite different in time-varying fields (i.e., electrical motors) or if the magnets are dropped on the floor.

The magnet circuit of Figure 9.16a illustrates how to use the demagnetization curve in the design of a simple device. A permanent magnet of length L_m and cross section area A_m with easy axis in the horizontal direction creates a field in an air gap of length L_g and area A_g. A soft iron flux return structure completes the magnetic circuit. We approximate the iron as an ideal conductor of magnetic flux with $\mu_r \gg 1$. Ignoring details of fringing and leakage fields, conservation of magnetic flux through the device implies that

$$A_g \, B_g = A_m \, B_\parallel . \tag{9.66}$$

The quantities B_g and $B_\parallel$ are the total magnetic fields in the gap and magnet. Introduction of an air gap into a permanent magnet circuit is equivalent to the application of a negative applied field to the material (see, for instance, Humphries, 1986). Ampere's law implies that the integral of applied field around the magnet circuit equals zero,

$$B_{o\parallel} \, L_m = -B_{og} \, L_g . \tag{9.67}$$

Taking $B_{o\parallel} = B_g$, we find that the total and applied fields in the magnet material are related by

$$B_\parallel = -\left[\frac{A_g \, L_m}{A_m \, L_g} \right] B_{o\parallel} . \tag{9.68}$$

The intersection of the line of Equation 9.68 with the demagnetization curve gives the *operating point* of the material in the circuit. We can then use the value of $B_\parallel$ to find the magnetic field in the gap. For example, the condition $A_m \approx A_g$ holds for the device of Figure 9.16a. The gap width is 0.5 cm and the total length of the two Alnico 5 magnets is 3.5 cm. The line $B_\parallel / B_{o\parallel} = 7$ intersects the curve of Figure 9.15 at $B_\parallel = 0.44$. Figures 9.16b and 9.16c show field lines in the device and a plot of horizontal field over the midplane. The average operating point and the peak gap field are close to the predicted value.

The target for magnetic field energy in the gap is set by the application requirements. We shall see in Section 10.2 that the gap field energy is given by

$$U_g \cong \frac{B_g \, B_{og}}{2\mu_o} A_g \, L_g . \tag{9.69}$$

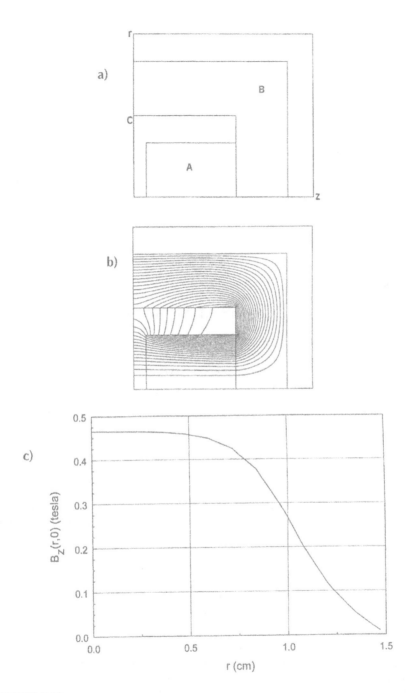

FIGURE 9.16
Design of a simple permanent magnet device. (a) Geometry, one half of system. A, Cylindrical Alnico 5 magnet; B, soft iron flux return structure; C, Neumann boundary. $r_{min} = 0.0$ cm, $r_{max} = 3.0$ cm, $z_{min} = 0.0$ cm, $z_{max} = 3.5$ cm. (b) Computed magnetic field lines separated by equal increments of flux. (c) Axial magnetic field vs. r at z = 0.0 cm.

Substitution from Equations 9.66 and 9.67 gives

$$U_g = \left[\frac{B_{\parallel} B_{o\parallel}}{2\mu_o} \right] A_m L_m .$$

(9.70)

Equation 9.70 shows that U_g is proportional to the magnet volume — we can achieve higher fields by buying larger magnets. We can also maximize the field energy by seeking an operating point that gives the highest value of $B_{\parallel} B_{o\parallel}$, thereby using the material most effectively. The bracketed quantity in Equation 9.70 is called the *energy product* and has units of J/m^3. The energy product is a measure of material quality. The maximum energy product for a ceramic magnet is about $10 \ kJ/m^3$ compared with values exceeding $100 \ kJ/m^3$ for neodymium-iron.

9.7   Magnetostatic Solutions with Permanent Magnets

Section 8.4 derived the finite-element equations for solutions on a conformal triangular mesh for linear anisotropic dielectrics. The anisotropy led to modified expressions for the coupling constants. With minor changes, the results apply to permanent magnets. The total and applied fields in an element can be resolved into components along the easy and normal axes: $B_{\parallel}$, $B_{o\parallel}$, $B_{\perp}$, and $B_{o\perp}$. The field components perpendicular to the easy axis are related by

$$B_{\perp} = \mu_{\perp}\left(\left|B_{\perp}\right|\right) B_{o\perp} .$$

(9.71)

The relative magnetic permeability $\mu_{\perp r}$ is written as a function of the normal component of total field. Because there is little data available on this quantity, we typically set it equal to unity. The equation for the field components along the easy axis is

$$B_{\parallel} = \mu_{\parallel r}\left(\left|B_{\parallel}\right|\right) \left(B_{o\parallel} - B_c\right).$$

(9.72)

The functional dependence of $\mu_{\parallel r}$ is given by the demagnetization curve. For an ideal material, $\mu_{\parallel r} = 1$.

Depending on the magnetic material, we can create numerical models of varying levels of complexity. We begin with the simplest case, a planar geometry with an ideal magnet where $\mu_{\perp r} = \mu_{\parallel r} = 1$. If the easy axis makes an angle θ with the x-axis, the applied field is related to the total field by

$$B_o = B + B_c = B + B_c \ \cos\theta \ \hat{x} + B_c \ \sin\theta \ \hat{y} .$$

(9.73)

Application of Ampere's law (Equation 9.43) around the path shown in Figure 9.12 gives a relationship between the vector potential at a point and its neighbors. New terms arise from the circuital integral of $\mathbf{B}_c$. Following the mathematics of Sections 9.4, the finite-element difference equation is

$$A_o = \frac{1}{\sum_i W_i}$$

$$\times \left[\sum_i W_i A_i + \sum_i \frac{\mu_o j_i a_i}{3} - \sum_i \frac{B_{ci}}{2} \left[\cos\theta_i \left(x_i - x_{i-1} \right) + \sin\theta_i \left(y_i - y_{i-1} \right) \right] \right].$$

(9.74)

The sum extends over the elements surrounding a vertex, which may represent vacuum, ferromagnetic materials, coils, or permanent magnets. The coupling constants are given by Equations 9.54. Equation 9.74 is the same as Equation 9.53 with the exception of the third term in brackets. This source term for permanent magnet elements is a specified spatial function and makes contributions similar to those from applied currents. We can understand the physical meaning of Equation 9.74 by considering a homogeneous permanent magnet where all elements have the same value of easy axis angle and coercive field. The sum of the third term in brackets equals zero at internal vertices surrounded by permanent magnet elements. Values are non-zero only for vertices on boundaries adjacent to both magnet and external elements. Therefore, the ideal magnet generates field by producing a surface current density. Equation 9.74 can easily be extended to cylindrical systems using the analogy between magnetization and applied current. Following the discussion of Section 9.4, no changes are necessary in the source terms. The difference is that the solution is carried out for the stream function and a factor of $1/r$ is introduced in the definition of coupling constants (Equation 9.59). In these solutions, is important to remember that the device magnetization must be either axial or radial to satisfy cylindrical symmetry.

Figure 9.17a is an example of a planar field calculation for a ring dipole magnet. Eight permanent magnet blocks are arranged as shown to generate a vertical magnetic field $B_y \approx -1.0$ tesla. The directions of magnetization are marked with arrows. Because of the system symmetry, it is only necessary to model the quadrant of the structure shown with a Dirichlet boundary on the left ($A_z = 0$) and Neumann boundary on the bottom ($dA_z/dy = 0$). The problem has four regions, the air region of the solution volume and three neodymium-iron blocks with $B_c = -1.12$ tesla. Figure 9.17b shows the resulting field lines. Finally, Figure 9.17c shows field lines in a periodic permanent magnet (PPM) array for beam focusing in traveling wave tubes. The annular Alnico magnets with axial magnetization produce a periodic variation of axial magnetic fields with a peak field of ±0.075 tesla on axis.

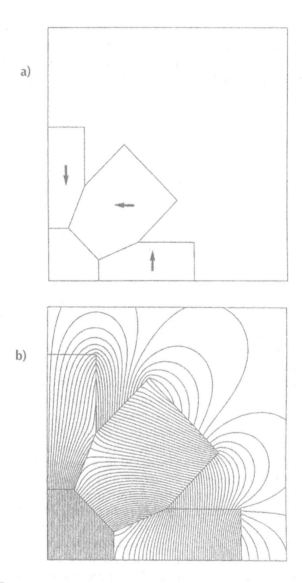

FIGURE 9.17
Examples of permanent magnet calculations. (a) Planar solution, one quarter of a 1 tesla permanent magnet ring dipole showing directions of magnetization. Neodymium iron, $x_{min} =$ 0.0 cm, $x_{max} = 12.0$ cm, $y_{min} = 0.0$ cm, $y_{max} = 12.0$ cm. (b) Calculated magnetic field lines for the ring dipole. (c) Magnetic field lines for a cylindrical PPM (periodic permanent magnet) structure for electron beam focusing. Annular Alnico 5 magnets separated by iron inserts. Peak field on axis: 0.083 tesla, $r_{min} = 0.0$, $r_{max} = 6.0$ cm, $z_{min} = 0.0$ cm, $z_{max} = 9.0$ cm.

The next level of difficulty is to handle permanent magnet materials that follow an approximately straight-line demagnetization curve that falls below the ideal. This approximation describes most neodymium-iron and samarium-cobalt magnets. Here, we must handle anisotropies because $\mu_\perp \approx 1$ while

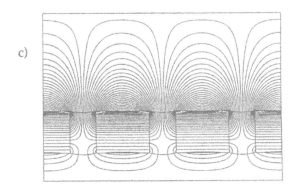

c)

FIGURE 9.17 (continued)

$\mu_\parallel > 1$. The equations for these materials are most conveniently expressed in terms of the inverse relative magnetic permeabilities, $\gamma_\perp = 1/\mu_\perp$ and $\gamma_\parallel = 1/\mu_\parallel$. The equations for the coupling constants are slightly different from those for anisotropic dielectrics (Section 8.4). Replacing the surface integral with the line integral of Ampere's law leads to the form

$$W_i = w_{ia} + w_{i+1,b} \tag{9.75}$$

where

$$W_{ia} =$$

$$\frac{\gamma_{yyi}y_i(y_i - y_{i-1}) + \gamma_{yxi}x_i(y_i - y_{i-1}) + \gamma_{xyi}y_i(x_i - x_{i-1}) + \gamma_{xxi}x_i(x_i - x_{i-1})}{4a_i} \tag{9.76}$$

and

$$W_{ib} =$$

$$\frac{-\gamma_{yyi}y_{i-1}(y_i - y_{i-1}) - \gamma_{yxi}x_{i-1}(y_i - y_{i-1}) - \gamma_{xyi}y_{i-1}(x_i - x_{i-1}) + \gamma_{xxi}x_{i-1}(x_i - x_{i-1})}{4a_i}. \tag{9.77}$$

The values of inverse permeability projected on the coordinate axes are given by

$$\gamma_{xx} = \gamma_\parallel \cos^2\theta + \gamma_\perp \sin^2\theta,$$

$$\gamma_{xy} = \gamma_{yx} = (\gamma_\parallel - \gamma_\perp)\cos\theta\sin\theta, \tag{9.78}$$

$$\gamma_{yy} = \gamma_\parallel \sin^2\theta + \gamma_\perp \cos^2\theta.$$

The final challenge is to treat standard magnetic materials like Alnico which are both anisotropic and nonlinear. The extra task is to find self-consistent values for $\gamma_\parallel$ and $\gamma_\perp$ in each element from the local magnetic field. As with any nonlinear boundary value problem, we must cyclically compute the fields and correct the values of the material parameters. The process is straightforward. We resolve the magnetic field vector in each element into amplitude components along the easy and normal axes. Numerical tables for each material give interpolated values for $\mu_\perp(|B_\perp|)$ and $\mu_\parallel(|B_\parallel|)$. Finally, we insert the values into Equations 9.78 to find γ values along the coordinates axes and then recompute coupling constants for affected vertices. The example of Figure 9.16 was solved with this method.

Chapter 9   Exercises

9.1. We measure the following components of current density in a flowing plasma: $j_x = 400$ A/m^2, $j_y = 550$ A/m^2, and $j_z = 350$ A/m^2. Find the total current through a circular loop lying in the x–y plane of radius 0.025 m.

9.2. Use Equation 9.6 to find the magnetic field at the center of a square current loop. The loop carries current $I = 1500$ A and has sides of length $a = 0.04$ m.

9.3. An electron with kinetic energy $T_e = 10$ keV (1.6×10^{-15} J) is injected normal to a uniform magnetic field $B_o = 0.010$ tesla. The electron moves in a circular orbit of radius r_g. Show by balancing centrifugal and magnetic forces that the radius is given by $r_g = \gamma m_e \beta_c / eB_o$, where m_e is the electron rest mass and c is the speed of light. The parameters β and γ are relativistic factors defined in Section 10.6.

9.4. Show that the on-axis field produced by a thin circular coil of radius R carrying current I is

$$B_z(0,z) = \frac{\mu_o\, I\, R^2}{2\left(R^2 + z^2\right)^{3/2}}.$$

9.5. A long solenoid magnetic coil with a large number of windings creates an internal magnetic field B_o. Use Equation 9.9 to show that the total force per unit area on the windings equals $B_o^2/2\mu_o$ newtons/m^2. For $B_o = 2$ tesla, compare the value to the force per unit area corresponding to 1 atm.

9.6. A large pulsed current $I = 2.5 \times 10^6$ A passes axially through the array of wires illustrated in Figure E9.1. The array with radius 0.035 m is composed of 12 wires of diameter 2.5×10^{-4} m.

(a) Find the force on a wire.

(b) If the wires are composed of aluminum with density 2700 kg/m³, find their initial acceleration.

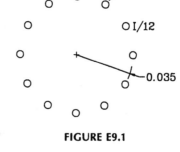

FIGURE E9.1

9.7. A long solenoid coil of radius R_1 is centered inside a perfectly conducting metal cylinder of radius R_2. The coil current per unit length has a step excitation to J_o at $t = 0$.

(a) Find the magnetic field in the regions $0 \le r \le R_1$ and $R_1 \le r \le R_2$.

(b) Find the vector potential A_θ as a function of radius.

9.8. Prove the relationship

$$\nabla \cdot (\nabla \times \mathbf{A}) = 0$$

by evaluating the partial derivatives in Cartesian coordinates.

9.9. Confirm that the vector potential around a long straight wire carrying current I is

$$A_z(r) = -\frac{\mu_o I}{2\pi} \ln\left(\frac{r}{R}\right) + A_o$$

where r is the distance from the wire center and A_o is the value on the wire surface. Find an equation for the magnetic field lines from two wires that carry currents $\pm I$ at positions $(-x_o, 0)$ and $(x_o, 0)$.

9.10. A current sheet with an x-directed current density $j_x = j_o \cos(\pi y / 2D)$ extends infinitely in the y and z directions. For the parameters $j_o = 1.5 \times 10^5$ A/m² and $D = 0.10$ m, use Ampere's law to find magnetic fields inside and outside the sheet.

9.11. Confirm the vector identity

$$\nabla \times (\nabla \times \mathbf{A}) = \nabla(\nabla \cdot \mathbf{A}) - \nabla^2 \mathbf{A}$$

by direct substitution in Cartesian coordinates.

9.12. Confirm that the total flux of B_r through a cylindrical surface of radius r between points (r,z_1) and (r,z_2) equals $2\pi r\ [A_\theta(r,z_2) - A_\theta(r,z_1)]$ (see Equation 9.39).

9.13. Construct diagrams of magnetic fields near current loops to confirm the following.

(a) The magnetic force on a loop in an applied field causes it to flip so that the field inside the loop is in the same direction as the applied field.

(b) The magnetic force on a loop in the field created by another flips the loop so that the external fields on the two loops are in opposite directions.

9.14. There are alternatives to Equation 9.42 to define the magnetic permeability, which depend on the application. For example, the quantity called the small signal magnetic permeability represents the effect of the material on small AC variations of field about a bias level. The quantity is given by $\mu_s = dB/dB_o$. The following table of values shows B_o as a function of B for a material. Add additional rows for $\mu(B)$ and $\mu_s(B)$. Calculate μ from Equation 9.42 and use the second-order polynomial fit of Exercise 8.4 to find μ_s.

B	B_o	B	B_o
0.00E+00	0.00E+00	6.00E-01	1.65E-03
1.00E-01	1.93E-04	7.00E-01	2.29E-03
2.00E-01	3.64E-04	8.00E-01	3.15E-03
3.00E-01	5.68E-04	9.00E-01	4.32E-03
4.00E-01	8.30E-04	1.00E+00	5.91E-03
5.00E-01	1.18E-03		

9.15. Consider a boundary between air and a ferrite with $\mu_r = 250$ along the line $y = 0$. The field components near the boundary inside the ferrite are $B_x = 0.055$ tesla and $B_y = 0.046$ tesla. Give values for the field components on the air side of the boundary.

9.16. Figure E9.2 shows a cross section of an H magnet that produces a vertical dipole field. The structure extends a long distance out of page. Each coil shown carries 7500 A. Assuming iron has very high relative permeability and for dimensions given, use Ampere's law to estimate the magnetic field in the gap.

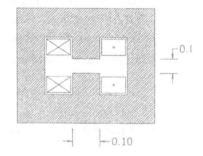

FIGURE E9.2

9.17. Figure E9.3 shows a cross section of a cylindrical laboratory magnet.

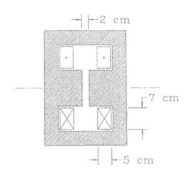

(a) What value of coil current creates a magnetic field of 0.8 tesla in the gap?

(b) We have a 50 A power supply that can drive the coils in series. How many turns should be in each coil?

(c) If the wires have a square cross section and fill the coil region, give the wire cross section area and total length of each winding.

FIGURE E9.3

(d) For copper wires with volume resistivity 1.72×10^{-8} Ω-m, what is the resistance of each winding?

(e) How much power does the magnet consume?

9.18. Show that the operating point of an ideal permanent magnet material for maximum energy product is $B_m = B_r/2$, $B_{om} = -B_r/2$.

9.19. Find the operating point parameters for maximum energy product for a material with a straight-line demagnetization curve that passes through the points ($B = B_r$, $B_o = 0$) and ($B = 0$, $B_o = -B_c$). Give a value for the maximum energy product for SmCo 22B with $B_r = 1.05$ tesla and $B_c = 0.78$ tesla.

9.20. Figure E9.4 shows a permanent magnet assembly to produce a dipole field in a particle spectrometer. The structure extends a long distance out of the page. Assume that (1) $A_g = A_m$, (2) $L_m = 0.075$ m, (3) the magnetic material is ideal with $B_r = -B_c = 1.1$ tesla, and (4) the iron return structure has effectively infinite magnetic permeability.

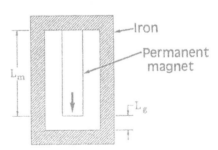

FIGURE E9.4

(a) Find the value of L_g that makes the most effective use of the magnet.

(b) What is the field in the gap for this choice?

10

Static Field Analysis and Applications

*To define it rudely but not inaptly, engineering is the art
of doing that well with one dollar, which any bungler can
do with two after a fashion.*

Arthur Mellen Wellington

This chapter covers techniques to apply solutions of previous chapters to the design of electric and magnetic field devices. Section 10.1 discusses volume and surface integrals over the regions of triangular meshes. These quantities can be used to determine inductances, capacitances, and forces. Volume integrals are useful when we know the distribution of current or space-charge over a spatial region. Surface integrals are necessary when field-dependent material charges or currents are concentrated on boundaries. Section 10.2 derives expression for magnetic field energy density and applies volume integrals to find field energy in regions of a solution space. Section 10.3 reviews capacitance calculations from electrostatic solutions including mutual capacitances between multiple electrodes. Section 10.4 describes methods to find self- and mutual inductances in systems with arbitrary sets of coils. Section 10.5 covers techniques to find electric and magnetic forces on rigid body regions.

The next two sections deal with charged-particle devices. Section 10.6 reviews particle orbit tracking in vacuum using numerical field solutions. The results apply to accelerators and electro-optical devices. Section 10.7 extends the methods to the design of high-current electron and ion guns where the particles make significant contributions to the field solution. These types of solutions are important for the design of high-power radio frequency (RF) and microwave tubes. Finally, Section 10.8 describes simulations of the response of Hall-effect probes, devices used for magnetic field measurements. The solutions illustrate the implementation of generalized Neumann boundary conditions in finite-element solutions.

10.1   Volume and Surface Integrals on a Finite-Element Mesh

In two-dimensional solutions, volume and surface integrals are equivalent to weighted summations over element areas or the lengths of vectors on boundaries. To illustrate the procedures, consider routines to find the volume and surface area of regions in a planar solution volume on a conformal triangular mesh. If NReg is the maximum number of regions, we set up two arrays, Volume(1:NReg) and SurfArea(1:NReg), with entries initially equal to zero. To find the volume, we cycle through all elements of the mesh, checking the region number NR and discarding elements with NR = 0. For a valid element with index i we determine the cross section area a_i using the routine of Section 7.1 and then increment Volume(NR) by the quantity. At the end of the procedure the array contains the volumes (per length in z) of the solution regions. Regions with zero volume are unfilled and correspond to boundaries or points. In cylindrical solutions the elements represent toroidal shapes that encircle the z-axis. In this case, the volume increment is approximately equal to

$$\Delta V = 2\pi \, R_i \, a_i , \tag{10.1}$$

where R_i is the average of the element vertex radii.

The calculation of surface areas involves a check of all mesh lines between vertices following the procedure of Section 7.4. We determine the region numbers of adjacent elements. If one number equals that of the target region and the other does not, the line is on a boundary. The corresponding component of the *SurfArea* array is incremented by the length of the line, Δl, for a planar solution. At the end, *SurfArea* contains the surface area (per length in z) of all valid regions. In cylindrical coordinates, the increment is

$$2\pi \, \frac{r_a + r_b}{2} \, \Delta I , \tag{10.2}$$

where r_a and r_b are the radii at the endpoints of the line.

10.2   Electric and Magnetic Field Energy

Section 3.1. showed that the work to assemble the charge distribution of an electrostatic solution can be expressed as a volume integral over the energy per volume,

$$u_e = \frac{\epsilon_r \, \epsilon_o \, E^2}{2} . \tag{10.3}$$

We can find the electrostatic energy for regions in a numerical solution by applying the volume integral technique of the previous section. After setting up an array Energy(1:NReg), we check each element to determine the region number NR and the electric field amplitude E_i (Section 7.3). In planar problems, the array components are incremented by

$$Energy(NR) = Energy(NR) + \frac{\epsilon_{ri} \, \epsilon_o \, E_i^2 \, a_i}{2}. \qquad (10.4)$$

In Equation 10.4, ϵ_{ri} is the relative dielectric constant of the element and a_i is the cross section area. The calculation yields the region energy per unit length in z. The energy increment in cylindrical coordinates is

$$Energy(NR) = Energy(NR) + \frac{2\pi \, R_i \, \epsilon_{ri} \, \epsilon_o \, E_i^2 \, a_i}{2}, \qquad (10.5)$$

where R_i is the average element radius.

To find the energy of magnetic fields, we must recognize that μ may change substantially during the creation of fields inside materials. Furthermore, time-dependent effects must be considered to find the work involved in generating a field distribution. The derivation requires Faraday's law (discussed further in Section 11.1). Consider the fixed closed path in space shown in Figure 10.1a. Application of a time-varying magnetic flux through the surface enclosed by the loop creates an electric field according to the relationship

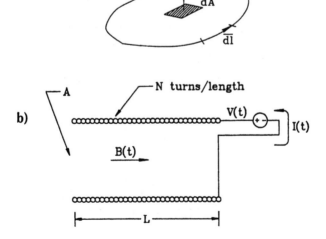

FIGURE 10.1
Calculation of field energy. (a) Geometry for the application of Faraday's law. (b) Configuration of a long solenoid coil.

$$\oint \mathbf{E} \cdot d\mathbf{l} = -\iint \frac{\partial \mathbf{B}}{\partial t} \cdot \hat{n} \, dA \, . \tag{10.6}$$

The circuital integral on the left-hand side of Equation 10.6 extends around the loop. The term on the right-hand side is the integral of the normal component of magnetic field over the enclosed area. We shall apply Faraday's law to the special geometry shown in Figure 10.1b, a solenoid coil of length L and cross section A with N turns per length wrapped around an isotropic magnetic material. The applied magnetic field is related to the current by

$$B_o = \mu_o \, NI \, . \tag{10.7}$$

The energy to create the field is the time integral of power from the drive circuit, the product of the current and voltage. Equation 10.6 implies that the voltage per turn is $V/(NL) = A(dB/dt)$. The work to create the field in the solenoid is

$$U = \int_0^t dt' \, B_o \, \frac{dB}{dt'} \, (AL) \, . \tag{10.8}$$

Dividing Equation 10.8 by the volume of the solenoid, AL, gives an expression for the energy density of magnetic fields,

$$u_m = \frac{1}{\mu_o} \int_0^B B_o(B') \, dB' \, . \tag{10.9}$$

Equation 10.9 applies to isotropic materials. The notation $B_o(B)$ denotes that the applied field can be expressed as a function of total field in the material during the field generation process. The integral in Equation 10.9 is taken over the material hysteresis curve (Section 9.3). We can write the energy density in an alternate form for soft magnetic materials where the applied field is a single-valued function of the total field magnitude,

$$u_m = \frac{1}{\mu_o} \int_0^B \frac{B' dB'}{\mu_r(B')} \, . \tag{10.10}$$

Given $\mu_r(B)$, we can construct a table of $u_m(B)$ by numerical integration.
Another form for energy density, similar to Equation 10.3, holds when μ_r is approximately constant over the range of field values,

$$u_m = \frac{B^2}{2\mu_o \mu_r} \, . \tag{10.11}$$

In air regions, the field energy density is $B_o^2/2\mu_o$.

As an example, consider the total field energy in a planar solution with several soft magnetic materials. The loop over elements is the same as in the electrostatic case. The increment of energy in a region corresponding to a magnetic material is

$$Energy(NR) = Energy(NR) + u_m\left(B_i\right) a_i . \tag{10.12}$$

In Equation 10.12 B_i is the magnitude of the total field in element i and a_i is the cross section area. The quantity $u_m(B_i)$ is determined by interpolation of a table that must be supplied to the program.

10.3 Capacitance Calculations

In a system with two electrodes the capacitance equals the magnitude of the charge induced on the surface of one electrode divided by the voltage between them,

$$C = Q/V . \tag{10.13}$$

If the system is originally charge neutral and the application of a voltage creates a charge $+Q$ on one electrode, the charge on the other electrode is $-Q$. We can show that the total field energy in the system is

$$U = \frac{CV^2}{2} . \tag{10.14}$$

One way to find the capacitance of a two-electrode system is to generate an electrostatic solution with an applied voltage difference of 1 V, determine the total field energy U using the method of Section 10.3, and then set C = 2U. For a planar solution, the procedure yields the capacitance per unit length.

Alternative approaches are necessary when there are three or more electrodes. For example, consider the voltage measurement illustrated in Figure 10.2. A cylindrical electrode inside a grounded vacuum chamber has applied voltage V(t). An isolated circular section of the chamber connected to ground through a current monitor acts as a probe. The *mutual capacitance* between the electrode and the probe, C_p, equals the charge induced on the probe divided by the electrode voltage. If the time scale for voltage changes is long compared to the electromagnetic propagation time over the system, the current in the probe circuit equals

$$i(t) = C_p \frac{dV}{dt} . \tag{10.15}$$

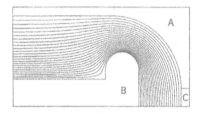

FIGURE 10.2
Mutual capacitance calculation for a voltage probe. Geometry and equipotential lines. A, Grounded chamber; B, high-voltage electrode; C, probe electrode.

If we know C_p, we can find the voltage from the time-integral of the probe current.

Mutual capacitance calculations require a routine to find the total surface charge on electrode regions. For this function we can combine Equation 2.25 with the surface integral technique of Section 10.1. Suppose we have a target region that corresponds to a fixed potential electrode. The routine identifies mesh lines, $L_i = (x_{i1}, y_{i1}, x_{i2}, y_{i2})$, that are adjacent to a single element of the target region. The points in each line vector are ordered so that they point in the direction of positive rotation with the electrode element on the inside. The other adjacent element must have a non-zero region number that does not correspond to a fixed potential. The routine finds the electric field E_i and dielectric constant ϵ_i in this element and increments the accumulated induced charge per length on the electrode by

$$dq_i(NR) = \frac{(L_i \times \hat{z}) \cdot E_i}{\epsilon_i}. \qquad (10.16)$$

Equation 10.16 uses the result from Section 2.7 that the unit vector pointing out of the electrode is given by $n = L_i \times z / |L_i|$. The charge increment for a cylindrical problem is

$$dQ_i(NR) = \frac{2\pi R_i (L_i \times \hat{z}) \cdot E_i}{\epsilon_i}, \qquad (10.17)$$

where R_i is the mean radius.

For the example of Figure 10.2 the procedure gives an induced surface charge on the probe of -2.058×10^{-13} coulombs for an applied voltage of 1 V. The mutual capacitance is therefore $C_p = 0.2058$ pF.

We must exercise caution applying surface integral methods when the calculated values of the electric field near the surface may be inaccurate. A common example is a rectangular electrode, where the field calculation in elements adjacent to the sharp edges may have large errors. An alternative for difficult shapes is to integrate the normal electric field around a surrounding

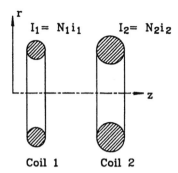

FIGURE 10.3
Inductance calculations for a simple transformer consisting of two circular coils. Coil 1 has a uniform density of N_1 windings, drive current i_1, and total current $I_1 = N_1 i_1$.

path that is well removed from the points of discontinuity. Gauss' law states that the integral gives the total enclosed charge. As an example, consider calculating the charge on the surface of the sharply curved central electrode of Figure 10.2. The surface integral routine yields a value 1.698×10^{-11} coulombs. This is significantly larger than the prediction of 1.387×10^{-11} coulombs from a calculation of the total field energy in the vacuum gap. In contrast, a Gauss' law integration around a surface approximately 1 cm from the electrode gives 1.386×10^{-11} coulombs.

10.4 Inductance Calculations

The field solutions discussed in Chapter 9 can be used to find the inductances of sets of coils in two-dimensional geometries. To illustrate the definition of inductance, consider the two circular coils of Figure 10.3. Coil 1 has a uniform density of N_1 windings and a driving current i_1. The total coil current is $I_1 = N_1 i_1$. The second coil has the parameters N_2, i_2, and I_2. Some of the magnetic flux generated by coil 1 will pass through coil 2. The flux generated by coil 1 enclosed within coil 2 is

$$\Phi_{12} = \int_0^{R_2} 2\pi r dr \, B_{z1} . \tag{10.18}$$

The quantity B_{z1} is the axial magnetic field generated by coil 1 at coil 2. Faraday's law (Equation 10.6) states that the magnitude of voltage across the leads of coil 2 is

$$V_2 = \frac{d}{dt} \left(N_2 \Phi_{12} \right) . \tag{10.19}$$

The flux linked by coil 2 is proportional to i_1,

$$N_2\, \Phi_{12} = M_{12}\, i_1 \,. \tag{10.20}$$

The constant of proportionality, M_{12}, is called the *mutual inductance* between coils 1 and 2. The unit of inductance is the henry. The flux generated by coil i_1 that links itself is

$$N_1\, \Phi_{11} = L_{11}\, i_1 \,. \tag{10.21}$$

The quantity L_{11} is the *self-inductance* of the coil. Given the inductances, we can write the circuit equation for the simple transformer as

$$V_1 = L_{11} \frac{di_1}{dt} + M_{21} \frac{di_2}{dt}\,,$$
$$V_2 = M_{12} \frac{di_1}{dt} + L_{22} \frac{di_2}{dt}\,. \tag{10.22}$$

Equation 10.22 can be extended to systems of three or more coils.

Although analytic methods give estimates of the inductances for simple coil geometries, we must turn to numerical methods when the coils have large or irregular cross sections. We can apply the solutions of Chapter 9 when the following conditions are satisfied.

- The relative spatial variations of time-varying fields are close to those of static fields
- Magnetic materials in the solution volume exhibit an approximately linear response
- The density of windings is uniform over the coil cross section areas

The first condition holds at low frequency, while the second assumption ensures that the total field is the superposition of fields from individual coils. As an illustration, consider calculating M_{21} for a two-coil system. We set up a field solution with $I_2 = 0$ and a current I_1 in coil 1 and find the stream function,

$$\psi_1(r, z) = r A_\theta(r, z)\,. \tag{10.23}$$

In the finite-element treatment, the cross section of coil 2 is divided into several triangles. The value of the stream function in element i, ψ_{12i}, is the average of values at the vertices. Following the discussion of Section 9.2, the axial magnetic flux inside the element radius is

$$\Delta\phi_{12i} = 2\pi\, \psi_{12i}\,. \tag{10.24}$$

If coil 2 has cross section area A_2 and the element has area a_{2i}, the total number of turns passing through the element is $N_2 a_{2i}/A_2$. Therefore, the total flux linked by the element is $2\pi\psi_{12i}N_2a_{2i}/A_2$. A sum over the elements of coil 2 gives the total flux linkage. The mutual inductance equals the flux divided by i_1,

$$M_{12} = \frac{2\pi N_2}{A_2 i_1} \sum_i \psi_{12i}\, a_{2i} = \frac{2\pi N_2 N_1}{A_2 I_1} \sum_i \psi_{12i}\, a_{2i}. \qquad (10.25)$$

In Equation 10.25 the quantity I_1 is the total amp-turns in coil 1, $I_1 = N_1 i_1$. Using similar reasoning, the self-inductance of coil 1 is

$$L_{11} = \frac{2\pi N_1^2}{A_1 I_1} \sum_i \psi_{11i}\, a_{1i}. \qquad (10.26)$$

The sum in Equation 10.26 includes all elements of coil 1. It is easy to extend Equations 10.25 and 10.26 to any collection of coils. Solutions with excitation of coil 1 give the quantities L_{11}, M_{12}, M_{13}, ... Similarly, a solution with an applied current in coil 2 gives M_{21}, L_{22}, M_{23}, ...

In planar systems we seek inductance per length in z. The process involves choosing a drive coil and finding the flux linked by one or more pickup coils. In the general case the coils can be tipped with respect to the axes and each other as in Figure 10.4. In the example, coil 1 is represented by two regions that carry currents of equal magnitude and opposite direction. Consider the calculation of flux linkage between the two regions of coil 2 (marked 5 and 6). Suppose we pick a reference point on the surface between the two regions. Following Section 9.2, the magnetic flux per axial length between the reference point and an element of region 2a equals the difference in vector potential, $A_{2ai} - A_{zo}$. A sum over the elements of the region gives the total flux per meter between the reference point and the coil. Similarly, the flux between the point and an element of region 2b is $A_{zo} - A_{2bi}$. Therefore, the total magnetic flux per meter between the regions of coil 2 is

$$\phi_{12} = N_2 \left| \sum_i^{RegA} \frac{A_{zi}\, a_i}{a_{2a}} - \sum_i^{RegB} \frac{A_{zi}\, a_{2bi}}{a_{2b}} \right|. \qquad (10.27)$$

In Equation 10.27 the quantities a_{2a} and a_{2b} are the areas of regions 2a and 2b. If coil 1 carries a total current I_1, the mutual inductance per meter is given by

$$m_{12} = \frac{N_2 N_1}{I_1} \left| \sum_i^{RegA} \frac{A_{zi}\, a_i}{a_{2a}} - \sum_i^{RegB} \frac{A_{zi}\, a_i}{a_{2b}} \right|. \qquad (10.28)$$

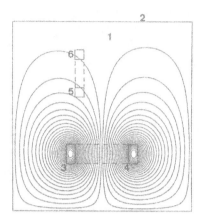

FIGURE 10.4
Inductance calculation in planar geometry for coils with arbitrary orientation. Region 1, solution volume (air); region 2, Dirichlet boundary; regions 3 and 4, drive coil; regions 5 and 6, pickup coil.

TABLE 10.1

Flux Integrals for the Example of Figure 10.4

Region No.	Inductance Factor (henry-A/m)
1	−1.10216E-07
3	4.18666E-04
4	−4.18672E-04
5	3.48825E-05
6	8.41368E-06

Similar expressions apply for the calculation of self-inductances. To illustrate, consider the system of Figure 10.4. The values in Table 10.1 are the sums of Equation 10.28 for each region with non-zero area for $I_1 = 1000$ A. The self-inductance of coil 1 equals the difference between the values for regions 3 and 4 multiplied by (N_1^2/I_1). Taking $N_1 = 100$ and $N_2 = 25$, the result is $l_{11} = 8.372$ mH/m. The mutual inductance equals the difference between the integrals of regions 5 and 6 multiplied by N_2N_1/I_1, or $m_{21} = 66.17$ μH/m.

10.5 Electric and Magnetic Forces on Materials

We next review methods to calculate force density on elements and volume-integrated forces on regions in two-dimensional solutions. To begin, consider magnetic forces in a planar geometry. Currents in the z-direction create field components B_x and B_y. The forces are normal to the current and lie in the x–y plane. The choice of procedure depends on whether the current density distribution is known (such as a specified coil) or whether it depends on the field solution. The latter option applies to surface current densities on magnetic materials.

Suppose that we know the applied current density j_o over the elements of a nonferromagnetic region. Applying Equation 9.9, the magnetic force components on element i are

$$f_{xi} = -j_{oi} \, a_i \, B_{yi},$$
$$f_{yi} = j_{oi} \, a_i \, B_{xi}. \tag{10.29}$$

Treating the region as a rigid body, the force per unit length equals the sum of the element components over the cross section area. In a cylindrical solution with azimuthal current, the element forces are

$$F_{zi} = \sum_i -2\pi \, R_i \, a_i \, j_{oi} \, B_{ri},$$
$$F_{ri} = \sum_i 2\pi \, R_i \, a_i \, j_{oi} \, B_{zi}. \tag{10.30}$$

where R_i is the average radius. The radial force in Equation 10.30 is the hoop stress on the region.

The torque on a body is the volume integral of force density weighted by distance from a rotation axis. If the axis is located at vector position x_c, then the torque equals

$$T = \iiint dV \, (x - x_c) \times f. \tag{10.31}$$

In two-dimensional calculations, torque is meaningful only in planar geometry. For an axis at position (x_c, y_c), the magnetic torque per unit length on an element is

$$t_{zi} = (x_i - x_c) \, j_{oi} \, a_i \, B_{xi} + (y_i - y_c) \, j_{oi} \, a_i \, B_{yi}, \tag{10.32}$$

where (x_i, y_i) is the center-of-mass of element i.

It is more challenging to determine forces on ferromagnetic regions. Here, the atomic current densities are mainly confined to a thin layer on the region surface. We can derive useful expressions if we convert volume integrals for forces to surface integrals. The total magnetic force on a body is

$$F = \iiint dV \, j \times B. \tag{10.33}$$

In Equation 10.33 the quantity j is the total current density in the material, the sum of applied and atomic contributions. Equation 9.21 implies that

$$j = \frac{1}{\mu_o} \nabla \times \mathbf{B} \tag{10.34}$$

Substituting in Equation 10.33, the body force is

$$\mathbf{F} = \frac{1}{\mu_o} \iiint dV \, (\nabla \times \mathbf{B}) \times \mathbf{B} \tag{10.35}$$

The component form of Equation 10.35 in planar geometry for elements of height Δz is

$$\mathbf{F} = \frac{\Delta z}{\mu_o} \iint dA \left[\hat{x} \left(-B_y \frac{\partial B_y}{\partial x} + B_y \frac{\partial B_x}{\partial y} \right) + \hat{y} \left(-B_x \frac{\partial B_x}{\partial y} + B_x \frac{\partial B_y}{\partial x} \right) \right]. \tag{10.36}$$

Equation 10.36 involves field gradients. It is not useful for finite-element calculations where fields are uniform in element volumes. To find an alternative form, consider the divergence of the following vector,

$$\mathbf{S}_x = \hat{x} \frac{B_x^2 - B_y^2}{2} + \hat{y} \, B_x B_y \,. \tag{10.37}$$

The result is

$$\nabla \cdot \mathbf{S}_x = B_x \frac{\partial B_x}{\partial x} - B_y \frac{\partial B_y}{\partial x} + B_y \frac{\partial B_x}{\partial y} + B_x \frac{\partial B_y}{\partial y} \,. \tag{10.38}$$

Applying $\nabla \cdot \mathbf{B} = 0$, the sum of the first and last terms equals zero. Note that the remaining terms are identical to the first term in parenthesis of Equation 10.36. We can therefore write the x-component for force per unit length as

$$f_x = \frac{1}{\mu_o} \iint dA \, \nabla \cdot \mathbf{S}_x = \frac{1}{\mu_o} \int dl \, \mathbf{S}_x \cdot \hat{n} = \frac{1}{\mu_o} \oint \mathbf{S}_x \times d\mathbf{l} \,. \tag{10.39}$$

The first alternate form in Equation 10.39 follows from the divergence theorem. We can express the force in terms of a surface integral over the region boundary. The final form follows from Equation 2.60. The quantity dl is a vector component of the region boundary pointing in the direction of positive rotation. If we denote the line segments constituting a region boundary as $dl_k = \Delta x_k x + \Delta y_k y$, the total x-directed force per unit length is given by the sum

$$f_x = \frac{1}{\mu_o} \sum_k \left[-\Delta x_k \, B_x B_y + \Delta y_k \, \frac{B_x^2 - B_y^2}{2} \right]. \tag{10.40}$$

To include the effect of atomic surface currents the field quantities in Equation 10.40 must be evaluated in elements adjacent to the boundary segments *outside* the material region.

Following similar reasoning, we can show that the y-directed force is given by

$$f_y = \frac{1}{\mu_o} \sum_k \left[\Delta x_k \, \frac{B_x^2 - B_y^2}{2} + \Delta y_k B_x B_y \right]. \tag{10.41}$$

The region forces in cylindrical coordinates are

$$F_r = \frac{2\pi}{\mu_o} \sum_k r_k \left[-\Delta r_k \, B_r B_z + \Delta z_k \, \frac{B_r^2 - B_z^2}{2} \right],$$

$$F_z = \frac{2\pi}{\mu_o} \sum_k r_k \left[\Delta r_k \, \frac{B_r^2 - B_z^2}{2} + \Delta z_k B_r B_z \right]. \tag{10.42}$$

A parallel treatment can be applied to electrostatic solutions by integrating the force density

$$\rho E = \epsilon_o \left(\nabla \cdot E \right) E \tag{10.43}$$

and applying the condition $\nabla \times E = 0$. The expressions are the same as those of Equations 10.40, 10.41, and 10.42 with the replacements $B_x \rightarrow E_x$, $B_y \rightarrow E_y$, $B_r \rightarrow E_r$, and $B_z \rightarrow E_z$.

10.6 Charged-Particle Orbits

Many electronic devices and industrial processes depend on the motion of charged particles through regions of vacuum or low-density gas. Examples include particle accelerators, X-ray generators, cathode ray displays, high-power microwave sources, and electron beam welders. The force on a particle with charge q with velocity v is given by the Lorentz expression,

$$\mathbf{F} = q\ \mathbf{E} + q\ \mathbf{v} \times \mathbf{B}. \qquad (10.44)$$

All quantities in Equation 10.44 are evaluated at the position of the particle, x. The equation of motion specifies how a force changes the momentum of a particle,

$$\frac{d\mathbf{p}}{dt} = \mathbf{F}. \qquad (10.45)$$

The special theory of relativity gives the following relationship between momentum and velocity

$$\frac{d\mathbf{x}}{dt} = \frac{\mathbf{p}}{\gamma m_o}, \qquad (10.46)$$

where

$$\gamma = \sqrt{1 + \frac{p^2}{m_o^2 c^2}}. \qquad (10.47)$$

The quantity p in Equation 10.47 is the magnitude of the momentum vector and m_o is the particle rest mass. The kinetic energy T is given by

$$T = (\gamma - 1)\ m_o c^2. \qquad (10.48)$$

If we know the electric and magnetic fields as a function of position and time, the calculation of a particle orbit involves a straightforward numerical integration of Equations 10.45, 10.46, and 10.47 over small intervals Δt. We can extend the two-step method of Section 4.2 to the set of nonlinear coupled differential equations. To start, we shall express the equations in a convenient form for numerical calculations. It is usually easy to visualize results when the particle position is expressed in meters and elapsed time in seconds. On the other hand, values of momentum are small numbers that do not appeal to the intuition. We can simplify the equations and generate quantities that are easier to interpret by using a dimensionless momentum that is on the order of unity,

$$P_x = p_x / m_o c,$$

$$P_y = p_y / m_o c, \qquad (10.49)$$

$$P_z = p_z / m_o c.$$

For nonrelativistic particles, the quantity γ is close to unity. In this case Equation 10.47 involves small differences between numbers. It is more useful to track the quantity

$$\delta\gamma = \gamma - 1. \tag{10.50}$$

The energy factor $\delta\gamma$ is a useful quantity for both nonrelativistic and relativistic calculations.

The full set of differential equations for particle motion in Cartesian coordinates are

$$\frac{dx}{dt} = \frac{cP_x}{1+\delta\gamma},$$

$$\frac{dy}{dt} = \frac{cP_y}{1+\delta\gamma},$$

$$\frac{dz}{dt} = \frac{cP_z}{1+\delta\gamma},$$

$$\frac{dP_x}{dt} = \left[\frac{q}{m_o c}\right]E_x + \left[\frac{q}{m_o(1+\delta\gamma)}\right]\left(P_y B_z - P_z B_y\right), \tag{10.51}$$

$$\frac{dP_y}{dt} = \left[\frac{q}{m_o c}\right]E_y + \left[\frac{q}{m_o(1+\delta\gamma)}\right]\left(P_z B_x - P_x B_z\right),$$

$$\frac{dP_z}{dt} = \left[\frac{q}{m_o c}\right]E_z + \left[\frac{q}{m_o(1+\delta\gamma)}\right]\left(P_x B_y - P_y B_x\right),$$

$$\delta\gamma = -1 + \sqrt{1 + P_x^2 + P_y^2 + P_z^2}.$$

In analytic orbit calculations it sometimes useful to express the equations of motion in cylindrical or spherical coordinates. Generally, non-Cartesian coordinate systems should be avoided in numerical calculations. Special coordinates give no accuracy advantages and they often lead to program errors. For example, in cylindrical coordinates there is a numerical discontinuity in the azimuth, radial velocity, and angular momentum when particles pass near the axis.

To demonstrate application of the two-step method to Equations 10.49, we shall write out expressions for components in the x direction. Consider advancing from t_n to t_{n+1}. First, we find estimates at the midpoint of the interval,

$$x_{n+\frac{1}{2}} = x_n + \frac{cP_{xn}}{1+\delta\gamma_n} \frac{\Delta t}{2},$$

$$P_{xn+\frac{1}{2}} = P_{xn} + \left[\left(\frac{q}{m_oc}\right)E_{xn} + \left(\frac{q}{m_o(1+\delta\gamma_n)}\right)\left(P_{yn}B_{zn} - P_{zn}B_{yn}\right)\right]\frac{\Delta t}{2}, \quad (10.52)$$

$$\delta\gamma_{n+\frac{1}{2}} = -1 + \sqrt{1 + P^2_{xn+\frac{1}{2}} + P^2_{yn+\frac{1}{2}} + P^2_{zn+\frac{1}{2}}}.$$

The second step of the integration is

$$x_{n+1} = x_n + \frac{cP_{xn+\frac{1}{2}}}{1+\delta\gamma_{n+\frac{1}{2}}}\Delta t,$$

$$P_{xn+1} = P_{xn}$$

$$+\left[\left(\frac{q}{m_oc}\right)E_{xn+\frac{1}{2}} + \left(\frac{q}{m_o(1+\delta\gamma_{n+\frac{1}{2}})}\right)\left(P_{yn+\frac{1}{2}}B_{zn+\frac{1}{2}} - P_{zn+\frac{1}{2}}B_{yn+\frac{1}{2}}\right)\right]\Delta t, \quad (10.53)$$

$$\delta\gamma_{n+1} = -1 + \sqrt{1 + P^2_{xn+1} + P^2_{yn+1} + P^2_{zn+1}}.$$

The field quantities are determined either from analytic expressions or interpolations of numerical data. The main task in orbit calculations on a conformal triangular mesh is to locate the element occupied by the particle (Section 7.1). A full mesh search is usually necessary to initiate the particle orbit. Thereafter, a local search of elements surrounding the previous position is sufficient if the time step is not too large.

Figure 10.5 shows an example of a numerical calculation for the orbit of an electron trapped in a magnetic mirror. The field, produced by two circular coils, has lower magnitude at the midpoint. Electrons follow an approximately circular orbit normal to the field while oscillating axially in the reduced field region. The plot is a projection of the three-dimensional orbit into the r–z plane of the cylindrically symmetric system.

More involved integration techniques such as the fourth-order Runge-Kutta method give better results, if we know field values with unlimited accuracy. This is true if the fields are given by analytic expressions. On the other hand, interpolations of numerical quantities introduce errors. Therefore, higher-order integration methods that require field interpolations at several intermediate locations may not give higher accuracy. Furthermore, the optimum time step is often set by the application rather than the integration scheme. One example is the assignment of space-charge (discussed in the next section). For this process, we choose Δt so that particles move about one element per time step.

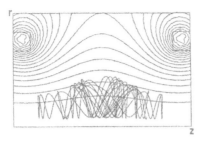

FIGURE 10.5
Numerical calculation of trajectories of electrons trapped in a magnetic mirror generated by two coils. Plot shows magnetic field lines and the orbits of five electrons with different pitch angles at the midplane.

10.7 Electron and Ion Guns

Models of beam devices become more difficult when the fields created by the particles are comparable to the applied fields. Applications of high-current beams include klystron guns, intense ion beam extractors, and plasma devices. The challenge is that the particle orbits simultaneously depend on and affect the total fields of the system. In the discussion of this section we shall limit attention to electric field contributions. Although beams create both electric and magnetic fields, magnetic forces are significant only when particles have relativistic energies. The following models apply to low-energy electron devices and ion extractors.

We can divide high-current beam solutions into initial value and boundary value types. An example of an initial value problem is the particle-in-cell simulation method applied in plasma physics. The strategy is to represent the behavior of large distributions with a limited number of model particles. These computational particles characterize the average behavior of the distribution. The model particles are injected into a solution volume and advanced in position over small time steps. At each step the charges, positions, and velocities are analyzed to estimate distributions of charge and current density, ρ and j. The source functions are then used to advance static or dynamic fields. The book *Plasma Physics Via Computer Simulation* (Birdsall and Langdon, 1985) gives a complete description of the process. In this book we shall limit attention to boundary value solutions that apply to steady-state beams. The category also encompasses pulsed beams when the duration is long compared to particle transit times through the device. We can separate solutions for steady-state high-current beam devices into two categories. In the first we know the characteristics of particles that enter the solution volume. The task is to calculate trajectories through the device using self-consistent fields. An

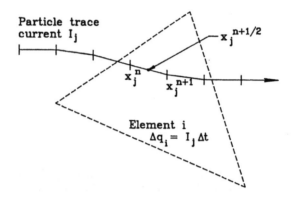

FIGURE 10.6
Particle orbit traces and assignment of space-charge on a triangular mesh.

example is a beam transport line where the entering beam parameters are determined by an external source. The second category applies to the design of injectors that create beams. Here, both the entering particle properties and downstream orbits depend on fields that are initially unknown.

We shall start with transport problems. In order to calculate electrostatic fields, we must know the space-charge density of the beam, $\rho(x)$. We can estimate this quantity by replacing the beam with a number of model particles. Because of the laminar nature of particle orbits in the absence of collisions, a model particle orbit represents the average behavior of many neighboring particles in phase space (see, for instance, Humphries, 1990). The procedure is to follow the model orbits as though they were single electrons or ions but to assign space-charge along the trajectory as though the particle carried a significant portion of the beam current. Suppose a model particle carries current ΔI_i and that we have an initial approximation to the total electric field. We apply the orbit integration technique of the previous section to find the model particle trajectory, advancing with a uniform time step Δt. The integration yields a set of line segments that define a particle streamline called the *trace* of the particle orbit (Figure 10.6). The coordinates of a segment for particle i from time t^n to t^{n+1} are $x_i[n\Delta t]$ and $x_i[(n+1)\Delta t]$. For continuous beams each segment represents a charge $\Delta q_i = I_i \Delta t$ centered at $x[(n+\frac{1}{2})\Delta t]$. To set the space-charge density on a mesh, we check each segment as it is generated and find which element contains the coordinate $x[(n+\frac{1}{2})\Delta t]$. The space-charge density of this element is incremented by

$$\frac{\Delta q_i}{\Delta V_i} = \frac{I_i \Delta t}{\Delta V_i}, \tag{10.54}$$

where ΔV_i is the element volume. After advancing through all model particles, the mesh contains a map of the space-charge density of the beam.

The problem is that we do not know the fields in advance. The method used in orbit tracing programs to resolve the conundrum is to apply cyclic correction. Initially, the applied field solution is used to find approximate particle trajectories. After depositing space-charge from the initial orbits, the field is recalculated with the extra contributions to $\rho(x)$. The corrected fields give better approximations for the orbits which, in turn, lead to improved field estimates. With appropriate charge averaging, the process usually converges to a self-consistent solution even for a high-intensity beam. Figure 10.7 shows a challenging example to illustrate the method, injection of an intense proton beam into a vacuum transport region. There is no applied field so the electrostatic potential results solely from the beam space-charge. A 250 keV slot beam with linear current density J = 1750 A/m and width 1.0 cm enters through the grounded left-hand boundary. The beam is represented by 100 model particles, each carrying 17.5 A/m. If the particles traveled in a straight line, the space-charge potential at the midplane of the transport volume would be 250 kV. Clearly this is inconsistent with propagation. Instead, the protons generate a virtual anode a short distance from the entrance foil. The particles slow almost to stopping and there is a transverse explosion of the beam. Figure 10.7a shows orbits generated by a numerical code on the second cycle. The particles moved in the field created by the particles of the first cycle. The first cycle particles crossed the solution volume in straight-line orbits because there was no applied field. They created a potential symmetric in x with a maximum value of 250 kV. Therefore, on the second cycle all orbits were driven to the side wall. After several iterations, the maximum potential dropped because of beam expansion, allowing a portion of the particles to propagate forward. Figures 10.7 b and c show the final self-consistent beam traces and the potential distribution.

Turning next to injectors, the values of current assigned to the model particles depends on electric field values near the source if the device operates at the *space-charge limit*. We can understand the basis of space-charge limited flow by considering the one-dimensional gap of Figure 10.8a. We shall treat electron flow; modifications for ions are straightforward. The gap has width d and an applied voltage V_o. In steady state, the time derivative of space-charge density is zero so that the current density must have a constant value j_o at all positions in the gap. Therefore, the charge density of nonrelativistic electrons is given by

$$\rho_e(x) = -en_e(x) = -\frac{j_o}{v_e(x)} = -\frac{j_o}{\sqrt{2e\phi/m_e}}. \tag{10.55}$$

The electron density, proportional to $1/v_e$, diverges at the cathode. The second form of Equation 10.55 follows from conservation of energy if the electrostatic potential is taken as zero at the source, $\frac{1}{2}m_e v_e^2 = e\phi$.

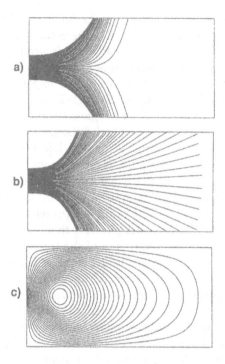

FIGURE 10.7
Self-consistent field calculation for a high-current beam. Injection of a slot beam into a metal-walled transport chamber. Beam linear current, 1750.0 A/m; beam kinetic energy, 250 keV; beam width, 1.0 cm; chamber width; 4.0 cm, chamber length; 8.0 cm. (a) Orbits of 100 model particles on second cycle. (b) Orbit distribution after nine cycles. (c) Self-consistent electrostatic potential distribution — maximum potential, 210 kV.

The one-dimensional Poisson equation can be written

$$\frac{d^2\phi}{dx^2} = \frac{j_o}{\epsilon_o \sqrt{\dfrac{2e\phi}{m_e}}} .$$
(10.56)

Equation 10.56 can also be written in terms of the dimensionless variables $\Phi = \phi/V_o$ and $\chi = x/d$ as

$$\frac{d^2\Phi}{d\chi^2} = \frac{\Gamma}{\sqrt{\Phi}} ,$$
(10.57)

where

$$\Gamma = \frac{j_o d^2}{\epsilon_o \sqrt{2e/m_e} \; V_o^{3/2}} .$$
(10.58)

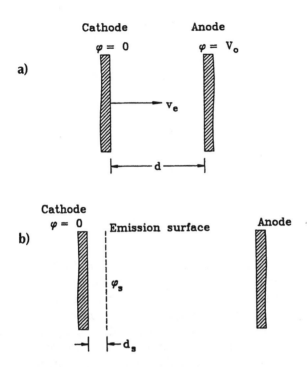

FIGURE 10.8
Space-charge-limited emission of electrons. (a) Geometry for Child law derivation. (b) Emission surface construction for a numerical solution.

We can solve Equation 10.57 for different values of Γ with the boundary conditions $\Phi(0) = 0$ and $\Phi(1) = 1$. For example, with no space-charge ($\Gamma = 0$) the solution is a linear variation of potential. In this case, the field at the cathode has the value $d\Phi(0)/d\chi = 1$. Increasing values of Γ give reduced values of the field at the cathode. The limiting case is where $d\Phi(0)/d\chi = 0$. Any higher value of Γ would correspond to a decelerating field at the source, inconsistent with extraction of zero energy electrons.

Solution of Equation 10.57 with the condition $\Phi'(0) = 0$ gives the maximum value $\Gamma = 4/9 = 0.444444$. The corresponding maximum current density is

$$j_o = \left(\frac{4\epsilon_o}{9}\right) \sqrt{\frac{2e}{m_e}} \frac{V_o^{3/2}}{d^2}. \tag{10.59}$$

Equation 10.59 is the Child law for space-charge limited flow. In this case, the electrostatic potential varies as $\phi = V_o (x/d)^{4/3}$. The electric field is zero at $x = 0.0$ and has the value $E_x = -1.333V_o/d$ at $x = d$.

The zero-field condition raises problems for numerical models of space-charge limited guns. It is impossible to start model particles at the source because they never advance. A common solution is to originate orbits at a hypothetical emission surface a short distance d_s from the source. Model particle currents are assigned by applying Child's law over the thin gap (Figure 10.8b). An iterative process gives the self-consistent value for ϕ_s, the absolute potential at the emission surface. Suppose we make an initial guess of ϕ_s. Model electrons reach the emission surface with velocity

$$v_e = \sqrt{\frac{2e\phi_s}{m_e}}$$

(10.60)

and carry current consistent with the current density

$$j_o = \left(\frac{4\epsilon_o}{9}\right)\sqrt{\frac{2e}{m_e}}\frac{\phi_s^{3/2}}{d_s^2}.$$

(10.61)

The current for a model particle is the product of the current density and cross section area occupied by the particle at the surface. The electrons have non-zero velocities in the region downstream of the emission surface. Therefore, we can find the associated space-charge from an integration of the relativistic equations of motion and the prescription of Equation 10.54.

The problem is how to assign charge to the elements between the source and emission surfaces. Direct calculations from the analytic Child law expressions are difficult to implement on an arbitrary mesh. An alternative is to project model electrons from the emission surface toward the source with a velocity given by the negative of Equation 10.60 and current derived from Equation 10.61. This procedure can lead to density divergences because the particle velocity approaches zero at the cathode. We can circumvent the problem with a trick. Instead of the density variation of Equation 10.55, we use an equivalent uniform density that gives the correct value of electric field at the emission surface, $E_x = 1.333\phi_s/d_s$. The difference between fields generated by the two distributions is negligibly small when $d_s \ll d$. The motivation is that we set a uniform charge density on a triangular mesh by projecting model particles backward at constant velocity. A solution of the one-dimensional Poisson equation shows that the desired density value is $\rho_o = 4\epsilon_o\phi_s/6d_s^2$. This density results if the model particle moves backward from the emission surface to the source with the current density of Equation 10.61 and a velocity equal to two thirds that of Equation 10.60,

$$v_x = -\left(\frac{2}{3}\right)\sqrt{\frac{2e\phi_s}{m_e}}.$$

(10.62)

This method is easy to implement on a two-dimensional mesh because the logic of particle tracking and charge assignment is the same for forward and backward motions. After several iterations, the emission surface potential in a planar gap approaches the value

$$\phi_s = V_o \left[\frac{d_s}{d} \right]^{4/3} \tag{10.63}$$

and the current density carried by model particles approaches the value of Equation 10.59.

Benchmark tests against analytic solutions show that the method predicts absolute values of emitted current with an absolute accuracy of about 0.1%. Figure 10.9a shows an application example, a converging beam electron gun for a traveling wave tube. The goal is to produce a narrow beam with relatively high current density. The empty space between the cathode and the plotted orbits corresponds to the emission layer. The gun produces a total current of 1.12 A that compresses to a radius of 1.2 mm at the entrance to the tube. Numerical calculations are essential in gun design because beam quality is sensitive to small changes in geometry. The calculation shows some overfocusing on the beam edge resulting in the hollow profile of Figure 10.9b.

10.8 Generalized Neumann Boundaries — Hall-Effect Devices

Hall-effect probes are solid-state devices used to measure static and time-varying magnetic fields. Figure 10.10 shows a schematic geometry. The probe consists of a thin sheet of metal or semiconductor deposited on an insulating substrate. Electrodes at the ends drive a current. A component of magnetic field normal to the sheet exerts a transverse force on the conduction particles. Displacement of the charge generates a transverse electric field that compensates the magnetic force, allowing particles to travel in relatively straight lines between the electrodes. Detection of the resulting voltage gives a measurement of the magnetic field. We shall use equations appropriate to an n-type material where electrons are the majority carriers. If field variations are slow compared with the electron transit time, a steady-state solution is sufficient. In this case, the governing equations are

$$\nabla \cdot \mathbf{J} = S \tag{10.64}$$

and

$$\mathbf{E} = \rho \mathbf{J} - K_H \left(\mathbf{J} \times \mathbf{B} \right). \tag{10.65}$$

Equation 10.64 expresses conservation of charge — the quantity S is the source current per volume added or extracted at a point. Equation 10.65 describes current flow in the presence of crossed electric and magnetic fields. The quantity ρ is the volume resistivity of the material and K_H is the Hall

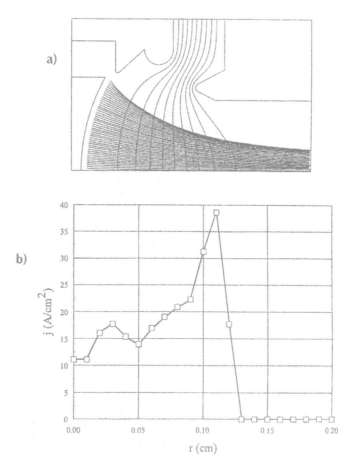

FIGURE 10.9

Cylindrical converging beam electron gun for a traveling wave tube. (a) Geometry showing equipotential lines and traces of 46 model electrons with current density assigned to give space-charge-limited emission. Length in z, 12.0 mm; width in r, 7.5 m; emission surface 0.3 mm from the cathode surface. A, Cathode (–10 kV); B, focusing electrode (–10 kV); C, anode and transport tube (0 kV). Current after 12 cycles, 1.12 A. (b) Current density near the exit boundary determined by assigning model electron current to radial bins.

coefficient, a negative number when the majority carriers are electrons. In the absence of a magnetic field, Equation 10.65 is equivalent to Ohm's law. With low resistivity, the equation expresses the balance of electrical and magnetic forces normal to the current.

Commercial Hall probes are usually thin compared to their transverse dimensions. Therefore, a two-dimensional planar solution is adequate. If the probe lies in the x–y plane and the magnetic field is in the z direction ($\mathbf{B} = B_o\mathbf{z}$) the component forms of Equation 10.65 are

$$\frac{E_x}{\rho} = J_x - \alpha J_y ,$$

(10.66)

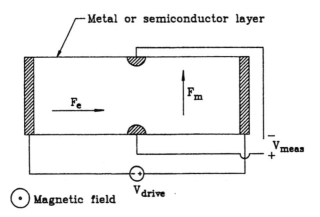

FIGURE 10.10
Schematic view of a Hall effect probe showing electric and magnetic forces.

$$\frac{E_y}{\rho} = J_y + \alpha J_x ,$$

(10.67)

where $\alpha = K_H B_o / \rho$. Solving for the current density components gives

$$J_x = \frac{E_x + \alpha E_y}{\rho(1+\alpha^2)} ,$$

(10.68)

$$J_y = \frac{E_y - \alpha E_x}{\rho(1+\alpha^2)} .$$

(10.69)

We can find finite-element equations for a Hall probe on a conformal triangular mesh by substituting Equations 10.68 and 10.69 into Equation 10.64 and taking volume integrals of both sides of the equation over the standard path of Figure 2.12a. The right-hand side equals one third of the sum of the current added to the surrounding elements,

$$\sum_{i=1}^{} \frac{S_i a_i}{3} .$$

(10.70)

Current sources can be used to represent the effect of circuit loading on contact pads. The left-hand side can be expressed as a surface integral. Following Equation 2.60 and 2.61, the contribution to the integral in triangle 2 is

$$W_2 \hat{z} \int J_2 \cdot \hat{n}_a = w_2 (J_2 \times L).$$

(10.71)

The quantity w_2 is the height of the element in z. This quantity is taken as an element characteristic in order to model Hall probes with varying thickness. Substituting into Equation 10.69 from Equations 2.61 and 2.62 and taking $E = -\nabla\phi$, the right-hand side of Equation 10.71 can be written

$$\frac{w_2}{\rho_2\left(1+\alpha_2^2\right)}\left[\left[\frac{\left(y_2-y_1\right)u}{2}-\frac{\left(x_2-x_1\right)v}{2}\right]+\alpha_2\left[\frac{\left(y_2-y_1\right)v}{2}-\frac{\left(x_2-x_1\right)u}{2}\right]\right], \quad (10.72)$$

where ρ_2 and α_2 represent material properties of element 2.

The first bracketed term in Equation 10.72 is similar to the expressions in the electrostatic treatment of Chapter 2. It leads to a form for the coupling coefficients similar to Equation 2.67, with the replacement of the element dielectric constant ϵ_i by

$$\gamma_i = \frac{w_i}{\rho_i\left(1+\alpha_i^2\right)}. \quad (10.73)$$

The second bracketed term in Equation 10.72 represents the effect of the magnetic field. Substituting for u and v from Equations 2.56 and 2.57 reduces the term to the expression

$$\alpha_2\frac{\phi_2-\phi_1}{2}. \quad (10.74)$$

Adding terms for the surrounding elements and solving for the potential at the enclosed vertex gives the final result,

$$\phi_0 = \frac{\displaystyle\sum_{i=1}^{6} V_i\,\phi_i + \sum_{i=1}^{6}\frac{S_i a_i}{3}}{\displaystyle\sum_{i=1}^{6} W_i}, \quad (10.75)$$

where

$$W_1 = \frac{\left(\gamma_2\cot\theta_{2b}+\gamma_1\cot\theta_{1a}\right)}{2},$$

$$\cdots \quad (10.76)$$

$$W_6 = \frac{\left(\gamma_1\cot\theta_{1b}+\gamma_6\cot\theta_{6a}\right)}{2},$$

and

$$V_1 = \frac{\left(\gamma_2 \cot\theta_{2b} + \gamma_1 \cot\theta_{1a} + \gamma_2 \alpha_2 - \gamma_1 \alpha_1\right)}{2},$$

$$\ldots \tag{10.77}$$

$$V_6 = \frac{\left(\gamma_1 \cot\theta_{1b} + \gamma_6 \cot\theta_{6a} + \gamma_1 \alpha_1 - \gamma_6 \alpha_6\right)}{2}.$$

Equation 10.75 describes the potential at internal vertices of the solution region with any spatial variation of magnetic field, resistivity, Hall coefficient, and thickness. With no modification, the equation also applies to generalized Hall boundaries with curvature and arbitrary variations of medium properties. The treatment parallels that of a Neumann boundary in electrostatics except that the condition is that the normal component of J is zero along the boundary. This condition implies that electric fields intersect the boundary at the Hall angle, $\theta_H = \tan^{-1}\alpha$. We can implement the Hall boundary simply by treating the computational region as though it were surrounded with a layer of elements with $\alpha_i = 0$ and $\gamma_i = 0$. In this way, all points are treated equivalently and all boundaries automatically assume the Hall condition unless they have a fixed potential (Dirichlet condition). A Hall boundary adjacent to elements with $\alpha_i = 0$ is identical to a specialized Neumann boundary.

The following construction gives insight into the physical meaning of Equation 10.75. Figure 10.11 shows a reference point on a Hall boundary for a homogeneous medium. Triangles 1, 2, and 3 are inside the solution volume while triangles 4, 5, and 6 are outside. We determine an equation that relates ϕ_o to ϕ_1, ϕ_2, and ϕ_3 by applying Gauss' law around the dashed line. Because there is no space charge, the integral of the normal field equals zero. If the values of α and γ are zero in the external triangles, then the integral around the curve in the internal triangles is $\phi_o(W_6 + W_1 + W_2 + W_3) - \phi_6 W_6 - \phi_1 W_1 - \phi_2 W_2 - \phi_3 W_3$. To do the integral along the boundary, note that the condition $J_\perp = 0$ implies that $E_\perp = \alpha E_\parallel$. The integral of parallel electric field from point a to point o is $(\phi_3 - \phi_o)/2$. Therefore, the integral of normal electric field over the boundary from point a to point b is

$$\int_a^b E_\perp \cdot ds = \frac{\alpha\left(\phi_3 - \phi_6\right)}{2}. \tag{10.78}$$

Gathering terms in the integral and solving for ϕ_o replicates the result of Equation 10.75. For a homogeneous medium terms of the form $(\gamma_{i+1}\alpha_{i+1} - \gamma_i\alpha_i)$ in the expression for V_i cancel in adjacent triangles except when the surrounding points are on a Hall boundary.

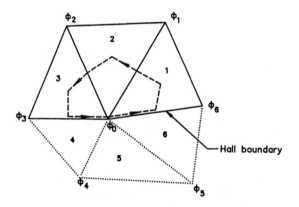

FIGURE 10.11
Integration path for the total current flux away from a reference point on a Hall boundary. Triangles 1, 2, and 3 are part of the probe, while triangles 4, 5, and 6 are external. The Hall boundary lies on the lines from vertex 3 to 0 and vertex 0 to 6.

Figure 10.12 illustrates an application of the model. The solution region represents a thin layer of homogeneous n-type material with a Hall angle of 16.7°. There are driving electrodes at the right and left boundaries. The circles at the top and bottom center are conducting contact pads. The mesh triangles (Figure 10.12a) conform to the boundaries and allow enhanced resolution near the pads. Figure 10.12b shows equipotential lines for a solution with the pads connected to an open circuit. There is a smooth transition from the Dirichlet condition on the side boundaries to Hall boundaries. The lines intersect open boundaries at the Hall angle, even in regions of strong curvature. Figure 10.12c shows the effect of circuit loading between the pads, approximately halving the voltage.

Chapter 10   Exercises

10.1.   Use calculus to find the expression for the volume of a torus of Figure E10.1 with circular cross section. The torus has major radius R and minor radius a.

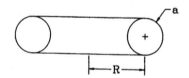

FIGURE E10.1

10.2.   Find the volume of a torus of Exercise 10.1 by splitting it into radial slices and summing the product of the slice area times $2\pi r$ using a spreadsheet. The torus has major radius R = 0.10 m and minor radius a = 0.04 m. Use 4, 10, and 20 radial slices. Compare the results to the prediction of Exercise 10.1 and a rough estimation from the formula $(\pi a^2)(2\pi R)$.

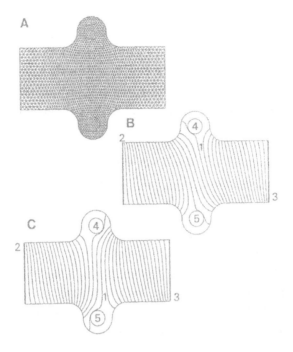

FIGURE 10.12
Example of a Hall probe model. Height in y direction: 40 μml length in x direction, 100 μm; thickness: 1.0 μm; $B_z = 0.01$ tesla; applied voltage, 10 mV; $K_H = -3.0 \times 10^{-2}$ V-m/tesla-A; $\rho = 0.001$ Ω-m. (a) Triangular mesh with 3600 elements. (b) Equipotential lines of solution at an interval of 0.25 mV, open circuit between contact pads at top and bottom. (c) Equipotential lines with 620 Ω load on contact pads.

10.3. For the table of μ(B) listed below, apply a spreadsheet to find the energy per unit volume of the material at B = 1.35 tesla. Compare the result to the prediction of Equation 10.11 using an average value of μ.

B (tesla)	μ	B (tesla)	μ
0.000	1074.918	1.279	456.734
0.214	1053.649	1.297	432.190
0.404	1010.527	1.312	409.998
0.565	949.611	1.326	389.986
0.703	881.153	1.339	371.899
0.820	817.114	1.351	355.533
0.917	764.000	1.363	340.677
0.999	713.144	1.374	327.534
1.066	666.011	1.384	314.371
1.121	622.707	1.393	301.819
1.166	583.062	1.401	291.348
1.203	546.843	1.406	284.198
1.233	513.866		
1.258	483.891		

10.4. Estimate the amount of field energy stored in the dipole magnets of a large synchrotron at a peak energy of 400 GeV. The ring radius is 1000 m and the peak magnetic field is 2.0 tesla. Assume that dipole magnets fill half the ring and have an air gap with cross section area 10^{-2} m².

10.5. A capacitive probe is used to measure time-variations of voltage on the center-conductor of a vacuum coaxial transmission line. The probe is a section of the outer conductor with a rectangular projection that extends 0.01 along z and 0.02 m along θ. The line has inner and outer radii $R_i = 0.075$ m and $R_o = 0.120$ m.

(a) What is the capacitance between the center conductor and the probe?

(b) The probe drives a 50 Ω terminated cable. What is the signal voltage on the center conductor if $V_o \cos(2\pi ft)$, where $V_o = 7.5 \times 10^4$ V and $f = 20$ MHz.

10.6. Figure E10.2 shows an array of cylindrical metal plates. The two central plates are suspended from the left and right plates by eight dielectric rods with radius $a = 0.025$ m and $\epsilon_r = 7.8$. For the dimensions given, estimate the mutual capacitances C_{14}, C_{12}, C_{13}, and C_{23}.

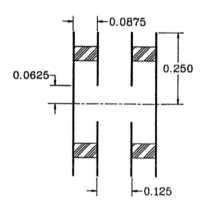

FIGURE E10.2

10.7. A biased spherical electrode suspended over a metal plate generates an electric field on the plate surface. The table below gives the normal field amplitude on the plate as a function of distance from the point closest to the sphere. The values were derived from a numerical solution for a sphere with radius $R = 0.05$ m and applied voltage $V_o = 25$ kV at a distance 0.03 m from the plate.

(a) Use a spreadsheet for a numerical integration to estimate the total surface charge on the plate.

(b) What is the capacitance between the sphere and plate?

R	EZ (R)	R	EZ (R)	R	EZ (R)
0.000000E-00	-6.719534E+05	6.800033E-02	-2.708070E+05	1.360002E-01	-7.705911E+04
4.000490E-03	-6.757886E+05	7.200032E-02	-2.500873E+05	1.400001E-01	-7.203613E+04
8.000480E-03	-6.665109E+05	7.600031E-02	-2.313666E+05	1.440001E-01	-6.783916E+04
1.200047E-02	-6.489993E+05	8.000030E-02	-2.124974E+05	1.480001E-01	-6.374920E+04
1.600046E-02	-6.270901E+05	8.400029E-02	-1.974112E+05	1.520001E-01	-5.995612E+04
2.000045E-02	-6.051807E+05	8.800028E-02	-1.823252E+05	1.560001E-01	-5.661727E+04
2.400044E-02	-5.796103E+05	9.200027E-02	-1.685106E+05	1.600001E-01	-5.327861E+04
2.800043E-02	-5.496437E+05	9.600026E-02	-1.564206E+05	1.640001E-01	-5.045857E+04
3.200042E-02	-5.222265E+05	1.000002E-01	-1.441157E+05	1.680001E-01	-4.770788E+04
3.600041E-02	-4.897924E+05	1.040002E-01	-1.344587E+05	1.720001E-01	-4.514091E+04
4.000040E-02	-4.598331E+05	1.080002E-01	-1.248018E+05	1.760001E-01	-4.286048E+04
4.400039E-02	-4.290173E+05	1.120002E-01	-1.159065E+05	1.800001E-01	-4.054404E+04
4.800038E-02	-3.981993E+05	1.160002E-01	-1.081671E+05	1.840000E-01	-3.864405E+04
5.200037E-02	-3.704800E+05	1.200002E-01	-1.002747E+05	1.880000E-01	-3.674392E+04
5.600036E-02	-3.432740E+05	1.240002E-01	-9.405205E+04	1.920000E-01	-3.496770E+04
6.000035E-02	-3.160672E+05	1.280002E-01	-8.782948E+04	1.960000E-01	-3.337416E+04
6.400034E-02	-2.936947E+05	1.320002E-01	-8.208208E+04	2.000000E-01	-3.175648E+04

10.8. A magnetic field is created by a current I_1 in a thin loop with N_1 turns and radius R_1. Use the result of Exercise 9.4 to estimate the flux linked by a smaller second loop with N_2 turns. The loops are coaxial and separated by a distance $D = 1.5R_1$. What is the mutual inductance M_{12} for the values $N_1 = 100$, $R_1 = 0.10$ m, $N_2 = 75$, and $R_2 = 0.06$ m?

10.9. Figure E10.3 shows a detector used to sense the passage of a metal projectile. A permanent magnet creates a solenoid field of 0.070 tesla inside a cylindrical shell 0.03 m in radius. The flux inside the shell is constant during the rapid passage of the bullet. A 10 turn loop of radius 0.015 m senses flux exclusion from the metal. Estimate the peak amplitude and waveform of the probe signal for a projectile of radius 0.003 m, length 0.02 m, and velocity 1000 km/s.

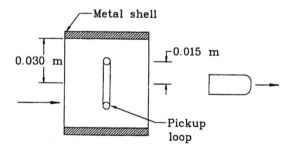

FIGURE E10.3

10.10. Find an expression for the self inductance of a solenoid coil with an internal ferrite slug. The coil has length L and radius R with n turns/length. The slug has $\mu_r = 50.0$ and a radius R/3.

10.11. A magnetic field parallel to a metal surface is turned on instantaneously at $t = 0$. If B_o is the field magnitude, the field distribution as function of depth in the material is given by $B(\zeta) = B_o \exp(-\zeta/\delta)$. The quantity δ is the skin depth given by $\delta = (2\rho t/\mu)^{\frac{1}{2}}$ (see Exercise 12.12).

(a) Give an expression for the current density in the metal.

(b) Integrate over the current sheet to show that the total force per unit area on the metal is $B_o^2/2\mu_o$.

(c) What is the force per unit area for $B_o = 2.5$ tesla?

10.12. The circular current loop of Figure E10.4 has radius R, carries current I, and pivots on an axis along z. There is a uniform magnetic field B_o in the x-direction. Give the torque on the loop as a function of the angle θ.

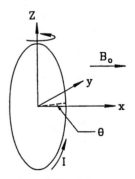

FIGURE E10.4

10.13. A proton beam moving in the z-direction with velocity $v_z = 7.5 \times 10^6$ m/s has the radial variation of current density $j(r) = j_o \exp(-r^2/R^2)$, where $j_o = 5.0 \times 10^3$ A/m² and $R = 0.01$ m. The beam moves in a magnetic field with vector components in tesla of **B** = (0.25,0.10,0.60). Find the components of net force on the beam.

10.14. Use a vector of the form

$$S_y = -\hat{x}\, B_x B_y + \hat{y}\, \frac{B_x^2 - B_y^2}{2}$$

to derive the formula for the y-directed force in a planar solution (Equation 10.41)

10.15. A uniform electric field accelerates electrons across a planar gap. The gap has width $d = 0.05$ m and applied voltage $V_o = 1.5$ MV. If the electrostatic potential has the value $\phi = 0$ at the source, then the electron kinetic energy at any position in the gap is $T_e = e\phi$. Consequently, the relativistic energy factor is $\gamma = 1 + e\phi/m_e c^2$. Set up a spreadsheet to estimate the relativistically correct transit time of an electron in the gap, Δt. Divide the space into 20 sections. Check the result by ensuring that $\Delta t > d/c$.

10.16. Investigate particle oscillations in a linear force $F_x = -Ax$ in the relativistic and nonrelativistic limits using two-step integration procedure on a spreadsheet. Use Equations 10.52 and 10.53.

10.17. Suppose we represent a uniform current-density cylindrical beam of radius 2.0 cm carrying 25 A by 19 model particles. The particles are distributed uniformly in radius with spacing Δr. Particle n starts at radius $(n-\frac{1}{2})\Delta r$. What current should be assigned to each model particle?

10.18. A high-current 10 MeV electron beam crosses a gap between wire grids. The gap has width $d = 0.05$ m and accelerating voltage $V_0 = 1$ MeV. We want to ensure that secondary electrons generated on the entrance grid are not accelerated. Applying a one-dimensional treatment, what is the beam current density such that the electric field on the entrance grid has a positive value and decelerates low-energy electrons?

10.19. Give a value for the Child limited current density of He$^+$ ions in a one-dimensional gap with width $d = 0.05$ m and applied voltage $V_0 = 5.0 \times 10^4$ V.

10.20. A simple Hall probe is a rectangular layer with thickness Δ, length L between the drive electrodes, and width W between the pickups. The material has Hall coefficient K_H and volume resistivity ρ. Assume that L » W.

(a) If the drive voltage V_0 is constrained, show that the ratio of output to drive voltage is given approximately by

$$\frac{V_H}{V_0} = \frac{K_H W}{L\rho} B.$$

(b) If the drive current I_0 is constrained, show that the ratio of output voltage to drive current is approximately

$$\frac{V_H}{I_0} = \frac{K_H}{\Delta} B.$$

10.21. In many references Hall effect characteristics of materials are given in terms of quantity R_H, defined by

$$E = \rho J - R_H \left(J \times H\right).$$

(a) Show that $K_H = R_H/\mu_0$.

(b) For a material with $R_H = 1.0 \times 10^{-10}$ m^3/A-s, $\rho = 1.0 \times 10^{-6}$ Ω-m, and $\Delta = 1$ μm, find the probe output voltage for a drive current of 1 mA and a magnetic field of 0.005 tesla. (Use the results of Exercise 10.20.)

11

Low-Frequency Electric and Magnetic Fields

Someone remarked to me once: "Physicians shouldn't say, I have cured this man, but, this man didn't die under my care." In physics too, instead of saying, I have explained such and such a phenomenon, one might say, I have determined causes for it the absurdity of which cannot be conclusively proved.

Lichtenberg

In this chapter we advance to numerical solutions for electric and magnetic fields with time variations. Section 11.1 reviews the physical implications of the Maxwell equations in integral and differential forms. Complete solutions lead to the phenomenon of electromagnetic radiation, the subject of Chapters 13 and 14. On the other hand, there are many applications where time variations of fields are important but where radiation is negligible. Rather than solve the complete Maxwell equations and deal with radiation boundary conditions and severe limitations on time step, it is often more efficient to simplify the equations. We shall cover such approximate solutions here and in Chapter 12.

Time-dependent solution methods fall into two classes: *frequency domain* and *time domain*. This chapter concentrates on frequency domain solutions. Here, we seek a steady state where all sources and fields vary harmonically at a given angular frequency ω. Continuous wave solutions are boundary value problems; therefore, we can adapt methods developed in previous chapters. It is important to note that the condition of a unique frequency is valid only in systems with linear materials. Nonlinear materials may lead to frequency conversion. For linear materials, we can apply Fourier analysis to find the fields resulting from nonharmonic periodic drives. These include step-function and sawtooth waveforms. The procedure is to find a solution for each Fourier component and then to combine solutions following the principle of superposition.

Time-domain calculations involve direct solutions of the field equations over small time steps for specified initial conditions. You might ask, if we can solve initial value problems with arbitrary time dependencies why bother with difficult boundary value problems? The answer is that the initial value approach is inefficient for many applications. For example, suppose we tried to calculate power loss as a function of frequency in an audio transformer by initiating a sinusoidal excitation at t = 0. Depending on the losses, it may take 10 to 100 cycles for transients to damp. It would therefore be necessary to make an extended run with the possibility of accumulated errors. Time-domain solutions are most effective to model devices with nonlinear materials or to calculate the transient behavior of isolated pulses.

The field equations that we shall derive in Sections 11.3 and 11.5 can be written succinctly in terms of complex numbers. In preparation, Section 11.2 reviews harmonic functions in complex exponential notation. Section 11.3 discusses equations for electric fields created by electrodes with harmonic voltages in imperfect dielectrics. The relationships hold at low to moderate frequencies. Applications include industrial radio frequency (RF) heating, electrosurgery, and plasma processing of microcircuits. Resistive materials carry both displacement and conductive current. We can represent the material response in terms of a complex dielectric constant. Section 11.4 covers methods to solve for the potential with finite-element techniques and to interpret the results. Section 11.5 covers the mathematical description of magnetic fields generated by harmonic applied currents in the presence of materials that may have both ferromagnetic and resistive properties. In this case, the electric fields associated with time-varying magnetic flux drive *eddy currents*. These current components may significantly affect the design of transformers, pulsed magnets, and magnetic recording devices.

11.1 Maxwell Equations

The Maxwell equations inspire awe because they convey so much information in such a succinct form. The four simple equations form the basis of the physical universe, the world economic system, and most advanced video games. On the negative side, summing the hours invested by decades of perplexed students, these relationships have made impressive contributions to the sum of human misery. Our goal is to exert some control over this puissant knowledge. Sections 2.3 and 9.1 discussed the static form of the equations in integral and differential form. In this limit the relationships for electric and magnetic fields are independent. Charges give rise to electric fields while charges in motion create magnetic fields. Clearly, this viewpoint is not totally consistent. The motion of a charge depends on the frame of reference. A correct theory should hold in any frame and must therefore couple electric and magnetic fields.

In this section, we shall complete the picture by including time variations of fields. The time-dependent Maxwell equations follow from two sources. The first is the experimental observation by Faraday that electric fields are

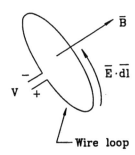

FIGURE 11.1
Illustration of Faraday's law, voltage induced on a wire loop by
a changing magnetic flux.

produced by time-varying magnetic fields. The second is the postulate of
Maxwell that time-varying electric fields create magnetic fields. The introduction of the *displacement current density,* proportional to the time-derivative
of local electric fields, unites classical electromagnetism.

To understand Faraday's law, consider a wire loop of arbitrary shape
immersed in a magnetic field (Figure 11.1). The voltage between the leads
equals the circuit integral of electric field around the loop,

$$V = \oint \mathbf{E} \cdot \mathbf{dl} . \tag{11.1}$$

By convention, the line integral follows the direction of positive rotation
given by the right-hand rule applied along the z-axis shown. The *magnetic
flux* linked by the loop is the integral of the normal component of magnetic
field over the surface bounded by the loop,

$$\Phi = \iint dA \ \mathbf{B} \cdot \hat{\mathbf{n}} . \tag{11.2}$$

The unit vector **n** is normal to the surface and points in the positive
z direction. The integral form of Faraday's law states that the voltage on the
loop equals the negative of the time rate of change of the flux,

$$\oint \mathbf{E} \cdot \mathbf{dl} = -\frac{d\Phi}{dt} = -\frac{d}{dt} \iint dA \ \mathbf{B} \cdot \hat{\mathbf{n}} . \tag{11.3}$$

In applications where elements may move, it is important to recognize that
there are several ways that the quantity Φ in Equation 11.3 can change.

- Time variations of a local magnetic field acting on a rigid and
  stationary loop
- Motion of a rigid loop in a constant magnetic field with spatial
  gradients
- Changes in the area of a stationary loop in a constant magnetic field

General flux changes may result from any combination of the above processes.

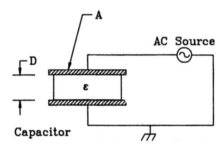

FIGURE 11.2
Illustration of displacement current — capacitor driven by an AC voltage source.

It is important to note that the differential form of Faraday's law is defined with respect to fixed surface elements in a stationary coordinate system. Therefore, flux changes result only from time variations of the local field. The derivation is similar to that of Sections 2.3 and 9.1. For example, the circuit integral in Equation 11.3 applied around an element $\Delta x \Delta y$ in the x–y plane leads to the z-component of the curl operator. The complete result is

$$\nabla \times E = -\frac{\partial B}{\partial t}. \tag{11.4}$$

The partial time derivative denotes a change of magnetic field with time at a constant position.

The circuit shown in Figure 11.2 illustrates the concept of displacement current. A source applies a time-varying voltage across a fixed parallel plate capacitor C. As discussed in Section 10.3, the charge stored in the capacitor at any time is $Q(t) = CV(t)$. Therefore, the source must supply a current

$$I = C\frac{dV}{dt}. \tag{11.5}$$

The circuit is continuous if we postulate that a current equal to that of Equation 11.5 flows across the capacitor gap. The capacitance between parallel plates with area A and spacing d is

$$C = \frac{\epsilon A}{d}. \tag{11.6}$$

Inserting Equation 11.6 in 11.5 and noting that the electric field in the gap equals V/d, the current density is

$$j = \frac{I}{A} = \frac{\epsilon}{d}\frac{dV}{dt} = \epsilon\frac{dE}{dt}. \tag{11.7}$$

Experiments show that current density in a capacitor is more than an abstraction. It generates magnetic fields the same way as a real current density. The general expression for the *displacement current density* is

$$j_d = \epsilon \frac{\partial E}{\partial t}.$$ (11.8)

The partial derivative denotes a time change of electric field at a given location.

Adding displacement current to the differential form of Ampere's law gives

$$\nabla \times \frac{B}{\mu} = j_o + \epsilon \frac{\partial E}{\partial t}.$$ (11.9)

In Equation 11.9, the quantity j_o is the applied current (i.e., specified coil current) and ϵ and μ are total values of local dielectric constant and magnetic permeability, $\epsilon = \epsilon_r \epsilon_o$ and $\mu = \mu_{r}\mu_o$. Again, the spatial variations of the quantity μ represent effects of atomic currents in isotropic magnetic media. The integral form of Equation 11.9 for media with constant properties is

$$\oint \frac{1}{\mu} B \cdot dl = \iint dA\, j_o \cdot \hat{n} + \epsilon \frac{d}{dt} \iint dA\, E \cdot \hat{n}.$$ (11.10)

Again, the time derivative in the last term may arise from changes in the local value of the electric field or time variations in the integration path. For reference, the full set of coupled Maxwell equations in differential form are

$$\nabla \cdot \epsilon E = \rho_o,$$ (11.11)

$$\nabla \cdot B = 0,$$ (11.12)

$$\nabla \cdot E = -\frac{\partial B}{\partial t},$$ (11.13)

and

$$\nabla \times \frac{B}{\mu} = j_o + \epsilon \frac{\partial E}{\partial t}.$$ (11.14)

11.2 Complex Numbers for Harmonic Quantities

The solutions discussed in Sections 11.3, 11.4, and 11.5 are relatively easy if we represent potentials as complex numbers. This section reviews the relationship

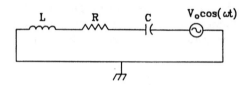

FIGURE 11.3
Circuit to illustrate complex number solutions. Series inductor, resistor, and capacitor driven by an AC voltage source.

between harmonic and complex exponential functions and shows how to extract real physical quantities from complex values. To illustrate the procedures, consider finding the current in the circuit of Figure 11.3. The circuit consists of a series combination of inductor, resistor, and capacitor driven by an AC voltage source. The condition of zero voltage around the circuit gives the differential equation,

$$\alpha \frac{di}{dt} + \beta\, i + \gamma \int idt = V_o\, \cos \omega t. \tag{11.15}$$

The solution of Equation 11.15 may involve a particular part that represents transients and a homogeneous solution that corresponds to the steady state. The current of the steady-state solution varies harmonically at angular frequency ω,

$$i = I_o\, \cos(\omega t + \phi). \tag{11.16}$$

The goal is to find the amplitude and phase of the current: I_o and ϕ. One approach is to substitute Equation 11.16 into Equation 11.15 and to manipulate the trigonometric functions, a laborious process for complicated circuits. We can simplify the mathematics by converting trigonometric functions to complex notation. It is important to remember that the answers to physical problems must be real numbers. Therefore, in the end it should be possible to group the complex numbers of a solution so that imaginary parts cancel.

The complex exponential function is related to trigonometric functions by Euler's formula,

$$\exp(j\omega t) = \cos(\omega t) + j\, \sin(\omega t), \tag{11.17}$$

where $j = (-1)^{1/2}$. The quantity j should not be confused with the current density. The inverse relationships are

$$\cos(\omega t) = \frac{\exp(j\omega t) + \exp(-j\omega t)}{2},$$

$$\sin(\omega t) = \frac{\exp(j\omega t) - \exp(-j\omega t)}{2j}. \tag{11.18}$$

We rewrite the right-hand side of Equation 11.15 as $V_o[\exp(j\omega t) + \exp(-j\omega t)]/2$ and express the current in the form

$$i(t) = A \, \exp(j\omega t) + B \, \exp(-j\omega t).$$ (11.19)

The coefficients A and B in Equation 11.19 are complex numbers that contain amplitude and phase information. We can find their values by substituting Equation 11.19 in Equation 11.15 and recognizing that terms involving $\exp(j\omega t)$ and $\exp(-j\omega t)$ must be individually equal.

$$A = \frac{j\omega V_o}{2} \Big/ \left(-\alpha\omega^2 + j\omega\beta + \gamma\right),$$

$$B = \frac{-j\omega V_o}{2} \Big/ \left(-\alpha\omega^2 - j\omega\beta + \gamma\right).$$ (11.20)

The complex conjugate of a number C is denoted $\overset{*}{C}$ and equals the value of C with $-j$ substituted for j. Inspection of Equation 11.20 shows that $B = A^*$.

Equations 11.20 give the mathematical answer; we must rewrite the results in terms of real numbers to extract the physical solution. Expressing Equation 11.16 in complex notation and setting the result equal to Equation 11.19, we find that

$$A \, \exp(j\omega t) + A^* \exp(-j\omega t) =$$
$$\frac{I_o}{2} \left[\exp(j\omega t) \, \exp(j\phi) + \exp(-j\omega t) \, \exp(-j\phi) \right].$$ (11.21)

Equation 11.20 implies that

$$A = \frac{I_o}{2} \, \exp(j\phi) = \frac{I_o}{2} \left[\cos(\phi) + j\sin(\phi)\right].$$ (11.22)

Multiplying Equation 11.22 by its complex conjugate gives an expression for the current amplitude,

$$I_o = 2\sqrt{A \, A^*} = \frac{\omega V_o}{\sqrt{\left(\gamma - \alpha\omega^2\right)^2 + \omega^2\beta^2}}.$$ (11.23)

The phase is given by

$$\phi = \tan^{-1}\left[\frac{\text{Im}(A)}{\text{Re}(A)}\right] = \tan^{-1}\left[\frac{\gamma - \alpha\omega^2}{\omega\beta}\right].$$ (11.24)

The term in brackets is the imaginary part of A divided by the real part.

Half the effort in the above solution was redundant. There was no need to find information on the complex conjugate of the current. We could have arrived at the same result with the following convention.

- For the drive term, substitute $V_o\exp(j\omega t)$ for $V_o\cos(\omega t)$
- Assume a current variation of the form $i = A\,\exp(j\omega t)$ and solve for the complex number A
- Determine the real values of the current amplitude and phase from the relationships

$$I_o = \sqrt{AA^*},$$

$$\phi = \tan^{-1}\left[\frac{Im(A)}{Re(A)}\right]. \tag{11.25}$$

The procedure can be adapted to any steady-state harmonic problem. We shall see how to interpret electrostatic solutions with complex potentials in Section 11.4.

11.3 Electric Field Equations in Resistive Media

In this section we shall discuss methods to find low-frequency electric fields in resistive media. Figure 11.4 shows an example to motivate the discussion, a three-phase 440-Hz, high-voltage transmission line. The outer boundary is a 6 in. diameter ground shield. The applied voltages on the three wires have 100 kV amplitude and phases of $-120°$, $0°$, and $+120°$. The assembly is immersed in a polyethylene dielectric with $\epsilon_r = 2.71$. The wires are surrounded by carbon-loaded conductive polyethylene sheaths (conductivity of 4.0×10^{-8} mhos/m) for field grading. The goal is to find the peak electric field amplitude and to evaluate power losses in the conductive sheaths.

For the parameters of the problem, we can reduce the Maxwell equations to a form similar to the electrostatic equations of Chapter 2. At low frequency, radiation effects are negligible. The criterion is that the system size L is much smaller than the wavelength of electromagnetic radiation, or

$$\frac{L\,\omega}{2\pi\,\sqrt{\epsilon\mu}} \ll 1. \tag{11.26}$$

For the example with $L = 0.15$ m, $\omega = 2.8 \times 10^3$, $\mu = \mu_o$ and $\epsilon = 2.71\epsilon_o$, the ratio L/λ equals only 2.3×10^{-6}. In this case, the electric field arises primarily from Equation 11.11 and

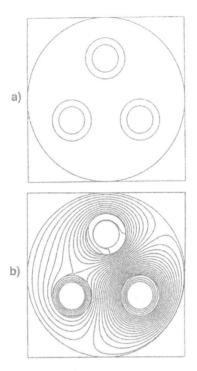

FIGURE 11.4
Cross section of a three-phase, 440-Hz, high-voltage transmission line. (a) Geometry. Outer shield is 6 in. in diameter. The applied voltages on the wires have amplitude 100 kV and phases of –120°, 0°, and +120°. The polyethylene dielectric has $\epsilon_r = 2.71$. The carbon-loaded wire sheaths have volume resistivity $\rho = 1.25 \times 10^7$ Ω-m. (b) Contour plot of the real part of the potential at 0° phase.

$$\nabla \times \mathbf{E} \cong 0. \tag{11.27}$$

Equation 11.27 implies that the electric field can be approximated as the gradient of a scalar potential, $\mathbf{E}(x,t) = -\nabla\phi(x)\exp(j\omega t)$. Finally, we assume that the properties of the medium do not change with time ($\partial\epsilon/\partial t = 0$) and that currents result primarily from conduction.

The conductive current density is related to the electric field by

$$\mathbf{j}_c = \sigma\, \mathbf{E}, \tag{11.28}$$

where σ is the material conductivity. In contrast to the discussion of Section 5.4, the time-varying conductive current generates local concentrations of space charge, ρ_c. The equation for the conservation of conductive current is

$$\frac{\partial\rho_c}{\partial t} = j\omega\, \rho_c = -\nabla\cdot\mathbf{j}_c = -\sigma\, \nabla\cdot\mathbf{E}. \tag{11.29}$$

Equation 11.11 implies that

$$\nabla \cdot (\epsilon \mathbf{E}) = \rho_c .$$ (11.30)

Combining Equations 11.29 and 11.30 gives

$$\nabla \cdot (\epsilon \mathbf{E}) = \nabla \cdot \frac{j\sigma}{\omega} \mathbf{E}$$ (11.31)

or

$$\nabla \cdot \left[\left(\epsilon - \frac{j\sigma}{\omega} \right) \nabla \phi \right] = 0 .$$ (11.32)

Equation 11.32 is identical to Poisson's equation if we interpret the quantity in brackets as a complex dielectric constant,

$$\epsilon = \epsilon' + j\epsilon'' = \epsilon_r \epsilon_0 - \frac{j\sigma}{\omega} .$$ (11.33)

The implication is that the finite element equations for low-frequency fields in resistive media are the same as those for electrostatics except that the quantities ϕ_i, ϵ_i, and W_i may be complex numbers. Physically, this means that there are phase differences in electric fields at different locations because the medium has both resistive and capacitive properties. Figure 11.4b shows the solution for the three-phase line. Contours of the real part of the potential are plotted at t = 0.0. In contrast to an electrostatic solution, the lines are skewed because of phase shifts in the conductive sheaths. The following section describes how to solve Equation 11.32 and how to understand results like those of Figure 11.4.

11.4 Electric Field Solutions with Complex Number Potentials

In this section we shall concentrate on electric field solutions in conductive media using a two-dimensional triangular mesh. The three main tasks are understanding boundary conditions, solving the simultaneous set of linear equations, and interpreting the results. For illustration, Figure 11.5 shows an example from bioengineering, namely heating of subcutaneous tissue by bipolar electrical probes pressed into the skin. By symmetry, only half the problem need be represented. Because of the poor conductivity of the epidermal layer, DC probe voltages would be useless for the application. On the other hand, the layer can conduct significant displacement current at high

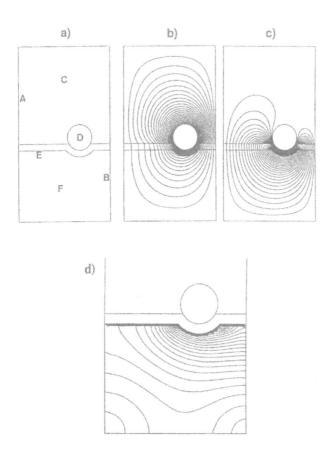

FIGURE 11.5
Heating of tissue of small bipolar electrical probes. (a) Geometry, x_{min}, −1.5 mm; x_{max}, 0.0 mm; y_{min}, −1.2 mm; y_{max}, 1.5 mm; A, grounded solution boundary; B, symmetry boundary, ground; C, air; D, probe, $V_0 = 60$, $\phi = -90°$; E, epidermal layer with $\epsilon_r = 60$ and zero conductivity; F, subcutaneous tissue with $\epsilon_r = 2$, $\rho = 1$ Ω-m. (b) Contour plot of the real part of the potential at $\phi = -90°$. (c) Potential contour plot at $\phi = 0°$. (d) Contour plot of power deposition.

frequency. To ensure that the special Neumann condition (Equation 2.72) applies on unspecified boundaries, we set up an indexing system so that all elements external to the solution region have zero real and imaginary parts of the dielectric constant. On fixed potential surfaces (such as the electrodes) the amplitude and phase of the voltage are given in terms of real and imaginary parts. Because we do not expect applied space-charge in the physical systems modeled, elements have only two material properties: the real part of the dielectric constant and the volume resistivity, $\rho[\Omega\text{-m}] = 1/\sigma[\text{mhos/m}]$. Given the frequency of the probe voltage, we can find the complex dielectric constant. For example, the value in element n which is part of material region k is

$$\epsilon_n = \epsilon_{rk}\epsilon_o - \frac{2\pi j}{f\,\rho_k}. \qquad (11.34)$$

Values of potential at vertices have the form

$$\phi(x, y, t) = \text{Re}\left[\Phi(x, y)\,\exp(j\omega t)\right]. \tag{11.35}$$

The complex number Φ contains information on amplitude and phase. We disregard the common factor of $\exp(j\omega t)$ in all source and field values and solve the boundary value problem for Φ. The complex potential is normally stored as a two-component record consisting of real and imaginary parts, $\Phi = [\Phi_r, \Phi_i]$. A grounded boundary has the fixed value $\Phi = [0,0]$. For a probe with amplitude V_0 and phase ϕ, the fixed potential is

$$\Phi = \left[V_0 \cos\phi, \quad V_0 \sin\phi\right]. \tag{11.36}$$

In the example of Figure 11.5, the probes have amplitude ± 60 V and phases $\pm 90°$, or $\Phi = [0,60]$ and $[0,-60]$. The calculation is performed inside a grounded box.

The coupling constants on a conformal triangular mesh are given by Equations 2.67 or 2.74. Equations 2.68 and 2.73 determine the potential values. The difference is that the quantities Φ, ϵ, and W are complex numbers. The equations can be solved with the same relaxation or matrix routines that we have developed simply by changing variables types to COMPLEX. In FORTRAN, we must also take care to change intrinsic functions to complex equivalents.

Regarding interpretation of the solutions, the main quantities of interest are interpolated values for the amplitude and phase of the potential and the two electric field components (E_x and E_y or E_z and E_r). Again, we can easily adapt existing routines. The linear three-point formula or least-squares fit procedure are applied separately to the real and imaginary parts of the complex potential to determine the quantities Φ_r, $\partial\Phi_r/\partial x$, $\partial\Phi_r/\partial y$, Φ_i, $\partial\Phi_i/\partial x$, and $\partial\Phi_i/\partial y$, where the subscripts r and i stand for real and imaginary parts. The amplitude and phase of the potential at the target point are

$$\phi_0 = \sqrt{\Phi_r^2 + \Phi_i^2},$$

$$\alpha = \tan^{-1}\left(\frac{\Phi_r}{\Phi_i}\right). \tag{11.37}$$

The real and imaginary parts of the electric field components are given in terms of the spatial derivatives of the complex potential. For example, assuming a form $E_{xo} \cos(\omega t + \alpha_x)$ for the x components of electric field, the amplitude and phase are given by

$$E_{xo} = \sqrt{\frac{\partial\Phi_r^2}{(\partial x)} + \frac{\partial\Phi_i^2}{(\partial x)}} \tag{11.38}$$

and

$$\alpha_x = \tan^{-1}\left[\frac{\partial \Phi_i/\partial x}{\partial \Phi_r/\partial x}\right]. \tag{11.39}$$

Amplitude and phase information for the fields can be plotted using the techniques of Chapter 7. A useful quantity for applications to RF heating is the time-averaged resistive power density in elements with non-zero conductivity,

$$p = \frac{\rho\left(E_{xo}^2 + E_{yo}^2\right)}{2} = \frac{E_{xo}^2 + E_{yo}^2}{2\sigma}. \tag{11.40}$$

The factor of two in the denominator results from averaging over the harmonic time variation. We can find the total power deposited in a solution region by multiplying the quantity of Equation 11.40 by the element volume and summing over all elements in the region.

In the probe example of Figure 11.5, the skin layer has zero conductivity and a relative dielectric constant of $\epsilon_r = 60$. The subcutaneous medium has $\epsilon_r = 2.0$ and $\rho = 1.0$ Ω-m. Figure 11.5b shows contours of the real value of potential at zero phase for $f = 100$ MHz. Although most of the electrical field is concentrated in the skin layer some energy is coupled to the region below. Figure 11.5c shows a contour plot of power density. An integral over elements gives a total power deposition of 1.46 kW/m. The power density information can be coupled to a thermal transport code (Chapter 12) to find temperature changes in the tissue.

11.5 Magnetic Fields with Eddy Currents

In the static magnetic field solutions of Chapter 9, we dealt with applied currents in coils and atomic currents that result from domain alignment. Time-varying magnetic fields can generate a third type of current. The electric field associated with changing magnetic flux drives currents through conductive materials. Eddy currents are of particular concern in magnetic materials with high conductivity like iron. They can significantly reduce the effectiveness of transformers and other devices.

To derive approximate equations we assume that real currents in the solution volume are much larger than displacement currents. The condition is equivalent to

$$f \ll \frac{1}{2\pi\epsilon\rho}. \tag{11.41}$$

In ferrites with high volume resistivity the limit of Equation 11.41 holds for frequencies below 1 MHz. The approximation holds for most frequencies of interest in iron. Dropping the displacement current, Equation 11.14 becomes

$$\nabla \times \frac{\mathbf{B}}{\mu} = \mathbf{J}_o + \mathbf{J}_c . \tag{11.42}$$

In Equation 11.42 the quantity $\mathbf{J}_o$ is the applied current density in coils and $\mathbf{J}_c$ is conductive current in metals and semiconductors. The magnetic field can be expressed in terms of a vector potential (Equation 9.24). We can rewrite Faraday's law as

$$\nabla \times \mathbf{E} = -\frac{\partial}{\partial t} (\nabla \times \mathbf{A}) . \tag{11.43}$$

If there are no free charges to create electrostatic-type fields, then Equation 11.43 implies that the total electric field is related to the vector potential by

$$\mathbf{E} = -\frac{\partial \mathbf{A}}{\partial t} . \tag{11.44}$$

The conductive current is therefore given by

$$\mathbf{J}_c = -\sigma \frac{\partial \mathbf{A}}{\partial t} . \tag{11.45}$$

Combining Equations 11.42 and 11.45 gives a differential equation for the vector potential,

$$\nabla \times \left(\frac{1}{\mu} \nabla \times \mathbf{A} \right) = \mathbf{J}_o - \sigma \frac{\partial \mathbf{A}}{\partial t} = \mathbf{J}_o - j\sigma\omega\mathbf{A} . \tag{11.46}$$

The final form applies to a system with steady-state harmonic drive currents at angular frequency ω.

Equation 11.46 is the same as the static equation for magnetic field with the exception of the term $j\sigma\omega\mathbf{A}$. We can apply the method described in Section 9.4 to convert the equation to the finite-element form on a conformal mesh, treating the eddy-current density the same way as the source current. For example, consider a planar solution determined by the vector potential A_z. The eddy current contributes a term

$$j \sum_i \frac{\sigma_i \omega a_i A_{zi}}{3} . \tag{11.47}$$

We must decide how to take the sum in Equation 11.47 over the vector potential in surrounding elements. If the local variation of A_z is approximately linear with x and y, contributions to the average from neighboring vertices cancel and it is sufficient to approximate the term as

$$\frac{j\omega A_{zo}}{3} \sum_i \sigma_i \, \alpha_i \,. \tag{11.48}$$

In Equation 11.48, the quantity A_{zo} is the value at the test vertex. By analogy with Section 9.4, the finite-element difference equation at a vertex is

$$\sum_i W_i \, A_{zi} - A_{zo} \sum_i W_i = -\sum_i \frac{J_{zoi} \, a_i}{3} + \frac{j\omega A_{zo}}{3} \sum_i \sigma_i \, \alpha_i. \tag{11.49}$$

Solving for A_{zo} gives a form suitable for a relaxation solution,

$$A_{zo} = \frac{\displaystyle\sum_i W_i \, A_{zi} + \sum_i \frac{J_{zoi} \, a_i}{3}}{\displaystyle\sum_i W_i - j\omega \sum_i \frac{\alpha_i \, \sigma_i}{3}}. \tag{11.50}$$

In contrast to the electric field solutions of Section 11.4, the coupling constants W_i are real numbers. The quantities that must be stored as complex numbers are the vector potential (or stream function) and the sum of applied current densities around a point,

$$\sum_i \frac{J_{zoi} \, a_i}{3}. \tag{11.51}$$

The complex values of J_{zoi} represent the amplitude and phase of element drive currents. The mechanics of the solution are similar to that for static field solutions. For example, in a relaxation solution we calculate a complex residual

$$R = \frac{\displaystyle\sum_i W_i \, A_{zi} + \sum_i \frac{J_{zoi} \, a_i}{3}}{\displaystyle\sum_i W_i - j\omega \sum_i \frac{a_i \, \sigma_i}{3}} \tag{11.52}$$

for each vertex and correct values according to

$$A'_{zo} = A_{zo} + \omega R. \tag{11.53}$$

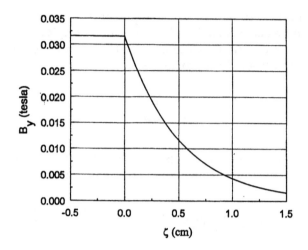

FIGURE 11.6
Benchmark eddy current solution, penetration of an AC magnetic field into a metal slab with $\mu/\mu_o = 1$, $\rho = 9.896 \times 10^{-5}$ Ω-m. Predicted skin depth at 1 MHz: $\delta = 0.5$ cm. Plot of the amplitude of $B_y(\zeta)$.

The quantity ω in Equation 11.53 is the over-relaxation factor. Determination of the amplitude and phase of magnetic fields from the complex vector potential is similar to the procedure described in Section 11.4. Again, resistive power dissipation per volume may be of interest for applications. The value in an element with vertices n_1, n_2, and n_3 is approximately

$$p = \frac{\omega^2}{2\sigma}\left(\frac{A_{z1}+A_{z2}+A_{z3}}{3}\right)^2. \tag{11.54}$$

Figure 11.6 shows a benchmark test to check the algorithm, penetration of an oscillating magnetic field into a metal surface. The analytic result (see, for instance Jackson, 1975) is that the magnetic field parallel to the surface varies inside the metal as

$$B = B_o \, \exp(-\zeta/\delta)\, \exp(j\zeta/\delta) = B_o \, \exp(-\zeta/\delta)\left[\cos(\zeta/\delta)+j\sin(\zeta/\delta)\right] \tag{11.55}$$

In Equation 11.55, B_o is the field at the surface, ζ is the distance from the surface, and δ is the skin depth,

$$\delta = \sqrt{\frac{\rho}{\pi\mu f}}. \tag{11.56}$$

The quantity ρ is the volume resistivity, μ is the magnetic permeability, and f is the field oscillation frequency. In the example, the choice of $\mu = \mu_o$, $\rho = 9.896 \times 10^{-5}$ Ω-m, and f = 1 MHz gives a skin depth of $\delta = 0.5$ cm. For the solution, we

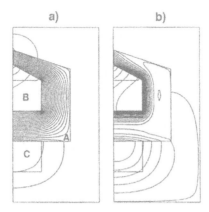

FIGURE 11.7
Frequency response of a recording head. x_{min}, 0.0 mm; x_{max}, 6.0 mm; y_{min}, –3.0 mm; y_{max}, 6.0 mm; A, ferrite with small gap, $\mu/\mu_o = 500$, $\rho = 10^{-3}$ Ω-m; B, top of drive coil; C, bottom of drive coil. (a) Geometric definitions and magnetic field lines at 1000 Hz. (b) Magnetic field lines at 2 MHz.

create a magnetic flux by assigning constant vector potential values of 0.0 tesla-m to the left boundary and 2.5×10^{-4} tesla-m to the right. Figure 11.6a shows contours of the vector potential amplitude at zero phase and Figure 11.6b shows the variation of $B_y(x)$. The field is constant in the air region and falls off exponentially with depth in the metal. The field amplitude at a depth of 0.5 cm is within 0.2% of the theoretical value and the phase difference is 55.6°, close to the predicted value of 57.3°.

Figure 11.7 shows an applications example, calculation of the frequency response of a recording head. The coil creates a high-frequency magnetic field carried through a ferrite to a narrow gap. The concentrated field magnetizes a tape passing over the gap. The solution region extends 6 mm in x and 9 mm in y. The top and bottom sections of the coil have equal amplitude and a 180° phase difference. The ferrite has $\mu/\mu_o = 500$ and $\rho = 1.0 \times 10^{-3}$ Ω-m, giving a skin depth of $\delta = 1$ mm at f = 505 kHz. Figure 11.7a shows magnetic field lines at low frequency (1 kHz). The result is indistinguishable from a static solution. The field is distributed throughout the ferrite with low values of fringing field near the gap and a small leakage flux. The results of Figure 11.7b at high frequency (2 MHz) are quite different. The field cannot penetrate the ferrite and is concentrated near the coils. Because of the increased leakage flux, the field magnitude at the gap drops by a factor of 3.5.

Chapter 11   Exercises

11.1.   A magnetic field is produced by loop of radius R = 0.3 m carrying I = 5.0×10^4 A. A smaller loop moves along the axis from z = $-\infty$ to z = ∞

at a constant velocity of 5 m/s. This loop, with 10 turns at radius a = 0.02 m, is connected to a high-impedance oscilloscope. Plot the measured voltage as a function of time.

11.2.   A long aluminum cylinder collapses radially toward the axis. At t = 0, the cylinder has radius R = 0.03 m, thickness δ = 0.002 m, and radial velocity –1000 m/s. At this time, the interior volume is filled with a solenoidal magnetic field of magnitude B_o = 0.1 tesla.

(a) Neglecting resistance, plot the magnetic field as a function of time and the acceleration history of the cylinder.

(b) What is the peak magnetic field at the turning point?

(c) Aluminum has the following parameters (at room temperature). Density: 2700 kg/m³, melting point: 660°C, volume resistivity 2.65 × 10⁻⁸ Ω-m, and specific heat 890 J/kg-°C. Make a rough estimate to determine if the cylinder reaches the turning point before melting.

11.3.   Figure E11.1 shows a circuit driven by an AC voltage source, $V_o\cos\omega t$.

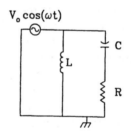

$V_o \cos(\omega t)$

C

L

R

FIGURE E11.1

(a) Find an expression for the current through the resistor.

(b) Find the amplitude and phase of the current for V_o = 12.5 V, L = 50.0 × 10⁻⁶ H, C = 2.5 × 10⁻⁶ F, R = 2.75 Ω and ω = 1.2 × 10⁴ s⁻¹.

11.4.   An AC voltage $V_o\cos(2\pi ft)$ is applied across circular parallel plates with spacing d and radius R.

(a) Assuming constant voltage over the plate areas and neglecting edge effects, find an expression for the magnetic field between the plates as a function of distance from the center.

(b) Find the magnetic field at r = R/2 if d = 0.015 m, R = 0.25 m, V_o = 1500.0 V, and f = 50 MHz.

(c) Is the assumption of constant voltage on the plates justified for the parameters of Part (b)?

11.5.   An AC electric field solution gives the following expression for the complex potential,

$$\phi(x,y) = 100.0 \left[1 + \cos(xy) - x^2\right] + 50.0j \left[\sin^2(y) - x + y^3\right]$$

Find numerical values for the amplitude and phase of ϕ, E_x, and E_y at the point (1.5,0.75).

11.6. Figure E11.2 shows a space of width d = 0.02 m between parallel plates of area 1 m². The space is divided into two layers of width 0.01 m. Layer 1 has parameters $\epsilon_r = 2.7$ and $\sigma = 0$, while layer 2 has $\epsilon_r \ll 1$ and $\sigma = 2.78 \times 10^{-3}$ mhos/m The right plate is grounded and a voltage $V_0 \cos\omega t$ is applied to the left plate. Use values $V_0 = 50$ V and $\omega = 6.283 \times 10^7$ s⁻¹.

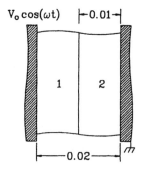

FIGURE E11.2

(a) Use the complex form of the Laplace equation (Equation 11.32) to find an expression for the complex value potential at the interface between the materials. Give values for the amplitude and phase of the potential.

(b) Layer 1 acts like a capacitor and layer 2 acts like a resistor. Find equivalent values of C and R for the layers.

(c) Solve the equivalent RC circuit and show that it leads to the same results as Part (a)

11.7. A medium with graded conductivity fills a space of width d = 0.075 between parallel plates. The right plate is grounded and the left plate has an AC applied voltage with amplitude $V_0 = 100$ V, phase $\phi = 60°$, and frequency f = 5.0 × 10⁴ Hz. The medium has uniform relative dielectric constant $\epsilon_r = 10.0$ and a spatially dependent conductivity $\sigma(x) = 10^5 (x/d)$.

(a) What are the values of complex potential on the two electrodes.

(b) Solve the one-dimensional Poisson equation for the complex potential with the given boundary conditions to find the voltage amplitude and phase as a function of position.

11.8. Take the curl of Equation 11.42 to show that one-dimensional diffusion of AC magnetic fields into a uniform material with no applied
current density is described by the equation,

$$\frac{d^2B_y}{dx^2} = j\sigma\omega\mu\ B_y\ .$$

Verify Equation 11.55 by substitution into the above equation and
confirm the skin depth expression of Equation 11.56.

11.9. An AC current $I_o\cos\omega t$ is applied to the long conducting rod of radius R
shown in Figure E11.3. The rod has uniform values of μ and σ. We
want to find the distribution of current density using a one-dimensional model. Confirm that Ampere's law gives the following relationship between the toroidal magnetic field and current density in
the rod.

$$\frac{dB_\theta}{dr} = r\ \mu\ j_z\ .$$

Faraday's law relates the two quantities as

$$\frac{dj_z}{dr} = j\omega\sigma\ B_\theta\ ,$$

so that the diffusion equation for axial current density is

$$\frac{d^2j_z}{dr^2} = j\omega\sigma\mu\ r\ j_z\ .$$

In the limit of high frequency, the equation implies that the current
density is confined to the surface of the rod within a distance equal to
the skin depth, $\delta = (2/\omega\mu\sigma)^{1/2}$. At low frequency, we expect that the
current density is uniform in radius.

(a) Estimate the transition frequency by a dimensional analysis of
the above equation.

(b) Estimate the transition frequency by setting $L/R = 1/\omega$, where
the inductance per unit length of the rod is about $L \sim \mu/2\pi$ and
the resistance per length is $R \sim 1/\sigma\pi R^2$.

(c) Alternatively, we could use the two equations to derive a diffusion equation for B_θ. What is the advantage of this approach?

11.10. A high-frequency transformer core is composed of laminations of thin silicon steel. The laminations have width 0.05 m, thickness 5×10^{-5} m, average magnetic permeability $\mu_r = 500$, and volume resistivity 50×10^{-8} Ω-m. What is the approximate limit on frequency such that the core behaves like an ideal inductor?

11.11. A transmission line has a solid copper inner conductor of radius $R_i = 0.010$ m and volume resistivity 1.7×10^{-8} Ω-m. The copper outer conductor has radius $R_o = 0.025$ m. The line carries an AC signal with current $I = 50.0\cos\omega t$. Estimate the power loss per meter in the inner and outer conductors at frequencies $f = 20.0$ MHz and 500 MHz.

12

Thermal Transport and Magnetic Field Diffusion

Scientists have odious manners, except when you prop up their theory; then you can borrow money of them.

Mark Twain

In this chapter we advance to fully time-dependent initial value solutions. We shall concentrate on diffusion solutions, an important class in all areas of physics and engineering. Applications include neutron transport in reactors, doping of semiconductors, dispersal of pollutants, and the distribution of ion species in plasma etching devices. As an introduction, the first three sections deal with thermal energy transport in solids. The goal is to find temperature variations in space and time in systems that consist of contiguous volumes of solid materials. An example is the transient temperature distribution in a nuclear fuel rod.

Section 12.1 applies the law of energy conservation to find the differential and integral equations of thermal transport. Section 12.2 reviews the conversion of the differential diffusion equation to a set of difference equations using finite-difference methods on a rectangular mesh. There are several options, some of which lead to severe constraints on the time step for numerical stability. The implication is that some methods may be mathematically correct but impractical. The section introduces the Dufort-Frankiel algorithm to solve diffusion problems. This technique is easy to implement in computer programs and is stable for any choice of Δt. Section 12.3 derives thermal transport difference equations from the finite-element viewpoint on a conformal mesh. A modified version of the Dufort-Frankiel method applies. It is relatively easy to extend solutions to complex materials where the thermal conductivity and specific heat vary with temperature. Section 12.4 discusses special problems of numerical stability on triangular meshes. Instabilities can be avoided by setting constraints on triangle geometry in the mesh generation process.

Section 12.5 shows that pulsed magnetic field penetration into conductive materials is governed by diffusion equations. Applications include high-field magnets, magnetic shaping of metal parts, pulse transformers, and high-power plasma devices. The section covers two-dimensional finite-element solutions with a single component of vector potential.

12.1 Thermal Transport Equation

The conduction of heat in one dimension between two positions in a solid is approximated by the relationship

$$q_x = -k \, \frac{T(x_2) - T(x_1)}{x_2 - x_1}. \tag{12.1}$$

The quantity $T(x)$ is the local temperature and q_x is the thermal flux in units of W/m^2. The quantity k is the material thermal conductivity with units $J/m\text{-}K$. We shall assume that k is independent of direction (isotropic material) but may vary with temperature. Equation 12.1 states that heat flows from locations of high temperature to those of low temperature. The flux is proportional to the temperature difference and inversely proportional to the distance between the locations. The three-dimensional form is

$$q = -k \, \nabla T. \tag{12.2}$$

Note that the thermal flux is a vector quantity.

A thermal equation with temperature as the only dependent variable can be derived by applying the principle of energy conservation to a small volume. We include the option of a volumetric heat source $U(x,y,z,t)$ with units of $J/(m^3\text{-}s)$. The source and thermal flux change the local energy density in the medium. In the limit that the material density is constant (no work by pressure forces), we can write the rate of change in energy density as

$$\rho \, C_P \frac{\partial T}{\partial t}. \tag{12.3}$$

In Equation 12.3, the quantity ρ is the material density in kg/m^3 and C_p is the specific heat at constant pressure in $J/kg\text{-}K$. Equating the expression of Equation 12.3 to the power contribution of sources and the incoming thermal flux gives the equation

$$\iiint dV \, \rho \, C_P \frac{\partial T}{\partial t} = -\iint q \cdot n \, dS + \iiint U \, dV . \qquad (12.4)$$

The first term on the right-hand side of Equation 12.4 is a surface integral of thermal flux. Here, the quantity dS is an area element of the surface and n is a unit vector pointing out of the volume normal to the surface. We can use the divergence theorem to convert the surface integral to a volume integral of $\nabla \cdot q$. Taking the limit at a point and substituting from Equation 12.2 leads to the differential form of the thermal transport equation,

$$\rho \, C_P \frac{\partial T}{\partial t} = \nabla \cdot (k \nabla T) + U . \qquad (12.5)$$

Section 12.2 shows how to solve thermal diffusion problems by converting Equation 12.5 to a set of difference equations on a rectangular mesh. Section 12.3 derives finite-element difference equations by applying Equation 12.4 to volumes on a conformal triangular mesh.

It is informative to work through a solution of Equation 12.5. The results are useful for estimating parameters for numerical solutions and creating benchmark tests. Consider a one-dimensional system that consists of a homogeneous material over the region $-L/2 \le x \le L/2$. The medium has no thermal sources and uniform values of k and C_P. The initial temperature distribution $T(x,0)$ has the uniform value T_0 over the range $-\alpha L/2 \le x \le \alpha L/2$, where $\alpha < 1$. The temperature at the boundaries has the constant value $T(-L/2) = T(L/2) = 0$.

Applying separation of variables, we assume a solution that consists of a set of Fourier spatial modes

$$T(x,t) = \sum_{n=1}^{\infty} T_{on} \, f_n(t) \, \cos\left(\frac{\pi n x}{L}\right) . \qquad (12.6)$$

We can find the temporal variations of each mode, $f_n(t)$ by substituting Equation 12.6 in Equation 12.5. The result is

$$\frac{\partial f_n}{\partial t} = -\left[\frac{k}{\rho C_P}\right]\left[\frac{\pi n}{L}\right]^2 f_n . \qquad (12.7)$$

If we assume that the temporal functions equal unity at $t = 0$, the solution of Equation 12.7 is

$$f_n(t) = \exp(-t/\tau_n) . \qquad (12.8)$$

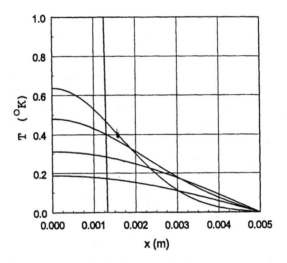

FIGURE 12.1
One-dimensional thermal diffusion. Step function initial temperature from x = 0.0 mm to x = 1.25 mm (L = 0.01 m, α = 0.25). Medium properties: $\rho = 1 \text{ kg/m}^3$, $C_p = 1 \text{ W/kg-K}$, $k = 1 \text{ W/m}^2\text{-s}$. Finite element calculation with 50 elements along the width. Neumann boundary on the left-hand side. Plots at 0, 1, 2, 5, and 10 μsec.

The mode time constant in Equation 12.8 is

$$\tau_n = \frac{\rho \, C_p \, L^2}{k \, \pi^2 \, n^2}.$$

(12.9)

We can find the mode amplitude T_{on} from a Fourier cosine series analysis of the initial temperature distribution. The result is

$$T_{on} = \frac{2 \sin(n\pi\alpha/2)}{n\pi}.$$

(12.10)

The sharp temperature boundary means that there is initially a substantial contribution from higher-order modes (n > 1). Inspection of Equation 12.9 shows that these modes decay rapidly. During an initial transient period diffusion solutions relax to the fundamental mode $T(x) \sim \cos(n\pi x/L)$. Thereafter, they decay at the rate given by τ_1. Figure 12.1 shows the temperature distribution for α = 0.25. The calculation, performed with a finite element code with 50 elements across the width, agrees with the prediction of Equation 12.6.

We can analyze Equation 12.9 to estimate the time step required for good accuracy in a numerical solution,

$$\Delta t < \frac{\rho \, C_p \, \Delta x^2}{k \, \pi^2}. \tag{12.11}$$

In Equation 12.11, ρ and C_p are local values of the thermal constants and Δx is a scale length for significant temperature variation. In the numerical solution of Figure 12.1, the quantity Δx is initially small because of the sharp temperature discontinuity. We cannot resolve variations to a length smaller than the element size; therefore, $\Delta x \approx L/50$. Therefore, the initial time step must be short. After the decay of higher-order modes, the scale length approaches $\Delta x \approx L/2$. The allowed time step increases by a factor of approximately 625. The time step was automatically adjusted during the solution of Figure 12.1 using the method described in Sections 12.2 and 12.3.

12.2 Finite-Difference Solution of the Diffusion Equation

As an introduction to numerical solutions of initial value problems we shall discuss the finite-difference representation of Equation 12.5 in a one-dimensional system with uniform material properties. The solution space is divided into uniform sections of width Δx and time into intervals Δt. The index i denotes the mesh point position $x_i = i\Delta x$ and n designates time, $t = n\Delta t$. Following Section 4.1, the difference expression for the second spatial derivative at x_i and t_n is

$$\frac{\partial^2 T(x_i, t_n)}{\partial x^2} \cong \frac{T_{i+1}^n - 2T_i^n + T_{i-1}^n}{\Delta x^2}. \tag{12.12}$$

We have several options for a difference expression in time. The simplest is the Euler method (Section 4.2) where the time derivative is expressed in terms of the temperature distribution at the current time. Substituting in Equation 12.5 and solving for temperature at the advanced time gives

$$T_i^{n+1} = T_i^n + \frac{\kappa \Delta t}{\Delta x^2} \left[T_{i+1}^n - 2T_i^n + T_{i-1}^n \right], \tag{12.13}$$

where $\kappa = k/\rho C_p$. As we saw in Section 4.2, the expression of Equation 12.13 is not time-centered. Therefore, it is accurate only to first order in Δt.

A second problem with Equation 12.13 is that it places strong limits on the time step for numerical stability. To show this we resolve the spatial variation of temperature into Fourier modes

$$T_k^n \exp(jkx). \tag{12.14}$$

The mode stability properties are given by the amplification factor

$$g_k = T_k^{n+1}/T_k^n . \tag{12.15}$$

The solution is stable only if the condition

$$|g_k| \le 1 \tag{12.16}$$

holds for all modes. Substitution of Equations 12.14 and 12.15 into Equation 12.13 gives the following relationship,

$$g_k = 1 + \frac{\kappa \Delta t}{\Delta x^2}\left[e^{ik\Delta x} - 2 + e^{-ik\Delta x}\right] = 1 + \frac{\kappa \Delta t}{\Delta x^2}\left[\cos(k\Delta x) - 2\right]. \tag{12.17}$$

The final term in brackets has the extreme value –4 when $k\Delta x = \pi$. In this case, the condition to satisfy Equation 12.16 is

$$\Delta t \le \Delta x^2/2\kappa . \tag{12.18}$$

A comparison of Equation 12.18 to Equation 12.9 shows that the time step for stability may be much shorter than the time step necessary for accuracy. At late times Δt must reflect the scale length of the mesh rather than of the system. Therefore, the computation time required with the Euler method may be 100 to 1000 times longer than that with more stable methods.

There are several time-centered alternatives to the Euler method that relax the stability condition of Equation 12.18. A good choice is the Dufort-Frankiel method. It requires a small number of operations per step, is easy to program, and is numerically stable for any choice of Δt. The price for these advantages is low — the method requires one additional array to store previous values of temperature. The one-dimensional expression to advance the temperature is

$$T_i^{n+1} = T_i^{n-1} + \frac{2\kappa \Delta t}{\Delta x^2}\left[T_{i+1}^n - T_i^{n-1} - T_i^{n+1} + T_{i-1}^n\right]. \tag{12.19}$$

The difference from the Euler expression is that Equation 12.19 represents an interval $2\Delta t$ from t_{n-1} to t_{n+1}. A combination of old and new values gives an expression for the second spatial derivative that is time-centered at t_n. Note that the contribution of the temperature at x_i is the average of past and future values. Solving Equation 12.19 for the future temperature gives the equation

$$T_i^{n+1} = \left(\frac{1-\alpha}{1+\alpha}\right) T_i^{n-1} + \left(\frac{\alpha}{1+\alpha}\right)\left(T_{i+1}^n + T_{i-1}^n\right), \qquad (12.20)$$

where

$$\alpha = \frac{2\kappa\Delta t}{\Delta x^2}. \qquad (12.21)$$

Implementation of Equation 12.20 on a computer requires two arrays: one to store the present values, $T(i)$, and one that holds both the past and future values, $TOldNew(i)$. In a time step, previous values at all positions are replaced by future values and then the arrays $T(i)$ and $TOldNew(i)$ are exchanged.

To analyze the stability properties of the Dufort-Frankiel method, we again consider a Fourier mode expansion. Substituting Equations 12.14 and 12.15 in Equation 12.20 gives

$$g_k^2 = \left(\frac{1-\alpha}{1+\alpha}\right) + \left(\frac{\alpha}{1+\alpha}\right) g \cos(k\Delta x). \qquad (12.22)$$

Applying the quadratic formula, the amplification factor is

$$g_k = \frac{\alpha \cos(k\Delta x) \pm \sqrt{1 - \alpha^2 \sin^2(k\Delta x)}}{1+\alpha}. \qquad (12.23)$$

We can understand the implications of Equation 12.23 by considering limiting cases. For small time step the following condition holds:

$$\alpha^2 \sin^2(k\Delta x) \preceq 1. \qquad (12.24)$$

In the limit of Equation 12.24, the amplification factor of Equation 12.23 is real and both roots have magnitude less than unity for any choice of k. For large time step, the amplification factor is complex with a magnitude given by

$$|g_k| = \frac{1-\alpha}{1+\alpha}. \qquad (12.25)$$

The expression of Equation 12.25 has a value less than unity. We can investigate intermediate values numerically to show that the Dufort-Frankiel method is stable for any choice of time step. It is straightforward to extend the derivation to two-dimensional calculations on a rectangular mesh with spacings Δx and Δy. The conclusion is that the calculation is stable for any choice of $\Delta y/\Delta x$ or α.

With stability assured, we can adjust the time step in the calculation to minimize computational time while maintaining accuracy. Automatic time-step adjustment is an essential feature of any practical diffusion program. To illustrate methods for picking Δt, consider a two-dimensional finite-difference solution with uniform k and C_p on a square mesh with spacing Δ. The Dufort-Frankiel algorithm is

$$T_{ij}^{n+1} = T_{ij}^{n-1} + \frac{2\kappa\Delta t}{\Delta^2}\left[T_{i+1,j}^n + T_{i-1,j}^n + T_{i,j+1}^n + T_{i,j-1}^n - 2T_{ij}^{n-1} - 2T_{ij}^{n+1}\right]. \quad (12.26)$$

We seek a criterion to pick the time step that will work with arbitrary spatial boundaries. One option is to take averages over all vertices that do not have fixed temperature. The following expression works well for most problems,

$$\Delta t = S_f \left[\frac{\rho C_p \Delta^2}{k}\right] \langle T_{ij}\rangle \frac{1}{\langle T_{i+1,j} + T_{i-1,j} + T_{i,j+1} + T_{i,j-1} - 4T_{ij}\rangle}. \quad (12.27)$$

The averaging symbols denote root-mean-squared averages. The quantity S_f is a safety factor with a value less than unity. A smaller time step or a different averaging technique may be necessary for problems with strong temperature gradients over small mesh regions.

12.3 Finite-Element Diffusion Solutions

It is easy to extend the concepts of the previous section to solutions on a conformal triangular mesh. We shall first consider planar solutions and then extend the results to cylindrical geometry. We take the volume and surface integrals of Equation 12.4 over the path of Figure 12.2 that encloses one third of the volume of each element. With the approximation that the temperature varies linearly in each element, the heat flux vector q is an element constant. We assume that the thermal properties (k, ρC_p, and S) are also element constants. If the system extends a unit distance in z, the volume integral on the left-hand side of Equation 12.4 is

$$\frac{\partial T_o}{\partial t} \sum_{i=1}^{6} \frac{\rho_i C_{pi} a_i}{3}, \quad (12.28)$$

where the quantities a_i are the element areas. The volume integral over heat sources is

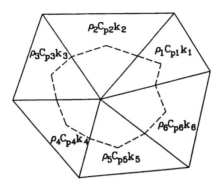

FIGURE 12.2
Application of energy conservation near a vertex of a conformal triangular mesh. The volume integral of thermal energy extends over the portion of the surrounding elements within the dashed line.

$$\sum_{i=1}^{6} \frac{U_i a_i}{3} . \tag{12.29}$$

By analogy with the Gauss's law derivation of Section 2.7, the integral of heat flux over the 12 facets of the surface is

$$\sum_{i=1}^{6} W_i T_i - T_o \sum_{i=1}^{6} W_i . \tag{12.30}$$

The coupling coefficients are given by

$$W_1 = \frac{k_2 \cot\theta_{2b} + k_1 \cot\theta_{1a}}{2},$$

$$W_2 = \frac{k_3 \cot\theta_{3b} + k_2 \cot\theta_{2a}}{2},$$

$$\cdots \tag{12.31}$$

$$W_6 = \frac{k_1 \cot\theta_{1b} + k_6 \cot\theta_{6a}}{2}.$$

We can incorporate Equation 12.30 into a difference equation that implements the Dufort-Frankiel method,

$$T_o^{n+1} =$$

$$T_o^{n-1} + \frac{6\Delta t}{\sum \rho_i C_{pi} a_i} \left[\sum W_i T_i^n - \frac{T_o^{n+1} + T_o^{n-1}}{2} \sum W_i + \sum \frac{U_i^n a_i}{3} \right] \quad (12.32)$$

Solving Equation 12.32 for T_o^{n+1} gives an equation to advance the temperature at each point.

We must address some preliminaries before beginning calculations. Following the discussion of Section 2.8, the condition on free boundaries in finite-element thermal solutions is that the normal derivative of the temperature equals zero. Therefore, an unspecified boundary represents an ideal thermal insulator. The alternative is a Dirichlet boundary where vertices are set to a temperature, either constant or varying in time. When the thermal conductivity varies with time or temperature, the coupling constants must be updated at each time step. The number of operations can be minimized using the methods described in Section 9.5. Ideally, the second term in brackets of Equation 12.32 should have the form

$$\frac{T_o^{n+1} \sum W_i^{n+1} + T_o^{n-1} \sum W_i^{n-1}}{2}, \quad (12.33)$$

where the coupling constant sums are evaluated at future and past times. Maintaining past, present, and future values of the coupling constants increases storage requirements and slows the calculation. In practice, material properties change little in one time step and it is sufficient to use only present coupling constants in Equation 12.32.

Figure 12.3 illustrates a finite-element solution of a challenging problem, propagation of a bleaching wave. The strongly nonlinear material has a sharp increase in thermal conductivity by a factor of 100 at 20°C. Thermal solutions in such a medium are qualitatively different from those described in Section 12.1. A shock-like transition from 0° to 20°C propagates at a velocity set by the heat capacity of the medium and conduction of heat through the high-conductivity material behind the front. In the system of Figure 12.3, the left boundary has a fixed time variation of temperature, a linear change from 0° to 25°C in 0.1 msec starting at $t = 0$. Heat conduction to the bleaching front through the high conductivity region roughly follows Equation 12.1. Because the propagation distance increases, the bleaching wave velocity decreases with time.

12.4 Instabilities in Finite-Element Diffusion Solutions

In contrast to diffusion solutions on a regular mesh, calculations on a conformal mesh may exhibit numerical instabilities for some choices of element

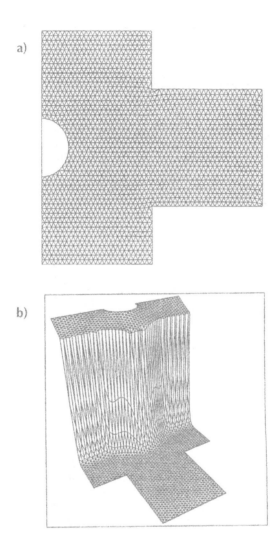

FIGURE 12.3
Thermal transport in a highly nonlinear medium — bleaching wave propagation. $x_{min} = 0.0$ cm, $x_{max} = 4.0$ cm, $y_{min} = 0.0$ cm, $y_{max} = 2.0$ cm. The fixed temperature on the left-boundary rises to 25°C in 0.1 msec. Thermal conductivity: $k = 1.0$ W/m²-s for $T < 19$°C and $k = 100$ W/m²-s for $T > 21$°C. (a) Geometry and computational mesh; (b) wire frame plot of temperature distribution at 3.5 msec showing a sharp thermal transition moving away from the left-hand boundary.

geometry. In this section, we shall discuss how to prevent such problems. Figure 12.4a shows the mesh for a benchmark calculation. The solution region has an initial temperature of 0°C, Neumann boundaries on the left and right, and fixed temperatures of 0° and 1°C at the bottom and top. Ultimately the temperature should relax to a linear variation in y. With the base of the triangular elements fixed at $\Delta x = 0.001$ m, we investigate solution stability for different choices of the height, Δy. The result is that the solutions are stable when $\Delta y > \Delta x/2$. When $\Delta y \leq \Delta x/2$ the solutions are unstable for any choice

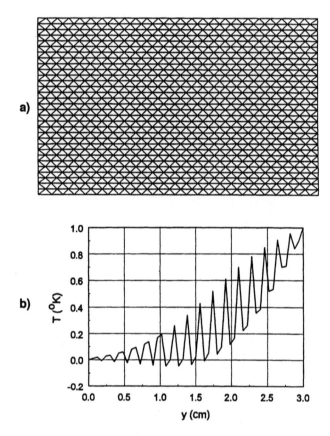

FIGURE 12.4
Instabilities of diffusion calculations on a conformal triangular mesh. (a) Geometry of benchmark test and computational mesh. Bottom boundary fixed at 0°C and top boundary at 1.0°C.
Initial temperature of 0°C on all other vertices. Neumann boundaries on the left and right. x_{min},
0 cm; x_{max}, 5.0 cm; y_{min}, 0.0 cm; y_{max}, 3.0 cm. k = 200.0 W/m²-s; p = 1000 kg/m3; Cp =
1.0 W/kg-°C. Isosceles triangle mesh with $\Delta y/\Delta x$ = 0.475. b) Temperature distribution at t = 0.2
msec.

of time step. The temperature contour plot of Figure 12.4b shows the problem
at an early stage for $\Delta y/\Delta x$ = 0.475. Note that the temperature deviations are
above and below the correct values on alternate rows.

We can investigate the stability of the Dufort-Frankiel algorithm on the uniform mesh of triangles shown in Figure 12.5. The angles in the figure are
given by

$$\beta = \tan^{-1}\left(\frac{2\Delta y}{\Delta x}\right),$$

$$\gamma = \pi - 2\beta. \tag{12.34}$$

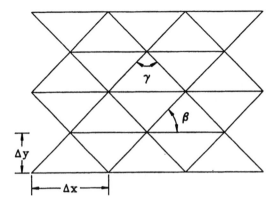

FIGURE 12.5
Simple mesh geometry to model diffusion instabilities.

The coupling coefficients to the six neighbors of a vertex are

$$W_1 = \cot(\beta), \quad W_2 = \cot(\beta), \quad W_3 = \cot(\gamma),$$
$$W_4 = \cot(\beta), \quad W_5 = \cot(\beta), \quad W_6 = \cot(\gamma). \tag{12.35}$$

Note that when $\Delta y \le \Delta x / 2$ the angle γ is larger than $\pi/2$ and the coupling coefficients W_3 and W_6 have negative values.

Under the conditions that there are no sources and that all elements have equal areas and thermal properties, Equation 12.32 reduces to

$$T_o^{n+1} = T_o^{n-1} + \alpha \left[\sum W_i T_i^n + \frac{T_o^{n+1} + T_o^{n-1}}{2} \sum W_i \right], \tag{12.36}$$

where

$$\alpha = \frac{\Delta t}{\rho \, C_p \, \alpha}. \tag{12.37}$$

Motivated by the results of Figure 12.4, we shall investigate a mode that is uniform in x but reverses sign on each row in y. This solution has the following values of temperature at vertices surrounding a test vertex with temperature T_o: $T_1 = -T_o$, $T_2 = -T_o$, $T_3 = T_o$, $T_4 = -T_o$, $T_5 = -T_o$, and $T_6 = T_o$. Further, because the instability is independent of time step we shall take $\alpha = 1$. Substitution in Equation 12.36 gives a relationship for the value of the temperature at the test vertex for the mode at three times,

$$T_0^{n+1} \left(1 + \frac{\sum W_i}{2} \right) + T_0^n \left(W_1 + W_2 - W_3 + W_4 + W_5 - W_6 \right)$$

$$- T_0^{n-1} \left(1 + \frac{\sum W_i}{2} \right) = 0. \tag{12.38}$$

Assume that the mode grows as $T_0^{n+1} = g\, T_0^n$. The solution is unstable if $|g| \geq 1$. Equation 12.38 implies that the growth constant is

$$g = \frac{-B \pm \sqrt{B^2 - 4AC}}{2A}, \tag{12.39}$$

where

$$A = 1 + \sum W_i/2,$$

$$B = \sum W_i - 2\left(W_3 + W_6 \right), \tag{12.40}$$

$$C = -1 + \sum W_i/2.$$

We can confirm the solution is unstable when $\beta \leq \pi/4$ by direct substitution from Equations 12.35.

The general conclusion is that the mesh instability occurs when there are extended regions of contiguous triangles that have one internal angle greater than 90°. This limit makes it difficult to generate meshes with fine resolution in y ($\Delta y < \Delta x/2$) under the mesh logic of Section 5.2. Figure 12.6 shows one solution. The logical mesh is initially composed of right triangles with neutral stability properties. We can create a logical mesh with all angles less than 90° by making small positive displacements in x of all vertices on odd rows and negative displacements on even rows (Figure 12.6). When the logical mesh is distorted to fit boundaries, some triangles may violate the stability criterion. The next issue is to determine how many connected bad triangles are necessary to drive the instability. Although there is no definitive answer for all meshes, we can make simple tests for guidance. In the mesh of Figure 12.7a, 12 triangles have been modified to include angles greater than 90° by moving four points on an odd row in the negative x direction. The temperature contours of Figure 12.7b confirm that there is a slow-growing instability in the region of modified triangles. The solution is stable if fewer points are shifted. Generally, a few connected triangles with large angles are not dangerous in diffusion calculations.

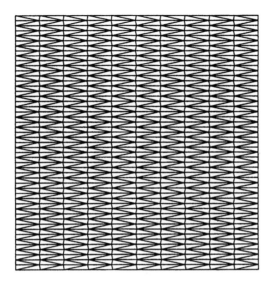

FIGURE 12.6
Stable triangular mesh with short elements ($\Delta y/\Delta x < 0.5$) and no obtuse triangles.

12.5 Magnetic Field Diffusion

In this section we shall discuss time-domain magnetic field solutions in the eddy current regime of Section 11.5. The goal is to generate equations for initial value problems with arbitrary time variations of driving currents. For stationary media, magnetic fields obey the diffusion equation. Therefore, the numerical techniques developed for thermal transport can be applied directly to magnetic fields. For convenience, we study the differential forms and then write the finite-element difference equations by analogy with previous sections.

There are three governing relationships. The first is Faraday's law

$$\nabla \times \mathbf{E} = -\frac{\partial \mathbf{B}}{\partial t}. \tag{12.41}$$

Ampere's law has the form

$$\nabla \times \left(\frac{\mathbf{B}}{\mu_r}\right) = \mu_o \, \mathbf{J}_c + \mu_o \, \mathbf{J}_o. \tag{12.42}$$

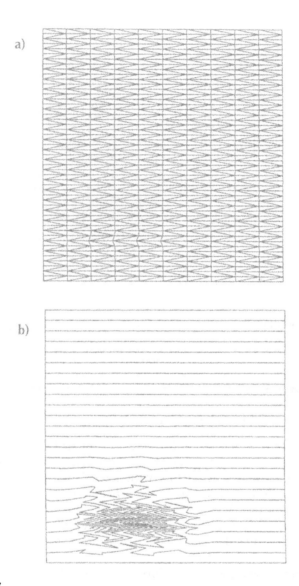

FIGURE 12.7
Investigation of the number of contiguous obtuse triangles to drive a diffusion instability.
(a) Destabilized mesh — displacement of four vertices (lower, left of center) to create 12 obtuse
triangles. (b) Temperature contours showing a slowly growing instability in the region.

In Equation 12.42, the quantity μ_r is the local value of relative magnetic permeability, J_c is the current density induced in materials by changing magnetic
fields, and J_o is the current density in driving coils. In most problems, the
drive current is a specified function of time. The final equation is Ohm's law,

$$J_c = \frac{E}{\rho},$$

(12.43)

where ρ is the volume resistivity in ohm-m.

The strategy for combining Equations 12.41, 12.42, and 12.43 into a single diffusion equation depends on the geometry. We shall limit the discussion to two-dimensional solutions and seek equations that involve a single scalar quantity. The magnetic field is given by

$$\mathbf{B} = \nabla \times \mathbf{A}. \tag{12.44}$$

Substituting Equation 12.44 into Equation 12.43 gives

$$\nabla \times \mathbf{E} = -\frac{\partial}{\partial t}(\nabla \times \mathbf{A}). \tag{12.45}$$

In the limit that there are no electrostatic fields, Equation 12.45 implies that

$$\mathbf{E} = -\frac{\partial \mathbf{A}}{\partial t}. \tag{12.46}$$

Substituting Equations 12.43 and 12.46 in Equation 12.42 gives the following relationship for the vector potential,

$$\frac{\partial \mathbf{A}}{\partial t} = \frac{\rho}{\mu_o}\left[\nabla \times \frac{1}{\mu_r}(\nabla \times \mathbf{A})\right] + \rho \mathbf{J}_o. \tag{12.47}$$

To begin, consider a system uniform in z with variations in x and y. As we saw in Chapter 9, the quantity A_z is the only non-zero component of vector potential. The drive current in the z direction generates field components B_x and B_y. For the two-dimensional geometry, the component form of Equation 12.47 is

$$\frac{\partial A_z}{\partial t} = \frac{\rho}{\mu_o}\left[\frac{\partial}{\partial x}\frac{1}{\mu_r}\frac{\partial A_z}{\partial x} + \frac{\partial}{\partial y}\frac{1}{\mu_r}\frac{\partial A_z}{\partial y}\right] + \rho J_{zo}. \tag{12.48}$$

The Dufort-Frankiel form of Equation 12.48 on a conformal triangular mesh is

$$A_{zo}^{n+1} = A_{zo}^{n-1}$$

$$+ \frac{2\Delta t}{\sum a_i/3\rho_i^n}\left[\sum W_i^n A_{zi}^n - \frac{A_{zo}^{n+1} + A_{zo}^{n-1}}{2}\sum W_i^n + \sum a_i J_{zoi}^n\right]. \tag{12.49}$$

The coupling constants have the form

$$W_1^n = \frac{\cot\theta_{2b}/\mu_{r2}^n + \cot\theta_{1a}/\mu_{r1}^n}{2\mu_0}.\tag{12.50}$$

The extension of Equations 12.49 and 12.50 to cylindrical coordinates closely follows the development of Section 9.4. The solution gives the time dependence of rA_θ from which B_z and B_r can be derived.

Equation 12.49 can be solved with specified driving currents and Neumann or fixed Dirichlet boundary conditions on A_z. Alternatively, some initial value solutions are most conveniently approached by setting a waveform for A_z on boundaries to specify a time-dependent flux. In the system of Figure 12.8a a current is driven through a shaped conductor with height h in z by applying a time-dependent voltage, V(t). The goal is to find the current density and magnetic field distributions in the material. These quantities are given, respectively, by time and space derivatives of A_z. If the solution boundary is distant from the object, the magnetic field lines are approximately parallel to the surface and the axial electric field is almost constant. Applying Equation 12.46, the time-variation of the vector potential on the boundary is

$$A_z(t) = \frac{\int_0^t V(t')\,dt'}{h}.\tag{12.51}$$

Figure 12.8a illustrates the condition of Equation 12.51. A pulsed axial current penetrates a stainless steel bus bar. One quarter of the geometry is shown. The bar has length 10 m and total cross section area 0.0016 m². The bottom and left sides are Neumann boundaries and the curved boundary has a given vector potential $A_{zo}(t) = -(0.1)\,t$. Equation 12.51 implies a step voltage pulse of 1 V along the bar length starting at $t = 0$. With a volume resistivity of 81×10^{-8} Ω-m, the bar resistance is $R = 5.1$ mΩ. The inductance of the bar volume is $L \approx 2$ μH. Therefore, the field penetration time is roughly $L/R \approx 0.4$ msec. For the simulation, the air space between the boundary and the bar has a resistivity of 0.001 Ω-m, about 1000 times that of the bar. This value is high enough to ensure rapid diffusion of vector potential to the bar surface. The expected current in the quarter section of the bar for long times ($t \gg L/R$) is $V/R = 49$ A. The current is smaller early in time because the current flows in only a fraction of the volume. Figure 12.8a illustrates that the magnetic field early in time (0.02 msec) is constrained to the surface. Figure 12.8b plots the time history of current density at two positions in the bar. The curves show the delay in current density moving into the bar and confirm that the average current varies approximately as $(1 - \exp[-t/(L/R)])$ with $L/R = 0.35$ msec. A spatial integral of current density in the bar at $t = 1.0$ msec gives 45.3 A, consistent with the long-term prediction.

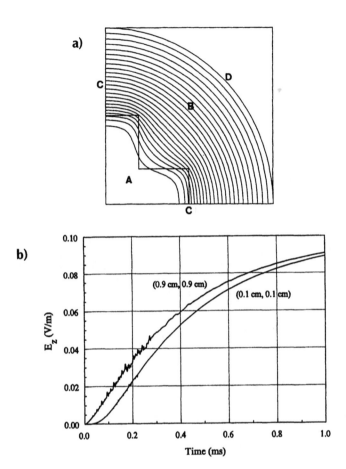

FIGURE 12.8
Magnetic field diffusion solution — pulsed current in a long bus bar. (a) System geometry and magnetic field lines, one quadrant shown. x_{min}, 0.0 cm; x_{max}, 5.0 cm; y_{min}, 0.0 cm; y_{max}, 5.0 cm; A, stainless steel bus bar with $\rho = 81 \times 10^{-8}$ Ω-m; B, vacuum; C, Neumann boundaries; D, fixed vector potential boundary to represent a constant applied axial field of 0.1 V/m, $A_{zo}(t) = -0.1t$. Magnetic field lines plotted at 0.02 msec. (b) Time history of current density near the surface and near the center of the bar.

Figure 12.9 shows an alternate two-dimensional geometry that applies to simulations of high-power plasma or metal pinches. Here, an axial current flows through a cylindrical system. Azimuthal symmetry implies that there is only one component of magnetic field, B_θ. Therefore, it is most convenient to cast the diffusion equation directly in terms of this quantity. For pinch applications we make two simplifying assumptions: the medium is nonmagnetic ($\mu_r = 1$) and there are no internal applied currents ($J_o = 0$). Equation 12.42 becomes

$$J_c = \frac{1}{\mu_o} \nabla \times B.$$ (12.52)

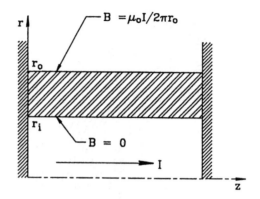

FIGURE 12.9
Geometry for a pinch-type field solution of $B_\theta(r,z,t)$. An annular metal shell carries a pulsed axial current $I(t)$. The Dirichlet boundary conditions at r_i and r_o are $B_\theta(r_i,z,t) = 0$, $B_\theta(r_o,z,t) = \mu_o I/2\pi r_o$. The Neumann boundary conditions at $z = 0$ and $z = z_o$ imply that $E_r = 0$.

Combining Equations 12.41, 12.43, and 12.52 gives

$$\frac{\partial \mathbf{B}}{\partial t} = -\frac{1}{\mu_o} \nabla \times \left(\rho \, \nabla \times \mathbf{B} \right). \tag{12.53}$$

The component form of Equation 12.53 is

$$\frac{\partial B_\theta}{\partial t} = -\frac{1}{\mu_o} \left[\frac{1}{r} \frac{\partial}{\partial r} \, \rho \, r \frac{\partial B_\theta}{\partial r} + \frac{\partial}{\partial z} \, \rho \frac{\partial B_\theta}{\partial z} \right]. \tag{12.54}$$

The finite-element difference equivalent of Equation 12.54 using the Dufort-Frankiel method is

$$B_{\theta o}^{n+1} = B_{\theta o}^{n-1} + \frac{3\Delta t \sum W_i}{\sum r_i \, a_i} \left[\sum W_i B_{\theta i} - \sum W_i \frac{B_{\theta o}^{n+1} + B_{\theta o}^{n-1}}{2} \right] \tag{12.55}$$

with coupling coefficients of the form

$$W_1 = \frac{\rho_2 r_2 \cot \theta_{2b} + \rho_1 r_1 \cot \theta_{1a}}{2}. \tag{12.56}$$

Figure 12.9 illustrates boundary conditions for a typical application of Equation 12.55. An axial large current is driven through a conducting cylindrical shell clamped to metal plates at the left and right. The condition on the inner cylindrical boundary (bottom) is that $B_\theta = 0$ because current cannot flow inside the shell. The top boundary (the outer radius of the solution volume) also has a Dirichlet condition. Following Equation 9.12 the toroidal field is a specified function of time related to the axial current supplied by an external circuit,

$$B_\theta\left(r_{out}, z, t\right) = \frac{\mu_o\, I_z(t)}{2\pi r_{out}}. \tag{12.57}$$

Finally, the condition on the conducting plates at the left and right is that $E_r = 0$. Applying Equation 12.44, the condition is equivalent to $\partial B_\theta / \partial z = 0$. Therefore, the boundaries satisfy the specialized Neumann condition.

Chapter 12 Exercises

12.1. A long bar of steel at temperature $T_0 = 500°C$ is suddenly dipped into a large vat of water. The water conducts heat rapidly because of convection and the generation of steam. We can approximate it as an ideal thermal reservoir at $T = 30°C$. The cross section of the bar is a rectangle with dimensions $0.02\ m \times 0.15\ m$. Use the following values for thermal conductivity, specific heat and density: $k = 80.2\ W/m\text{-}K$, $C_p = 449\ J/kg\text{-}K$, $\rho = 7874\ kg/m^3$.

(a) Use the theory of Section 12.1 to find an expression for the temperature at the center of the bar as a function of time.

(b) What is the temperature at $t = 5\ s$?

12.2. Use a spreadsheet to investigate the stability of the Euler difference method for the solution of the one-dimensional thermal diffusion equation. Columns represent the spatial distribution of temperature at different times. Consider a system that extends from $x = -0.5\ m$ to $0.5\ m$ with boundaries at fixed temperature $T = 0°C$. At $t = 0$, the region from $x = -0.25$ to 0.25 has initial temperature $T_0 = 1°C$. Take $\kappa = 1.0$. Apply symmetry at $x = 0$ to model half the system. Use a uniform vertex spacing $\Delta x = 0.05\ m$. Program the spreadsheet cells in subsequent columns to advance according to Equation 12.13. Pick time steps above and below the value of Equation 12.18 to confirm the stability limit.

12.3. Modify the spreadsheet of Exercise 12.2 to advance temperature by the Dufort-Frankiel method. Confirm that the calculation is stable for any choice of Δt and experiment with the effect of time step on accuracy.

12.4. Modify the spreadsheet of Exercise 12.3 to include automatic time step adjustment, applying Equation 12.27 to the values in one column to determine the time step to the next column.

12.5. Derive finite-difference equations for two-dimensional thermal diffusion in the Dufort-Frankiel form. Assume a uniform regular mesh with spacings Δx and Δy. Apply the treatment of Section 12.2. to confirm that the method is stable for any choice of Δx, Δy, and Δt.

12.6. Find finite-difference equations for thermal diffusion in a cylindrical system with uniform mesh spacings Δr and Δz. Use the Dufort-Frankiel form.

12.7. Find an expression for the velocity of a thermal bleaching wave in a one-dimensional system. The uniform medium has an abrupt change to conductivity k_o from a value $k \ll k_o$ at the critical temperature T_o. The medium with density ρ and specific heat C_p is initially at ambient temperature ($T = 0$). The temperature of a boundary at $x = 0$ changes abruptly from $T = 0$ to $T = T_s$ at $t = 0$. Take $T_s > T_o$. Show that the position and velocity of the wave front are approximately given by,

$$v = \frac{k_o}{\rho C_p}\left[\frac{T_s}{T_o} - 1\right].$$

12.8. Verify that Equation 12.35 gives the coupling constants for the mesh of Figure 12.5.

12.9. Fill in the steps to convert Equation 12.54 to the finite-element representation of Equation 12.55.

12.10. Use Ampere's and Faraday's laws to show that the one-dimensional diffusion of magnetic field into a homogeneous conducting medium with magnetic permeability μ and volume resistivity ρ is governed by the equation,

$$\frac{\partial B_y}{\partial t} = \frac{\rho}{\mu}\frac{\partial^2 B_y}{\partial x^2}.$$

If a step function in magnetic field to amplitude B_o is applied at the surface, show by substitution that the following function is a solution to the differential equation,

$$B(x,t) = B_o \exp(-x/\delta).$$

The time-dependent skin depth is

$$\delta = \sqrt{2\rho t/\mu} \, .$$

12.11. A large laboratory magnet has a solid iron flux return yoke with a pole cross section of dimensions 0.6 m × 0.6 m. The iron has a volume resistivity $\rho = 10 \times 10^{-8}$ Ω-m and average relative magnetic permeability $\mu_r = 1000$. Estimate how long it takes after activating the drive coils for the fields to reach operating conditions where the field at the center is within 0.01% of its steady-state value.

13

Electromagnetic Fields in One Dimension

A specialist is someone who does everything else worse.

Ruggiero Ricci

Electromagnetic solutions divide into two types. The first is *the time-domain* solution which describes the generation and propagation of electromagnetic pulses. Here, we deal with time-centered difference expressions of the Maxwell equations to represent initial value problems. The second type is the *frequency-domain* solution, which we can divide into two classes. The first class is the *scattering* solution. It generally involves traveling-wave propagation in open systems. The associated techniques have application to radar and communications. For a given frequency, we solve a boundary value problem using matrix inversion. The second class is the *resonant solution* for waves in closed structures such as resonant cavities and waveguides. Here, we seek modes where waves constructively interfere. The process involves a search using several boundary value solutions at different frequencies.

This chapter introduces numerical electromagnetism through one-dimensional pulse and wave propagation. The mathematics is straightforward and the physical foundations of the solutions are easy to visualize. This groundwork will allow a smooth transition to the two- and three-dimensional solutions in the following chapter. Section 13.1 reviews the analytic theory of one-dimensional electromagnetism. Topics include wave propagation in free space, reflection and transmission at boundaries, and absorption by resistive media. Sections 13.2 through 13.4 cover one-dimensional finite-element models. The solutions exhibit a diversity of phenomena that make them considerably more interesting than one-dimensional electrostatics. Section 13.2 concentrates on time-domain solutions for transverse electromagnetic (TEM) pulses. Here, the electric field and magnetic intensity vectors are normal to the direction of propagation. We shall derive finite-element equations from the integral form of Maxwell's equations. They advance either E_x or H_y for propagation along z. Section 13.3 covers the mechanics of solutions. The time step is constrained by the Courant condition, an upper limit for physical validity and numerical stability. The section discusses open- and short-circuit boundaries and addresses the important topic of the ideal absorbing boundary. This condition, which is essential for simulations of wave interactions in unbounded volumes, is analogous to a matched termination on a transmission line.

Section 13.4 proceeds to one-dimensional scattering solutions for steady-state systems. Here all quantities have the harmonic time dependence exp[jωt]; therefore, time derivatives can be replaced with the quantity jω. The finite-element relationships reduce to a set of linear equations with complex variables to represent the amplitude and phase of field quantities. The equations can be solved by extensions of the matrix methods discussed in Chapter 6. If we take ϵ and μ as complex numbers, the imaginary part gives frequency-dependent loss processes in materials. Section 13.5 reviews the theoretical basis complex-number material properties and the definition of perfectly absorbing boundaries for free-space simulations. Section 13.6 covers resonant solutions. These solutions apply to closed systems where reflection of waves at boundaries gives rise to constructive interference at particular frequencies. The resulting solutions are the *resonant modes* of the system. Determination of modes involves a search for interference conditions and requires several solutions at different frequencies. The section reviews the theory of driven circuits with inductance, capacitance, and resistance to illustrate resonance criteria and summarizes numerical root-finding methods to minimize the steps in the search.

13.1 Planar Electromagnetic Waves

To derive the properties of plane waves we shall express the Maxwell equations in a form that is suitable for most radio frequency and microwave calculations. One simplification is that the solution volume has no externally generated space-charge. We shall limit the discussion to materials that are isotropic and linear. The magnetic permeability μ and dielectric constant ϵ are constant in time and uniform over element volumes. We include the option for time-dependent source current density (J_o) to represent drive structures like coupling loops and capacitive probes. We shall also consider the possibility of currents driven by electric fields in materials with conductivity σ. This feature enables modeling of pulse attenuation and implementation of absorbing boundaries. Section 13.3 generalizes the treatment of losses for frequency-domain solutions. In this regime we can represent high-frequency dissipation in materials by complex values of ϵ and μ.

Under the limiting conditions the differential Maxwell equations are

$$\nabla \times \frac{\mathbf{B}}{\mu} = \epsilon \frac{\partial \mathbf{E}}{\partial t} + \mathbf{J}_o + \sigma\, \mathbf{E},$$

$$\nabla \cdot \mathbf{B} = 0,$$

(13.1)

$$\nabla \times \mathbf{E} = -\frac{\partial \mathbf{B}}{\partial t},$$

$$\nabla \cdot \epsilon \mathbf{E} = 0.$$

The equations are more symmetric if we write them in terms of the *magnetic field intensity,*

$$\mathbf{H} = \frac{\mathbf{B}}{\mu}. \tag{13.2}$$

The quantity $\mathbf{H}$ is proportional to the quantity that we have called the *applied field,* $\mathbf{H} = \mathbf{B_0}/\mu_0$. It has units of A/m. Substituting Equation 13.2 in Equations 13.1 gives

$$\nabla \times \mathbf{H} = \epsilon\frac{\partial \mathbf{E}}{\partial t} + \mathbf{J_0} + \sigma\mathbf{E}, \tag{13.3}$$

$$\nabla \cdot \mu\mathbf{H} = 0, \tag{13.4}$$

$$\nabla \times \mathbf{E} = -\mu\frac{\partial \mathbf{H}}{\partial t}, \tag{13.5}$$

and

$$\nabla \cdot \epsilon\mathbf{E} = 0. \tag{13.6}$$

To begin, consider wave propagation in a uniform medium without sources and with no dissipation. In this case ϵ and μ are constant in space, $\sigma = 0$, $\mathbf{J_0} = 0$, and all quantities vary in time as $\exp(j\omega t)$. Taking the curl of Equation 13.5 and substituting from Equation 13.3 gives

$$\nabla \times \nabla \times \mathbf{E} = -\mu\epsilon\frac{\partial}{\partial t}(\nabla \times \mathbf{H}) = -\mu\epsilon\frac{\partial \mathbf{E}^2}{\partial t^2} = \omega^2\mu\epsilon\mathbf{E}. \tag{13.7}$$

The condition of Equation 13.6 implies that in Cartesian coordinates we can rewrite the double curl operator in terms of the Laplacian operation,

$$\nabla^2\mathbf{E} + \omega^2\mu\epsilon\mathbf{E} = 0. \tag{13.8}$$

We seek plane wave solutions to Equation 13.8 where quantities vary only in z. The condition that $\nabla \cdot \mathbf{E} = 0$, implies that there is no component $\mathbf{E}_z$. We choose a coordinate system with the electric field along x, so that

$$\frac{\partial^2 E_x}{\partial z^2} + \omega^2 \mu \epsilon \, E_x = 0 . \tag{13.9}$$

The function

$$E_x(z,t) = E_o \, \exp\left[j(\omega t \pm kz)\right] \tag{13.10}$$

is a general solution of Equation 13.9 if

$$k = \pm \, \omega \, \sqrt{\mu \epsilon} . \tag{13.11}$$

Equation 13.10 represents a traveling wave, a harmonic function of space that moves in the positive or negative z direction. The spatial wavelength is $\lambda = 2\pi/k$ and the velocity of a point of constant phase is

$$v_{phase} = m \, \frac{\omega}{k} = m \, \frac{1}{\sqrt{\mu \epsilon}} . \tag{13.12}$$

The phase velocity depends on the properties of the medium. In vacuum where $\epsilon = \epsilon_o$ and $\mu = \mu_o$, the phase velocity equals the speed of light, $c = 2.997925 \times 10^8$ m/s.

Equation 13.5 implies that there is also a magnetic intensity associated with the wave that is perpendicular to both the electric field and the direction of wave propagation. For negative k (wave propagation in the positive direction), the relationship is

$$H_y = \frac{j}{\omega \mu} \frac{\partial E_x}{\partial z} = \frac{k}{\omega \mu} E_x = \frac{E_x}{\sqrt{\mu/\epsilon}} . \tag{13.13}$$

The cross-product $\mathbf{E} \times \mathbf{H}$ points in the direction of propagation. Noting that the dimension of E_x is V/m and H_y is A/m, the quantity in the denominator on the right-hand side of Equation 13.13 has units of ohms. This quantity is called the *impedance* of the medium. In vacuum, the value is

$$Z_o = \sqrt{\frac{\mu_o}{\epsilon_o}} = 377.3 \, \Omega . \tag{13.14}$$

Note that we could have derived an equation for H_y similar to Equation 13.9 and then calculated E_x. In one-dimensional solutions, both choices lead to the same results.

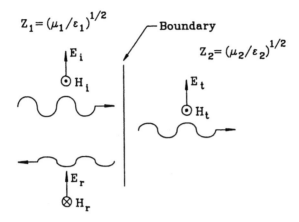

FIGURE 13.1
Traveling wave incident on a boundary between media with different impedances.

As an illustration, consider a wave traveling in the positive z direction incident on a plane boundary at $z = 0$ between two media. Figure 13.1 shows the labeling conventions. The incident wave generates a reflected wave traveling backward in material 1 and a transmitted wave moving forward in material 2. We denote the amplitudes of the incident, reflected, and transmitting waves as E_i, E_r, and E_t. From the discussion of Section 2.4, the parallel electric field is continuous across the boundary, or

$$E_i + E_r = E_t. \tag{13.15}$$

The common factor of $\exp[j\omega t]$ has been suppressed in Equation 13.15. The applied magnetic field, parallel to the interface, must also be continuous (Section 9.3). Therefore, the amplitudes of the magnetic intensity are related by

$$H_i - H_r = H_t. \tag{13.16}$$

Note the minus sign in Equation 13.16. A wave with positive E_x traveling in the negative z direction must have negative H_y. Substituting from Equation 13.13, we can rewrite Equation 13.16 as

$$\frac{E_i}{Z_1} - \frac{E_r}{Z_1} = \frac{E_t}{Z_2}. \tag{13.17}$$

Combining Equations 13.15 and 13.17 gives reflection and transmission coefficients for the electric field and magnetic intensity,

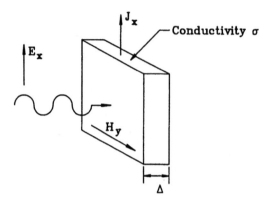

FIGURE 13.2
Traveling wave incident on a matched resistive layer.

$$R_E = \frac{E_r}{E_t} = \frac{Z_2 - Z_1}{Z_2 + Z_1}, \quad T_E = \frac{E_t}{E_i} = \frac{2Z_2}{Z_2 + Z_1} \qquad (13.18)$$

and

$$R_H = \frac{H_r}{H_i} = \frac{Z_1 - Z_2}{Z_2 + Z_1}, \quad T_H = \frac{H_t}{H_i} = \frac{2Z_1}{Z_2 + Z_1}. \qquad (13.19)$$

Note that conditions of Equations 13.18 and 13.19 are independent of frequency. Therefore, they apply to pulses or continuous waves.

Consider a pulse incident on a material with low impedance ($Z_2 \ll Z_1$) such as a dielectric with high ϵ_r. The *short-circuit* boundary condition gives total reflection of the wave with inversion of the electric field. At the other extreme, the *open-circuit* condition is $Z_2 \gg Z_1$. The pulse is again totally reflected but with positive electric field and inverted magnetic intensity. The cancellation of H near the interface is analogous to the condition of zero current at an open circuit. The special case where $Z_1 = Z_2$ is called an *impedance match*. Here the boundary has no effect and the wave is totally transmitted. A related solution that is important for numerical calculations is the termination of a wave by a lumped element resistor. Figure 13.2 shows the geometry. We introduce a resistive layer with conductivity σ and thickness Δ adjacent to an open-circuit boundary of a material with ϵ and μ. In the limit that $\Delta \ll \lambda$, the electric field is approximately uniform over the layer depth. There is no reflected wave if the resistor maintains the same conditions on E_x and H_y as an infinite extension of the medium. We can derive the correct value of σ by noting that the field E_x creates a linear current density of

$$J_x = \sigma \Delta \ E_x. \qquad (13.20)$$

The quantity $1/\sigma\Delta$ with units of ohms is called the *surface resistance* of the termination layer. Assuming zero magnetic field on the right-hand side of the resistor, the value of magnetic intensity on the left-hand side is $H_y = J_x$. Substituting in Equation 13.13 gives the condition for a matched termination,

$$Z_o = \sqrt{\frac{\mu}{\epsilon}} = \frac{1}{\sigma\Delta}. \tag{13.21}$$

Another example that we shall apply in benchmark tests in the following sections is wave attenuation in a lossy material. The equation for electric field in a uniform, isotropic medium with conductivity σ is

$$\frac{\partial^2 E_x}{\partial z^2} + \omega^2 \mu\epsilon\, E_x = j\omega\sigma\mu\, E_x. \tag{13.22}$$

For a wave moving in the positive z direction substituting $E_x = E_o \exp[j(\omega \pm \gamma z)]$ in Equation 13.22 gives an expression for the propagation constant,

$$\gamma^2 = \mu\epsilon\omega^2 - j\omega\mu\sigma. \tag{13.23}$$

When damping is weak the second term on the right-hand side of Equation 13.23 is small. In this case we can apply the binomial equation to find the square root,

$$\gamma \cong -\omega\mu\epsilon\left(1 - \frac{j\sigma}{2\omega\epsilon}\right). \tag{13.24}$$

Equation 13.24 implies that the electric field varies as

$$E_x(z,t) \cong E_o \exp(-\alpha z)\, \exp[j(\omega t - kz)], \tag{13.25}$$

where k is given by Equation 13.11 and

$$\alpha = \frac{\sigma}{2}\sqrt{\frac{\mu}{\epsilon}}\ (m^{-1}). \tag{13.26}$$

13.2 Time-Domain Electromagnetism in One Dimension

This section covers finite-element equations for plane electromagnetic pulses. We shall develop a model with options for variable-resolution meshes and

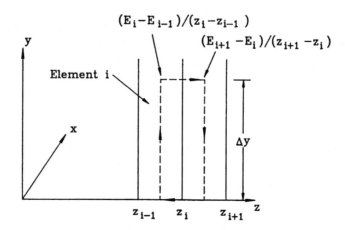

FIGURE 13.3
Computational mesh for one-dimensional finite-element equations of time-domain electromagnetics. Dashed line shows the path of the circuital integral around vertex I.

arbitrary variations of material properties. The approach is to apply Equations 13.3 and 13.5 on the computational mesh of Figure 13.3. Here a bounded region along the z-axis is divided into small elements with indices i = 1 to I. The element properties are the dielectric constant ϵ_i, magnetic permeability μ_i, source current J_{oi}, and conductivity σ_i. Field values are defined at vertices marked with indices i = 0 to I. Following Section 13.1, we can describe pulses either in terms of E_x or H_y. To derive the E-type equations, we combine Equations 13.3 and 13.5 to eliminate the magnetic intensity,

$$-\nabla \times \frac{1}{\mu} \nabla \times \mathbf{E} = \frac{\partial}{\partial t} \left(\nabla \times \mathbf{H} \right) = \epsilon \frac{\partial^2 \mathbf{E}}{\partial t^2} + \frac{\partial J_o}{\partial t} + \sigma \frac{\partial \mathbf{E}}{\partial t}. \qquad (13.27)$$

Using Stoke's theorem, we can rewrite Equation 13.27 as

$$-\oint \frac{1}{\mu} \nabla \times \mathbf{E} = \iint d\mathbf{S} \left[\epsilon \frac{\partial^2 \mathbf{E}}{\partial t^2} + \frac{\partial J_o}{\partial t} + \sigma \frac{\partial \mathbf{E}}{\partial t} \right]. \qquad (13.28)$$

The surface integral on the right-hand side extends over the region bounded by the circuit integral on the left.

Only derivatives in the z direction have non-zero values. Waves with field components E_x and H_y are generated by source currents J_{ox} in the x direction. We assume that the quantity $\partial J_{ox}/\partial t$ is a known function of space and time and take the integration path near vertex i along the dashed line in Figure 13.3. The enclosed space extends an arbitrary distance Δy along y and encloses half the volumes of adjacent elements. The left-hand side of Equation 13.28 is

$$-\oint \frac{1}{\mu} \frac{\partial E_x}{\partial z} = -\Delta y \left[\frac{1}{\mu_i} \frac{E_i - E_{i-1}}{z_i - z_{i-1}} - \frac{1}{\mu_{i+1}} \frac{E_{i+1} - E_i}{z_{i+1} - z_i} \right], \qquad (13.29)$$

where $E_i = E_x(z_i)$. To evaluate the area integrals on the right-hand side we note that although the quantities ϵ, σ, and $\partial J_{ox}/\partial t$ may vary significantly from element to element, changes in electric field must extend over several elements for an accurate calculation. Therefore, it is sufficient to approximate the electric field time derivatives by their value at the enclosed vertex i. The right-hand side of Equation 13.28 is approximately equal to

$$\Delta y \frac{\partial^2 E_i}{\partial t^2} \left[\frac{\epsilon_{i+1}(z_{i+1} - z_i) + \epsilon_i(z_i - z_{i-1})}{2} \right]$$

$$+ \Delta y \left[\frac{(\partial J_{i+1}/\partial t)(z_{i+1} - z_i) + (\partial J_i/\partial t)(z_i - z_{i-1})}{2} \right] \qquad (13.30)$$

$$+ \Delta y \frac{\partial E_i}{\partial t} \left[\frac{\sigma_{i+1}(z_{i+1} - z_i) + \sigma_i(z_i - z_{i-1})}{2} \right],$$

where $\partial J_i/\partial t = \partial J_{ox}/\partial t$ in element i.

Equation 13.30 applies to both time-domain and frequency-domain solutions — the difference lies in the expressions for the time derivatives. For time-domain solutions we shall use the time-centered forms for the first and second derivatives discussed in Section 4.1. Inserting the difference expressions and dropping the common factor of Δy, Equation 13.30 becomes

$$\frac{E_i^{n+1} - 2E_i^n + E_i^{n-1}}{\Delta t^2} \left[\frac{\epsilon_{i+1}(z_{i+1} - z_i) + \epsilon_i(z_i - z_{i-1})}{2} \right]$$

$$+ \left[\frac{\partial J_{i+1}^n}{\partial t} \frac{z_{i+1} - z_i}{2} + \frac{\partial J_i^n}{\partial t} \frac{z_i - z_{i-1}}{2} \right] \qquad (13.31)$$

$$+ \frac{E_i^{n+1} - E_i^{n-1}}{2\Delta t} \left[\frac{\sigma_{i+1}(z_{i+1} - z_i) + \sigma_i(z_i - z_{i-1})}{2} \right].$$

The superscript n denotes quantities measured at time $t = n\Delta t$. Combining Equations 13.29 and 13.31 and solving for E_i^{n+1} gives a time-centered finite-element equation to advance electric field values,

$$E_i^{n+1} =$$

$$\frac{-E_i^{n-1}\left(A_i - C_i\right) + E_i^n\left(2A_i - W_{ui} - W_{di}\right) + E_{i+1}^n W_{ui} + E_{i-1}^n W_{di} - S_i}{A_i + C_i}. \qquad (13.32)$$

The quantities in Equation 13.32 are

$$A_i = \frac{\epsilon_{i+1}\left(z_{i+1} - z_i\right) + \epsilon_i\left(z_i - z_{i-1}\right)}{2\Delta t^2}, \qquad (13.33)$$

$$C_i = \frac{\sigma_{i+1}\left(z_{i+1} - z_i\right) + \sigma_i\left(z_i - z_{i-1}\right)}{4\Delta t}, \qquad (13.34)$$

$$S_i = \frac{\partial J_{i+1}^n}{\partial t}\frac{z_{i+1} - z_i}{2} + \frac{\partial J_i^n}{\partial t}\frac{z_{i+1} - z_i}{2}, \qquad (13.35)$$

$$W_{di} = \frac{1}{\mu_i}\frac{1}{z_i - z_{i-1}}, \qquad (13.36)$$

$$W_{ui} = \frac{1}{\mu_{i+1}}\frac{1}{z_{i+1} - z_i}. \qquad (13.37)$$

The final two terms are the coupling constants to the neighboring vertices. Equation 13.32 is an impressively succinct relationship. We shall see in the following section that it encompasses all one-dimensional pulse propagation phenomena.

We can also describe one-dimensional pulses in terms of the magnetic field intensity. Dividing Equation 13.3 by the dielectric constant and taking the curl gives

$$\nabla \times \frac{1}{\epsilon}\left(\nabla \times H\right) = \frac{\partial}{\partial t}\left(\nabla \times E\right) + \nabla \times \frac{J_o}{\epsilon} + \nabla \times \frac{\sigma E}{\epsilon}. \qquad (13.38)$$

We proceed in the same way as the electric field equation, eliminating the electric field by using Equation 13.5 to replace $\nabla \times E$, taking surface integrals and applying Stoke's theorem to terms that involve the curl operation. One potential problem is that the final term in Equation 13.38 is not proportional to $\nabla \times E$. This is easily resolved in a linear finite-element treatment where σ, ϵ, and E are constant over element volumes. The surface integral of the term is

$$\iint dS\, \nabla \times \frac{\sigma E}{\epsilon} = \iint dS\, \nabla\left(\frac{\sigma}{\epsilon}\right) \times E + \iint dS\, \frac{\sigma}{\epsilon}\nabla \times E. \qquad (13.39)$$

In summing over elements the gradients of σ/ϵ in the first term on the right-hand side are zero. Therefore, we can eliminate the electric field by substitution from Equation 13.5.

For polarization H_y the integral form of the one-dimensional equation for electromagnetic pulse propagation is

$$-\oint \frac{1}{\epsilon} \frac{\partial H_y}{\partial z} \hat{x} \cdot d\mathbf{l} = -\iint dS \, \mu \frac{\partial^2 H_y}{\partial t^2} + \oint \frac{J_x}{\epsilon} \hat{x} \cdot d\mathbf{l} - \iint dS \frac{\sigma\mu}{\epsilon} \frac{\partial H_y}{\partial t}. \quad (13.40)$$

We can apply a procedure similar to that for E-type pulses to express the circuit and surface integrals on the mesh of Figure 13.3. Here, the surfaces lie in the x–z plane. At the vertex i the surface extends a distance Δx along x and from $(z_{i-1} + z_i)/2$ to $(z_i + z_{i+1})/2$ along z. The finite-element equation for one-dimensional H-type pulses is

$$H_i^{n+1} =$$

$$\frac{-H_i^{n-1}(A_i - C_i) + H_i^n(2A_i - W_{ui} - W_{di}) + H_{i+1}^n W_{ui} + H_{i-1}^n W_{di} + (J_{i+1}^n - J_i^n)}{A_i + C_i}. \quad (13.41)$$

The coefficients in Equation 13.41 are

$$A_i = \frac{\mu_{i+1}(z_{i+1} - z_i) + \mu_i(z_i - z_{i-1})}{2\Delta t^2}, \quad (13.42)$$

$$C_i = \frac{(\sigma_{i+1}\mu_{i+1}/\epsilon_{i+1})(z_{i+1} - z_i) + (\sigma_i\mu_i/\epsilon_i)(z_i - z_{i-1})}{4\Delta t}, \quad (13.43)$$

$$W_{id} = \frac{1}{\epsilon_i} \frac{1}{z_i - z_{i-1}}, \quad (13.44)$$

$$W_{iu} = \frac{1}{\epsilon_{i+1}} \frac{1}{z_{i+1} - z_i}. \quad (13.45)$$

13.3 Electromagnetic Pulse Solutions

We shall next discuss some of the solutions embodied in Equation 13.32 or 13.41. The results are useful because the mathematics of plane pulses is identical to that of transverse electromagnetic pulses in transmissions lines. With the model we can address the full range of transmission line problems, including resistive terminations, mismatched connections, dissipation, and reactive loads.

The choice of time step in an electromagnetic solution is important. For numerical stability, it must satisfy the *Courant condition*,

$$\Delta t < minimum\left[\sqrt{\epsilon_i \mu_i}\ (z_{i+1} - z_i)\right] = minimum\left[\frac{z_{i+1} - z_i}{v_i}\right]. \quad (13.46)$$

The quantity v_i is the speed of light in element i and the function *minimum* implies the smallest value on the mesh. Equation 13.46 states that Δt must be shorter than the time for an electromagnetic disturbance to propagate across the smallest element. If the Courant condition is violated, information can propagate through the mesh faster than the speed of light, violating the principle of causality. Therefore, the first step in a simulation is to scan the elements to find an acceptable value of Δt. In contrast to the diffusion solutions of Chapter 12, Equation 13.46 is a rigid constraint — the limit on time step applies through the full simulation.

As in previous discussions of numerical models we shall first find how to advance internal points of the solution volume and then deal with boundaries. The application of Equation 13.31 to internal points is simple. We maintain two arrays for the electric field, $E_x(i)$ and $E_x OldNew(i)$. At the beginning of step n+1, $E_x(i)$ contains values at t^n and $E_x OldNew(i)$ contains values at t^{n-1}. We loop through the array in any order, calculating values of t^{n+1}. These can be stored in $E_x OldNew(i)$ because values at t^{n-1} are no longer needed. Next, the time reference is shifted by exchanging the arrays.

To illustrate implementation of boundary conditions and interpretation of solutions, we shall use a baseline solution region of width L = 1 m with 250 uniform elements. Pulses follow a Gaussian waveform,

$$f(t) = \exp\left[-\left(\frac{t - 0.667}{0.133}\right)^2\right], \quad (13.47)$$

where the time is given in nanoseconds. The pulse of Equation 13.47 has a full width at half-maximum of 0.222 nsec and a peak at t = 0.667 nsec. In vacuum the spatial width is $w_p = 0.067$ m and the time to cross the system is 3.336 nsec.

To begin, consider E pulse propagation in a uniform vacuum region with metal boundaries. We set $\epsilon = \epsilon_o$, $\mu = \mu_o$, and $\sigma = 0$ in all elements. The conditions $E_x(0) = 0.0$ and $E_x(I)$ represent the short-circuit boundaries. We excite a pulse of 1 V/m amplitude by setting $E_x(0)$ equal to the function f(t) in Equation 13.47 for a few nanoseconds and then clamp the value to zero. From the discussion of Section 13.1 we expect the pulse to propagate at the speed of light with no change in amplitude, reversing polarity at each wall reflection. The results shown in Figure 13.4 closely follow this behavior. The figure

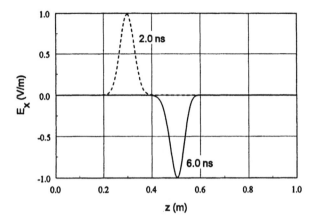

FIGURE 13.4
Numerical calculation of pulse propagation in vacuum. Solution width: 1 m. Uniform medium with $\epsilon = \epsilon_o$, $\mu = \mu_o$, and $\sigma = 0$. The drive boundary on the left generates a Gaussian pulse with full width at half-maximum of 0.222 nsec. Shorted boundaries on the left and right.

plots the pulse shortly after entry and after one reflection. The velocity equals the speed of light to within the accuracy set by the mesh resolution. The electrical field energy per unit cross section area in the pulse equals

$$U = \sum_1^I \frac{\epsilon_i E_i^2}{2} (z_i - z_{i-1}).$$
(13.48)

In checking energy conservation, we must take care not to apply Equation 13.48 near a reflected wall where the pulse energy is carried mainly in H_y. Extending the calculation of Figure 13.4 to 30 nsec shows that Equation 13.31 conserves energy to better than 1 part in 10^7.

The condition $dE_x/dz = 0$ holds on an open-circuit boundary. The Neumann boundary represents a symmetry plane where a traveling pulse is superimposed on a mirror pulse moving in the opposite direction. To implement a Neumann boundary at point I, we set $W_{uI} = 0$ and calculate A_I and C_I with values $\epsilon_{I+1} = 0$ and $\sigma_{I+1} = 0$. Figure 13.5 shows pulse reflection from an open boundary. The wave amplitude is doubled at the point of reflection (solid line). At this instant, the full pulse energy resides in the electric field. Application of Equation 13.48 gives an electric field energy twice that of the incident wave — the pulse amplitude doubles but the volume is halved. The dashed line shows the reflected wave with positive electric field polarity.

Numerical simulations represent continuous physical systems with discrete equations. We expect to find imperfections in the model. One problem is apparent if we extend the calculation of Figure 13.5 to a long propagation distance. Figure 13.6a shows the pulse profile at 30 nsec after traveling 9 m (2200 elements). Although energy conservation is almost perfect, the pulse exhibits noticeable distortion. This effect arises from *numerical dispersion*. The

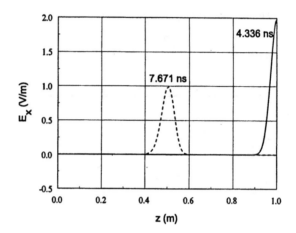

FIGURE 13.5
Numerical calculation of pulse reflection from an open boundary.

Gaussian pulse can be represented as a spectrum of Fourier modes of the form $\exp[j(2\pi z/\lambda - \omega t)]$. The pulse shape is preserved if all modes move at the same phase velocity. In the discrete approximation there are shifts in phase velocity for modes with λ comparable to the element width. Therefore, we expect more distortion for narrow pulses or larger elements. Figure 13.6b illustrates declining fidelity with increasing element size. In the three cases the pulse width was 0.067 m and the propagation distance was 2.5 m. (Note that small time shifts have been introduced to separate the pulses.) The plots correspond to different values of the element width. As a general rule, dispersion is small if the distance of propagation is less than or comparable to $2w_p^2/\Delta z$, where Δz is the element width.

We can represent nonuniform media simply by assigning different values for ϵ_i, μ_i, and σ_i. Figure 13.7a shows pulse reflection and transmission at a boundary between materials with different impedances. The region to the left of the boundary (dashed line) is vacuum, while the region to the right is a dielectric with $\epsilon = 2.5\epsilon_o$. The corresponding characteristic impedance is $Z_2 = 238.6\ \Omega$ and the propagation velocity is $v = 1.8961 \times 10^8$ m/s. Equation 13.18 predicts a reflected pulse amplitude of -0.2251 V/m and a transmitted amplitude of 0.7749. Figure 13.7a shows a numerical solution 0.667 nsec after the pulse peak strikes the interface. The amplitudes are in good agreement. Note that the reduced propagation velocity results in a narrowed transmitted pulse. The benchmark calculation of Figure 13.7b shows damping by a distributed conductivity. For $\epsilon = \epsilon_o$, $\mu = \mu_o$, and $\sigma = 5.306 \times 10^{-3}$ mhos/m, Equation 13.26 gives $\alpha = 1$ m^{-1}. The behavior of the pulse in Figure 13.7b is in agreement with the prediction of Equation 13.25. The small change in pulse shape is physically correct — the high value of α marginally satisfies the approximation of weak damping.

We can model pulse generation from distributed sources by assigning time-dependent J_o or $\partial J_o/\partial t$ to elements. For example, a current density $J_o f(t) =$

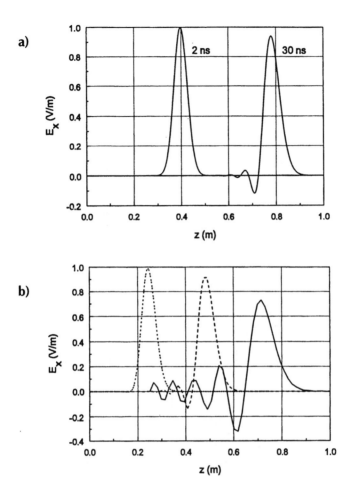

FIGURE 13.6
Numerical dispersion in time-domain electromagnetics. Gaussian pulse in vacuum with full-width at half-maximum of 0.067 m. (a) Pulse shapes after 0.6 and 9 m of propagation for an element width of 0.004 m. (b) Pulse shapes after 2.5 m of propagation as a function of element width. Ratio of pulse width to element width: 16.6 (dotted), 8.3 (dashed), and 4.2 (solid).

$1250 \ A/m^2$ in a central element of the baseline geometry gives a peak linear current density of $J_o\Delta z f(t) = 5.0 \ A/m$. Equation 13.41 implies that the pulsed current should create two Gaussian H pulses that travel away from the source element with amplitude $H_o = \pm 2.5 \ A/m$. We can check the correspondence of the models by using a source function $\partial J_o/\partial t = 1250.0 \ df/dt$ in an E-pulse solution. The result is two positive polarity waves moving away from the source with amplitude $E_o = 943.3 \ V/m$. The amplitude is close to the predicted value of $E_o = 377.3H_o$.

In principle, we could have obtained the previous solutions with finite-difference methods if we made some allowances for the fuzziness of material boundaries. The advantage of the finite-element approach is apparent when

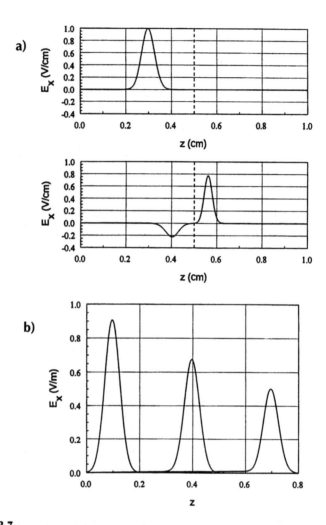

FIGURE 13.7

Numerical calculations of pulse propagation. (a) Reflection and transmission at a boundary between materials with different characteristic impedances (dashed line shows boundary). $Z_1 = 377.3 \ \Omega$, $Z_2 = 238.6 \ \Omega$. Top: incident pulse. Bottom: pulses 0.667 nsec after the incident pulse strikes the boundary. (b) Pulse attenuation in a uniform conducting medium, 1 V amplitude at the left-hand boundary. Medium properties: $\epsilon = \epsilon_o$, $\mu = \mu_o$, $\sigma = 5.306 \times 10^{-3}$ mhos/m, and $\alpha = 1 \ \text{m}^{-1}$.

we need advanced boundary conditions like ideal absorbers. Implementing the matched termination condition in finite-difference formulations with staggered field quantities is complex. The most common approach in finite-difference time-domain calculations is the *look-back method*. Here, time-delayed field values are assigned to a set of mesh points outside the solution region based on field values at points just inside the boundary (see, for instance, Kunz and Luebbers, 1993). If the time delay equals the propagation time for waves between the two points, the boundary simulates open space. The look-back method has severe drawbacks. It is straightforward to apply

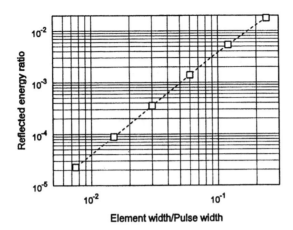

FIGURE 13.8
Performance of a matched terminating boundary layer. Relative reflected energy as a function of the ratio of layer thickness to pulse spatial width (Gaussian pulse full width at half-maximum).

only at planar boundaries of a regular mesh with uniform elements. Curved and sloped boundaries are a major challenge, and the method fails when portions of the boundary are adjacent to materials with different propagation velocities. In contrast, the finite-element matched-termination that we shall discuss is effective and easy to apply in any geometry.

To represent an absorbing boundary in a one-dimensional finite-element model we leave the boundary unspecified (open circuit) and assign an appropriate conductivity to a single adjacent element. Following Equation 13.21, if the element width is Δz then the conductivity for a matched termination is

$$\sigma = \frac{1}{\Delta z \sqrt{\mu/\epsilon}}. \tag{13.49}$$

The values ϵ and μ apply to the termination element and the adjacent medium. The method works for both E- and H-type pulses. Figure 13.8 shows the performance of a single-element termination boundary for a normally incident pulse. The baseline Gaussian pulse enters from a vacuum region. The figure plots the reflected pulse amplitude as a function of the termination element width Δz. At the baseline element width of 0.004 m the termination layer absorbs 99.86% of the pulse energy. Reducing the element thickness gives better absorption because the layer behavior approaches that of an ideal lumped resistor. On the other hand, because of the Courant condition, thin elements lead to short time steps and extended run times.

A useful feature of termination layers in finite-element calculations is the option to construct generalized reactive boundaries. This is physically equivalent to transmission line terminations with resistive, capacitive, and inductive components. To illustrate, consider the reflection of a pulse from the end of a vacuum transmission line terminated with a capacitor C. If the system

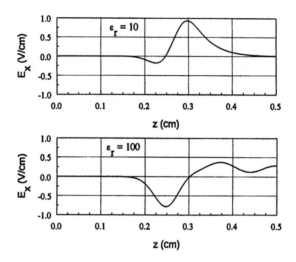

FIGURE 13.9
Demonstration of a reactive boundary layer — capacitive termination. Gaussian pulse with full width at half-maximum of 0.067 m. Layer thickness: 0.004 m. (a) Low capacitance, C = 0.35 pF. (b) High capacitance, C = 35.4 pF.

extends an arbitrary distance L in the x and y directions, the capacitance of an element of thickness Δz adjacent to an open boundary is

$$C = \frac{\epsilon L \Delta z}{L} = \epsilon \Delta z . \qquad (13.50)$$

To represent an ideal capacitor, we set $\sigma = 0$ and $\epsilon = C/\Delta z$. The choice $\mu = 1/\epsilon v = \Delta z/Cv$ (where v is the speed of light in the adjacent material) maintains continuity of the pulse velocity. As an example, we expect that a capacitive termination will strongly affect a pulse of width Δt_p on a transmission line with impedance Z_0 if $C_0 = \Delta t_p/Z_0$. For the baseline pulse in vacuum the values $\Delta t_p = 0.222$ nsec and $Z_0 = 377.3$ Ω correspond to $C_0 = 6.0 \times 10^{-13}$. If $C \ll C_0$ the termination acts like an open circuit while the behavior approximates a short circuit when $C \gg C_0$. Figure 13.9 shows the expected pulse reflection properties in the two regimes.

13.4 Frequency-Domain Equations

Frequency-domain solutions represent a steady state where all sources and fields have a harmonic time variation with angular frequency ω. As discussed in Section 11.2, we shall represent harmonic variations with complex numbers and eliminate the common factor of $\exp(j\omega t)$. The remaining complex field quantities give amplitude and phase as a function of position. Taking a time derivative is equivalent to multiplication by $j\omega$. In frequency-domain solutions we can generalize the treatment of losses in materials. At microwave frequencies dissipation may result from lagging response of molecular dipoles in dielectrics or domains in ferrites as well as common resistive processes. We can include these effects conveniently by adding complex parts to the dielectric constant and magnetic permeability. The following section reviews the theoretical basis of the representation and the definition of perfectly absorbing boundaries.

We shall concentrate on E-type waves — the extension to H waves is straightforward. The governing equation is

$$-\nabla \times \left(\frac{1}{\mu} \nabla \times E\right) = -\epsilon\omega^2 E + j\omega J_o . \tag{13.51}$$

The complex quantity E contains information on the amplitude and phase of the electric field. In a one-dimensional finite-element representation, we integrate both sides of Equation 13.51 over an area surrounding a vertex (Figure 13.3) and apply Stoke's theorem to transform the curl terms. The expressions for the left-hand side are similar to those of Section 13.2. The main difference is that W_{iu} and W_{id} may be complex if μ includes energy loss. The right-hand side of Equation 13.51 at a vertex is

$$-\frac{\omega^2 E_i}{2}\left[\epsilon_{i+1}(z_{i+1} - z_i) + \epsilon_i(z_i - z_{i-1})\right] + \frac{j\omega}{2}\left[J_{i+1}(z_{i+1} - z_i) + J_i(z_i - z_{i-1})\right]. \tag{13.52}$$

Combining results and solving for E_i gives a set of linear equations for the electric field,

$$E_i = \frac{W_{iu} E_{i+1} + W_{id} E_{i-1} - j\omega\left[J_{i+1}(z_{i+1} - z_i) + J_i(z_i - z_{i-1})\right]}{W_{iu} + W_{id} - A_i} . \tag{13.53}$$

Expressions for the coupling coefficients are the same as Equations 13.36 and 13.37. The quantity A_i is given by

$$A_i = \omega^2 \frac{\epsilon_{i+1}(z_{i+1} - z_i) + \epsilon_i(z_i - z_{i-1})}{2} . \tag{13.54}$$

We can understand the physical implication of Equation 13.53 by considering wave solutions in a uniform material region with no damping on a uniform mesh. For mesh spacing Δz the coefficients in the equation are $W_{iu} = W_{id} = 1/\mu\Delta$ and $A_i = \omega^2 \epsilon \Delta z$. Equation 13.53 reduces to

$$E_i = \frac{E_{i+1} + E_{i-1}}{2 - \mu\epsilon\omega^2 \Delta z^2} . \tag{13.55}$$

We can show that Equation 13.55 leads to the traveling wave solutions of Equations 13.10. Substituting Equation 13.10 into Equation 13.55 and canceling the factor of $\exp(j\omega t)$ gives

$$\frac{\exp[j\omega\Delta z/v] + \exp[-j\omega\Delta z/v]}{2 - \mu\epsilon\omega^2 \Delta z^2} = \frac{\cos(\omega\Delta z/v)}{1 - \mu\epsilon\omega^2 \Delta z^2/2} = 1 . \tag{13.56}$$

A binomial expansion of the cosine in the limit $\omega\Delta z/v = \Delta z/\lambda \ll 1$ confirms the identity.

Relaxation techniques do not give successful solutions of Equation 13.51. Instead, we can modify the method of back-substitution to solve tridiagonal equations with complex numbers. The implementation of short-circuit ($E_x = 0$) and open-circuit ($dE_x/dz = 0$) boundary conditions follows from the discussion of Section 4.4. We can initiate waves either by specifying the amplitude and phase of current sources in one or more elements or by setting the magnitude and phase of E_x on a boundary. The definition of absorbing or reactive boundaries with termination elements is the same as that for time-domain solutions.

A solution of Equation 13.51 consists of real and imaginary parts of E_x at mesh vertices. Transmission-line solutions are generally expressed in terms of standing and traveling waves. For reference, we shall document the correspondence between the viewpoints. Given a complex value of the electric field at a point z, $E_x = [E_{xr}, E_{xi}]$, the physical electric field is given by

$$E_x(z, t) = \text{Re}\left[\left(E_{xr} + jE_{xi}\right) \exp(j\omega t)\right] = E_{xr}(z) \cos(\omega t) - E_{xi}(z) \sin(\omega t) . \tag{13.57}$$

Alternatively, we can express the field in terms of single-frequency waves moving in the $\pm z$ directions,

$$E_+ \cos(kz - \omega t + \phi_+) + E_- \cos(kz + \omega t + \phi_-) . \tag{13.58}$$

Expanding Equation 13.58 gives

$$E_+\left[\cos(kz + \phi_+)\cos(\omega t) + \sin(kz + \phi_+)\sin(\omega t)\right]$$
$$+ E_-\left[\cos(kz + \phi_-)\cos(\omega t) - \sin(kz + \phi_-)\sin(\omega t)\right] . \tag{13.59}$$

Comparison with Equation 13.57 gives the correspondence between the complex number and traveling wave viewpoints,

$$E_{xr}(z) = E_+ \cos(kz + \phi_+) + E_- \cos(kz + \phi_-),$$
$$E_{xi}(z) = -E_+ \sin(kz + \phi_+) + E_- \sin(kz + \phi_-). \tag{13.60}$$

The values $[E_{xr}, E_{xi}]$ at two points in the solution space define four equations that determine the components E_+, ϕ_+, E_-, and ϕ_-. To illustrate the inverse conversion, a traveling wave with amplitude E_o moving in the +z direction in a uniform medium with $\phi_+ = 0$ is equivalent to the complex components $E_{xr} = E_o \cos(kz)$ and $E_{xi} = -E_o \sin(kz)$. A standing wave with $\phi_+ = \phi_- = 0$ corresponds to $E_{xr} = E_o \cos(kz)$ and $E_{xi} = 0$.

13.5 Scattering Solutions

At high frequencies material responses may lag behind driving fields. In this case, the fields created by shifts of dielectric charge or reorientation of magnetic domains may not be in phase with applied fields. This phase difference leads to energy losses in the material. We can represent phase differences with complex values of dielectric constant and magnetic permeability. The standard notation is

$$\epsilon = \epsilon' + j\epsilon'',$$
$$\mu = \mu' + j\mu''. \tag{13.61}$$

As an illustration, Figure 13.10 shows the variation of ϵ'/ϵ_o and ϵ''/ϵ_o in purified water. At low frequency the medium is an ideal dielectric with $\epsilon/\epsilon_o = 81$. At high frequency inertial effects in the reorientation of polar molecules cause a drop in the real part of the dielectric constant and increasing losses.

Poynting's theorem describes conservation of electromagnetic energy flow,

$$-\nabla \cdot (E^* \times H) = E^* \cdot \epsilon \frac{\partial E}{\partial t} + H^* \cdot \mu \frac{\partial H}{\partial t} + E^* \cdot J_c . \tag{13.62}$$

Equation 13.62 is derived in most introductory texts, including *Field and Wave Electromagnetics*, (Cheng, 1992). The quantity in parentheses on the left-hand side is the Poynting vector, equal to the flux of electromagnetic power in watts/m². Quantities on the right-hand side are volumetric power losses. The

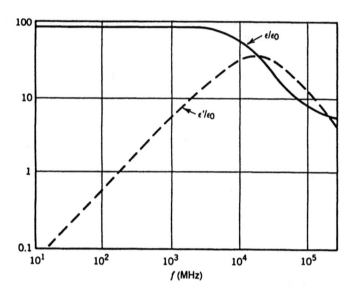

FIGURE 13.10
Real and imaginary parts of the relative dielectric constant of purified water as a function of frequency. (Adapted from Halstead, 1972.)

quantity J_c in the last term is the conductive current driven in resistive materials. The asterisk denotes complex conjugates. Assuming a harmonic field variation, applying Equation 13.62 and substituting $E = E_r + jE_i$, $H = H_r + jH_i$, the right-hand side of Equation 13.62 becomes

$$2j\omega \frac{\left[\epsilon'\left(E_r^2 + E_i^2\right) + \mu'\left(H_r^2 + H_i^2\right)\right]}{2}$$

$$-\omega\epsilon''\left(E_r^2 + E_i^2\right) - \omega\mu''\left(H_r^2 + H_i^2\right) + \sigma\left(E_r^2 + E_i^2\right). \tag{13.63}$$

The field quantities in Equation 13.63 represent time averages. The first term in Equation 13.63 is the time derivative of the total field energy (Equations 3.17 and 10.11). The remaining three terms represent power loss in materials. In the electric-field terms we can combine contributions of material response and resistivity into a single expression for the complex part of the dielectric constant,

$$\epsilon'' \to \epsilon'' - \frac{\sigma}{\omega}. \tag{13.64}$$

Equation 13.64 agrees with the derivation of Section 11.3 (Equation 11.34).

We can apply Equation 13.63 to determine the properties on an ideal absorbing layer for a frequency-domain solution. Consider first an E-type solution where a plane wave moves from a nonabsorbing medium (material 1) into an absorbing layer material of thickness Δz (material 2). Using Equation 13.13 to evaluate the Poynting vector, the power per unit area entering the latter is $Z_1(E_r^2 + E_i^2)$. The characteristic impedance of medium 1 is given by $Z_1 = (\mu_1'/\epsilon_1')^{1/2}$. If we express dissipation in the layer with a complex dielectric constant, the power absorbed per unit area is given by $-\omega\epsilon_2''(E_r^2 + E_i^2)\Delta z$. The condition for total absorption of E waves is

$$\epsilon_2'' = -\frac{1}{\omega Z_1 \Delta z}. \tag{13.65}$$

The real parts of the material properties for medium 2 are matched to those of medium 1, $\epsilon_2' = \epsilon_1'$ and $\mu_1' = \mu_2'$. The procedure for H-type waves is to use a complex magnetic permeability. The condition for perfect absorption is

$$\mu_2'' = -\frac{Z_1}{\omega \Delta z}. \tag{13.66}$$

Figure 13.11 shows benchmark solutions for wave propagation in a uniform vacuum of length L = 1 m with no damping. In both cases the left drive boundary has the fixed value $E_x = [1,0]$. In Figure 13.11a the right boundary is a short circuit, $E_x = [0,0]$. The frequency of 6.0×10^8 Hz (L = 2λ) gives a standing wave with $E_{xi} = 0$ at all points. We can create a traveling wave solution by adding a termination and an open-circuit condition on the right-hand boundary. For a width $\Delta z = 0.004$ m, the element on the right-hand side has $\epsilon = \epsilon_o (1 - 19.86j)$ and $\mu = \mu_o$. Figure 13.11b shows the result, a wave moving in the positive z direction with no reflection.

Finally, Figure 13.11c shows an interesting solution that we can compare to transmission line theory. Here, the right-hand half of the system consists of a dielectric with $\epsilon/\epsilon_o = 2.5$. The absorbing cell on the right-hand boundary has $\epsilon = \epsilon_o (2.5 - 31.40j)$ to match the characteristic impedance of 238.63 Ω. There are transmitted and reflected waves at the interface. An inspection of the real and imaginary parts of the solution shows a mixture of standing and traveling waves on the left-hand side and a pure traveling wave on the right-hand side. Equation 13.18 gives an electric field reflection coefficient of $R_E = -0.2252$. The predicted standing wave ratio is $(1+|R_E|)/(1-|R_E|) = 1.581$, consistent with Figure 13.11c.

13.6 One-Dimensional Resonant Modes

Resonators are closed systems with trapped traveling waves. Certain values of frequency give constructive interference, resulting in high field values for relatively weak excitation. The strategy to find resonant modes is to set up frequency-domain solutions (following the methods of Section 13.4) at several

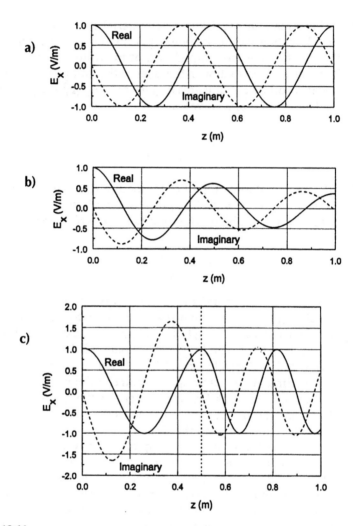

FIGURE 13.11

Benchmark frequency domain solutions. System length: L = 1 m. Element width: 0.004 m. Medium properties: $\mu = \mu_o$, $\epsilon = \epsilon_o$, and $\sigma = 0$. Left boundary: $E_x = [1,0]$. (a) Standing wave solution, right boundary Ex = [0,0], f = 600 MHz. (b) Traveling wave solution. Right boundary is an open circuit and the final element is a matched termination with $\sigma = 0.6626$ mhos/m. (c) Mixed standing and traveling wave solutions. Transition to $\epsilon/\epsilon_o = 2.5$ at z = L/2. Final element is an absorbing layer with $\sigma_l = 1.0477$ mhos/m.

frequencies, looking for characteristic signs of resonance. Although the modes of one-dimensional resonators are straightforward, the tools we develop in this section are directly relevant to the two- and three-dimensional systems of Chapter 14.

Figure 13.12 shows the simplest one-dimensional resonator, a vacuum region between two metal boundaries separated by a distance L. The boundary

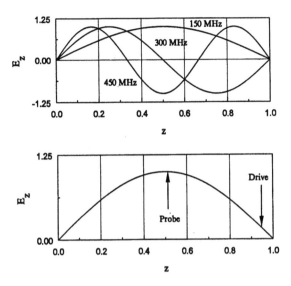

FIGURE 13.12
Lower-order E wave modes of a one-dimensional vacuum resonator. For L = 1 m, the predicted resonant frequencies are f_1 = 149.8962 MHz, f_2 = 299.7925 MHz, and f_3 = 449.6887 MHz.

condition for E-type waves is $E_x(0) = E_x(L) = 0$. The condition for constructive interference of traveling waves with wavelength λ is

$$\lambda = \frac{2L}{n}, \quad n = 1, 2, 3, \ldots . \tag{13.67}$$

The wavelength is related to frequency by $\lambda = c/f$. For L = 1 m the frequencies of the first three modes are f = 149.8962 MHz, 299.7925 MHz, and 449.6887 MHz. Figure 13.12 shows the corresponding variations of E_x. The figure also illustrates a common method to excite an E-type mode, a drive current near the expected location of maximum magnetic field. To check for a resonance, we sweep the drive through a range of frequencies and monitor the signal on a probe near the point of maximum electric field. The probe response rises sharply near the resonant frequency.

We can understand some features of resonance response by studying the behavior of the circuit shown in Figure 13.13. The circuit is a lumped-element model for the lowest frequency E-wave mode. Here, the capacitive energy (electric field) is highest near the center of the cavity and the inductive energy (magnetic field) at the edges. The drive current couples into a portion of the cavity inductance denoted L_1. The total system inductance is $L = L_1 + L_2$. If a harmonic current I_0 is applied at the drive point (with the sign convention shown) we can find the drive voltage by calculating the complex circuit impedance. The impedances of the individual circuit components are $Z_1 =$

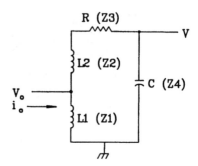

FIGURE 13.13
Driven LCR resonant circuit.

$j\omega L_1$, $Z_2 = j\omega L_2$, $Z_3 = R$, and $Z_4 = 1/j\omega C$. The total impedance at the drive is Z_1 in parallel with $(Z_2 + Z_3 + Z_4)$, so that

$$V_o = -\frac{Z_1 (Z_2 + Z_3 + Z_4)}{Z_1 + Z_2 + Z_3 + Z_4} i_o. \tag{13.68}$$

We measure the voltage across the capacitive region of the cavity, as shown. The probe voltage V is given in terms of V_o by the law of voltage division, $V = V_o Z_4/(Z_2 + Z_3 + Z_4)$. Alternatively, we can write

$$V = -\frac{Z_1 Z_4}{Z_1 + Z_2 + Z_3 + Z_4} i_o. \tag{13.69}$$

Inserting expressions for the component impedances in Equation 13.63 gives

$$V = i_o \frac{-j\omega L_1}{1 - \omega^2 LC + j\omega RC}. \tag{13.70}$$

We can generalize Equation 13.70 by expressing it in terms of the circuit resonant frequency ω_o, the resonator characteristic impedance Z_o, the quality factor Q, and the dimensionless frequency $\Omega = \omega/\omega_o$. The quantities are given by

$$\omega_o = \frac{1}{\sqrt{LC}}, \quad Z_o = \sqrt{L/C}, \quad Q = \frac{Z_o}{R}. \tag{13.71}$$

The parameter Q equals the ratio of stored electromagnetic energy U in the resonator multiplied by ω and divided by the average resistive power dissipation P:

$$Q = \frac{\omega U}{P}. \tag{13.72}$$

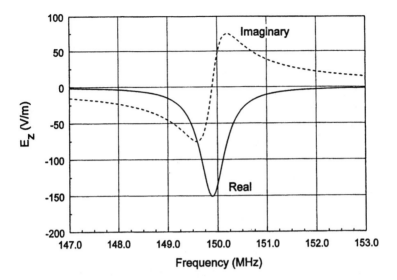

FIGURE 13.14
Resonant circuit, variation of V as a function of Ω near resonance. Drive current $I_o = [1,0]$, Q = 250.

The quantity $2\pi/Q$ is the approximate fraction of stored energy lost per cycle. A practical resonator has Q » 1.

Equation 13.70 takes the form

$$V = i_o \left[z_o \frac{L_1}{L} \right] \frac{\left[-\Omega^2/Q + j\Omega(\Omega^2 - 1) \right]}{\left[(1 - \Omega^2) + \Omega^2/Q^2 \right]}. \tag{13.73}$$

The resonance condition is that $\Omega = 1$. Figure 13.14 plots the variation of the real and imaginary parts of V as a function of Ω near resonance. In the plot the drive current has 0° phase ($i_o = [1,0]$) and the circuit has low damping (Q = 250). The imaginary part of the probe voltage has a sign change at resonance. At this point, the real part of the voltage has maximum amplitude $V_m = -QZ_o(L_1/L)$io. Substituting $\Omega = 1 + \Delta\Omega$ and making binomial expansions of terms in Equation 13.70, we find that the amplitude of the real part of V drops to half its maximum value at $\pm\Delta\Omega = 1/2Q$. This implies that the Q factor equals the reciprocal of the frequency difference between the half amplitude points.

The above discussion suggests a numerical technique to find the properties of damped and undamped resonators. As an example, consider E-wave solutions in the system of Figure 13.12. We set up a frequency-domain solution with a current source [1,0] in an element near the wall and monitor the real and imaginary parts of E_x near the expected field maximum. For the fundamental mode (n = 1), the best position is z = L/2. The next step is to pick upper and lower frequencies that bracket the anticipated value of resonant frequency, f_u and f_d. Initial solutions are performed to check that the imaginary part of the probe response is positive at f_u and negative at f_d. This condition guarantees that there is a resonance in the range. Subsequently, a frequency search for the sign change of the imaginary part gives the resonant frequency $f_o = 2\pi\omega_o$. Damping of the fields may result from volume resistivity in the medium, wall resistivity, or imperfect materials. To find the effect of dissipation, we can make several solutions near f_o to determine the frequency spread between the half amplitude points of the real part of the probe signal. The total quality factor is given by $Q = 2|f-f_o|/f_o$.

The open circles on the plot of Figure 13.14 shows numerical results for the system of Figure 13.12 with 250 elements. The predicted frequency of the n = 1 mode is $f_o = 149.89623$ MHz. A uniform conductivity σ_o gives damping. The stored energy in an element is $U = \epsilon_o E_x^2/2$ and the time-average power loss is $P = \sigma_o E_z^2$. Therefore, the predicted quality factor is $Q = 2\pi f_o \epsilon_o/\sigma_o$. A value $\sigma_o = 3.33 \times 10^{-5}$ mhos/m gives Q = 250. The points in Figure 13.14 are taken at increments of 40 kHz. Note that the imaginary part reverses sign at f_o and that the full width at half-maximum of the real part is 0.6 MHz as expected. Far from the resonance the probe electric field is purely imaginary, 90° out of phase from the drive current. The sign of the imaginary part is negative at low-frequency implying that the system acts like a capacitor. The impedance is imaginary and positive at high frequency, implying inductive behavior.

For most applications we want to determine the resonant frequency to high accuracy. Each point in a search involves a solution of finite-element equations. Solutions can be time-consuming in two- and three-dimensional problems. The implication is that we should seek methods to find the zero-crossing that minimize the number of steps. Figure 13.15a illustrates the *bisection method*. Initially, the frequency values f_d and f_u define a search range. Bisection of the range gives $f_1 = (f_u + f_d)/2$. If the probe response at f_1 is less than 0, we use f_1-f_u as the new interval; otherwise, the interval is f_d-f_1. The bisection continues either for a maximum number of steps or until the frequency width of the bisection region drops below a target frequency width. If Δf is the target width and Δf_o is the initial range, the maximum number of steps is $n = \log_2(\Delta f_o/\Delta f)$. The advantage of the bisection method is that it never fails to converge — the decreasing intervals always bracket the root. For well-behaved functions, alternative methods can achieve a target accuracy in fewer steps. Figure 13.15b shows the *false position method* for root finding. Starting again from points f_u and f_d that bracket the root, we interpolate the frequency as shown and calculate the corresponding value of E_x to find f_1. We choose the range f_d-f_1 or f_1-f_u that encloses the root and repeat the interpolation. The bisection method was used for the baseline calculation of Figure 13.14 with an initially broad range of 120 to 180 MHz. The search converged in 16 steps to the value 149.89496 MHz. The accuracy of 1 part in 1.2×10^5 was limited by the mesh size. Table 13.1 shows predicted resonant frequencies and numerical results for the first five modes of the one-dimensional resonator. The error is approximately proportional to the ratio $\Delta z/\lambda_n$ where Δz is the element size and λ_n is the mode wavelength.

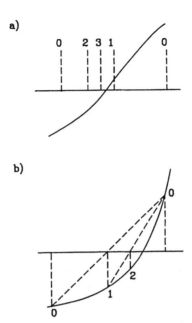

FIGURE 13.15
Numerical determination of zero crossing point of a function. (a) Bisection method, (b) false position method.

TABLE 13.1

Resonant Frequency Calculation: 250 Elements

Mode	f (Predicted) (MHz)	f (Numerical) (MHz)	Error
1	149.89525	149.89623	0.0007%
2	299.78457	299.79246	0.0026%
3	449.66204	449.68869	0.0059%
4	599.52185	599.58492	0.0105%

Chapter 13   Exercises

13.1.   Find the attenuation distance for a 100 MHz electromagnetic wave in the following aqueous media.

(a) Highly purified water with effectively zero conductivity (use the data of Figure 13.10).

(b) Distilled water with volume resistivity $\rho = 5 \times 10^3$ Ω-m.

(c) Seawater, with $\rho = 0.25$ Ω-m.

13.2. A plane electromagnetic wave with electric field amplitude $E_o = 10^5$ V/m moves through a uniform medium with $\epsilon_r = 4.5$ and $\mu_r = 10.8$. Find the amplitudes of the magnetic intensity H_o and magnetic field B_o of the wave.

13.3. An entering plane wave has electric field amplitude $E_i = 3.3 \times 10^6$ V/m in vacuum. The wave is normally incident on a ferrite with $\mu_r = 50$ and $\epsilon_r = 10$.

(a) Apply Equation 13.62 to find the time-averaged power density of the incident wave.

(b) Find the amplitudes of electric field and magnetic intensity in the transmitted and reflected waves.

(c) Give the time-averaged power densities of the transmitted and reflected waves.

13.4. Use a spreadsheet to investigate the stability of a one-dimensional time-domain electromagnetic calculation.

(a) For an E-type wave, write simplified forms of Equations 13.32 through 13.37 for a uniform mesh with spacing Δx. Assume the medium is homogeneous with no damping or internal current sources.

(b) Let each column of the spreadsheet represent a point in time. Model a region of width 1.0 m with 10 elements and 11 vertices. Set all initial values of E_z to zero.

(c) Maintain the fixed value $E_z(11) = 0$ to represent a wall. The first vertex has the fixed value $E_z(0) = 1$ to represent a step-function excitation. The other vertices advance according to the equations of Part a. Investigate solutions for time steps just above and just below the Courant limit of $\Delta t = 0.1667$ nsec.

13.5. Find the resonant frequencies of a dielectric-filled one-dimensional resonator. A distance $d = 0.5$ m separates conducting plates. The region $0 \le x \le d/2$ is occupied by a dielectric with $\epsilon_r = 2$, while the remaining volume is vacuum.

13.6. Figure E13.1 shows a simple resonant circuit driven by an AC voltage.

(a) Find the complex impedance of the circuit. Express the answer in terms of the parameters $\omega_0 = 1/(LC)^{1/2}$, $\beta = R/(L/C)^{1/2}$, $\Omega = \omega/\omega_0$.

(b) Plot the ratio of current through the capacitor to input current as a function of Ω.

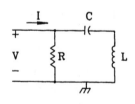

FIGURE E13.1

13.7. An AC voltage source with frequency f drives a transmission line of length L. The line is filled with a uniform dielectric with material characteristics ϵ and μ.

(a) Show that the wavelength of electromagnetic radiation in the line is $\lambda = 1/f(\epsilon\mu)^{\frac{1}{2}}$.

(b) A resonance occurs in a driven undamped circuit when the input impedance diverges to infinity. If the transmission line is shorted at the opposite end, explain why the resonance condition is $L = \lambda/4$.

(c) What is the resonance condition if the end of the transmission line has an open-circuit termination?

(d) Give the resonant frequency of a shorted polyethylene-filled line ($\epsilon_r = 2.7$) of length $L = 5$ m.

13.8. We want to define a resistive layer to absorb pulses propagating in a uniform medium with $\epsilon = \epsilon_0$ and $\mu = 5\mu_0$.

(a) If the temporal width of pulse is 2.0 nsec, find the spatial width.

(b) Suppose the absorber thickness equals 1/20 of the spatial pulse width. Find the layer conductivity for an ideal absorber.

13.9. Using the Fourier transform, a temporal Gaussian pulse with form $f(t) = \exp(-t^2/2\Delta t^2)$ can be written as an integral of frequency components,

$$f(t) = \int_0^\infty \exp\left[-\frac{(\omega\Delta t)^2}{2}\right] \cos(\omega t).$$

(a) Assuming propagation in a vacuum, convert the above equation to an expression for the spatial variation of fields $f(x)$ written in terms of the wave number k of the Fourier component.

(b) Find a value of the wave number k_o such that the integral of Fourier component amplitudes in the range $k_o \le k \le \infty$ is 1% of the integral from 0 to ∞. (Use tabulations of the error function.)

13.10. Use the results of Exercise 13.9 to investigate factors affecting the performance of an absorbing layer.

(a) Find the value of layer thickness in Figure 13.2 that gives a reflected pulse amplitude of 0.01, equivalent to a reflected energy fraction of 10^{-4}.

(b) Find the wave number k_o for the baseline Gaussian pulse of Equation 13.47, and compare $1/k_o$ to the layer thickness of Part a. What is the physical interpretation of this result?

13.11. We make two measurements of electric field amplitude and phase for a wave with frequency 200 MHz in a transmission line at positions z =

0 and z = 0.59 m. The line has a polyethylene dielectric with ϵ_r = 2.7. The results are as follows. Position 1: E_o = 17.748 V/m, ϕ = 8.184°. Position 2: E_o = 10.127 V/m, ϕ = –3.142°.

(a) Calculate the real and imaginary parts of the electric field at the two positions.

(b) Resolve the wave into positive and negative traveling waves by calculating the field amplitude E^+ and E^- following the discussion of Section 13.3.

13.12. A pillbox resonant cavity has radius R = 0.30 and length d = 0.06. The cavity is filled with purified water.

(a) Neglecting conductivity and dielectric losses, find the resonant frequency using the dielectric properties in Figure 13.10.

(b) If E_o is the electric field amplitude on axis, show that total electromagnetic field energy in the cavity at any time is U = $(\pi R^2 d)(\epsilon' E_o^2/2)J_1^2$ (2.405). The quantity ϵ' is the real part of the dielectric constant at the frequency of Part a. The first-order Bessel function has the value J_1(2.405) = 0.52.

(c) Find the power deposited in the water through dielectric losses.

(d) At what value of σ are conductive losses equal to dielectric losses?

(e) Under the condition of Part d, what is the Q factor of the resonator?

13.13. For the data set given below, find the value of x corresponding to f(x) = 0 using the two methods described in Section 13.6. Carry out the search step by step using a spreadsheet. The values are derived from the function $f(x) = \sin(3x/2) + x/2 + 1$, with a root at –0.54373094.

x	–2.0000	–1.5000	–1.0000	–0.5000	0.0000	0.5000	1.0000	1.5000	2.0000
f(x)	–0.1411	–0.5281	–0.4975	0.0684	1.0000	1.9316	2.4975	2.5281	2.1411

14

Two- and Three-Dimensional Electromagnetic Simulations

We shall not cease from exploration
And the end of all our exploring
Will be to arrive where we started
And know the place for the first time.

T. S. Eliot

This chapter is the culmination of the book. If you have a firm understanding of material in the previous chapters, the solution of the full Maxwell equations will be surprisingly easy. If you don't, the material will be one more hurdle of obfuscation. The good news is that it's the last one.

Section 14.1 extends the time-domain methods of Section 13.2 to two-dimensional solutions on a triangular mesh. In planar geometry, pulses propagate in the x–y plane. We can divide solutions into two classes: E or H. E-type solutions involve finite-element equations for E_z, which we shall call the *primary field component*. Gradients of the primary field give the *subsidiary* quantities H_x and H_y. The E-type solution is also called a TE (*transverse electric*) pulse because the electric field is always normal to the direction of propagation. The other class is the H-type solution or TM (*transverse magnetic*) pulse. Here, H_z is the primary field component and the subsidiary fields are E_x and E_y. Section 14.2 covers extensions of methods from the previous chapter to handle two-dimensional solutions. We must consider how to apply the Courant condition on a triangular mesh and how the performance of absorbing boundaries changes for pulses that are not normally incident. We shall also review equations to cover cylindrical systems. A useful application of H-type cylindrical solutions is propagation of transverse electromagnetic (TEM) pulses in coaxial transmission lines with properties that vary along z.

Section 14.3 addresses two-dimensional equations for scattering-type solutions. Most of the relevant physics was discussed in Section 13.4. The main challenge here is organizing the different types of propagation modes and finding expressions for the subsidiary field quantities. Section 14.4 covers solution techniques and several examples. Section 14.5 extends the frequency-domain theory to planar and cylindrical resonators. The resonance search techniques of Section 13.5 are directly applicable. Section 14.6 covers methods to calculate radio frequency (RF) power dissipation in finite-element simulations. One application is the determination of the Q factor of resonators.

To conclude, Sections 14.7 and 14. 8 cover electromagnetic pulse propagation in three dimensions on regular meshes. For reference, Section 14.7 covers finite-difference time-domain calculations. This popular approach has been intensively applied. It is based on time- and space-centered difference equations referenced to field components defined at six locations on a unit box cell. Section 14.8 introduces an element-based alternative. In addition to the material properties, the electric or magnetic field components are viewed as element properties. The solutions are easier to interpret and give a better representation of material discontinuities. In particular, the termination layer method can be used to simulate free-space boundaries. The section reviews practical considerations for implementing three-dimensional electromagnetic simulations on personal computers.

14.1 Time-Domain Equations on a Conformal Mesh

Extending the theories of Section 13.2 to two-dimensional time-domain solutions is straightforward. For illustration, consider the planar geometry where quantities vary in x and y but not in z. Electromagnetic disturbances propagate in the x–y plane. It is convenient to divide solutions into two classes.

- E pulse or transverse electric (TE) solutions where the electric field is transverse to the plane of propagation. The primary field component is E_z with subsidiary fields H_x and H_y.
- H pulse or transverse magnetic (TM) solutions where the magnetic field intensity is transverse to the plane of propagation. The primary field component is H_z with subsidiary fields E_x and E_y.

Following Section 13.2, the governing equation for E-pulse solutions is

$$-\oint \frac{\nabla \times E_z}{\mu} \cdot d\mathbf{l} = \iint dS \left[\epsilon \frac{\partial^2 E_z}{\partial t^2} + \frac{\partial J_{oz}}{\partial t} + \sigma \frac{\partial E_z}{\partial t} \right]. \tag{14.1}$$

In time-domain calculations the quantities E_z, ϵ, and μ are real. The only modification for a two-dimensional treatment is to take the circuit integral of Equation 14.1 over the 12-sided path of Figure 2.12 and to evaluate the surface integrals over the enclosed area. The result is

$$-E_{zo} \sum_i W_i + \sum_i W_i E_{zi} = \frac{\partial^2 E_{zo}}{\partial t^2} \sum_i \frac{\epsilon_i a_i}{3} + \sum_i \frac{a_i}{3} \frac{\partial J_{oi}}{\partial t} + \frac{\partial E_{zo}}{\partial t} \sum_i \frac{\sigma_i a_i}{3}. \quad (14.2)$$

The subscript o denotes the electric field at the test vertex and i refers to neighboring vertices or surrounding elements. The quantities a_i are element areas in the x–y plane. In Equation 14.2 the average of the field time derivative over the surrounding elements is approximately equal to the value at the test vertex. Following the notation of Section 2.7, the coupling constants are

$$W_i = \frac{1}{2} \left[\frac{\cot \theta_{b,i+1}}{\mu_{i+1}} + \frac{\cot \theta_{a,i}}{\mu_i} \right]. \quad (14.3)$$

Substituting time-centered derivatives as in Section 13.2, we find an equation to advance the axial electric field

$$E_{zo}^{n+1} = \frac{\left[E_{zo}^n \left(2A_o - \sum_i W_i \right) - E_{zo}^{n-1} (A_o - C_o) + \sum_i W_i E_{zi}^n - S_o^n \right]}{\left[A_o + C_o \right]}. \quad (14.4)$$

The quantities in Equation 14.4 are

$$A_o = \frac{1}{3\Delta t^2} \sum_i \epsilon_i a_i, \quad (14.5)$$

$$C_o = \frac{1}{6\Delta t} \sum_i a_i \sigma_i, \quad (14.6)$$

and

$$S_o^n = \sum_i \frac{a_i \partial J_{oi}^n / \partial t}{3}. \quad (14.7)$$

The quantity of Equation 14.7 is the time derivative of the drive current density, a specified function of time and position.

The treatment of H pulses in planar geometry is similar to the E pulse development. The governing equation is

$$-\oint \frac{\nabla \times H_z}{\epsilon} \cdot dl = \iint dS \left[\mu \frac{\partial^2 H_z}{\partial t^2} + \frac{\sigma \mu}{\epsilon} \frac{\partial H_z}{\partial t} \right] - \oint \frac{J_o}{\epsilon} \cdot dl. \quad (14.8)$$

The main difference is that the source current lies in the x–y plane. By continuity of current, the source generates displacement current and therefore creates field components E_x and E_y. To evaluate the curl term in Equation 14.8, suppose that source elements have a current density with amplitude J_{oi} inclined at an angle β_i relative to the x-axis. Following the discussion of Section 9.6, the curl can be expressed as a sum over elements surrounding a vertex,

$$S_o = \sum_i \frac{J_{oi}}{2\epsilon_i} \left[\cos\beta_i \left(x_i - x_{i-1} \right) + \sin\beta_i \left(y_i - y_{i-1} \right) \right]. \qquad (14.9)$$

The finite-element equation for H pulses is

$$H_{zo}^{n+1} = \frac{\left[H_{zo}^n \left(2A_o - \sum_i W_i \right) - H_{zo}^{n-1} \left(A_o - C_o \right) \sum_i W_i H_{zi}^n - S_o^n \right]}{\left[A_o + C_o \right]} \qquad (14.10)$$

The terms in Equation 14.10 are given by Equation 14.9 and the following expressions:

$$A_o = \frac{1}{3\Delta t^2} \sum_i \mu_i a_i , \qquad (14.11)$$

$$W_i = \frac{1}{2} \left[\frac{\cot\theta_{b,i+1}}{\epsilon_{i+1}} + \frac{\cot\theta_{a,i}}{\epsilon_i} \right], \qquad (14.12)$$

and

$$C_o = \frac{1}{6\Delta t} \sum_i a_i \frac{\sigma_i \mu_i}{\epsilon_i} . \qquad (14.13)$$

The source terms for E and H pulses are analogous to the input devices used to excite waveguides and resonators with power from transmission lines. For E pulses the source acts like the coupling loop illustrated in Figure 14.1a. The loop generates magnetic fields parallel to its surface. In turn, the changing magnetic flux creates electric fields. These fields are locally parallel to the exciting current with magnitude proportional to the time derivative of current. The source term for H pulses acts like the capacitive probe of Figure 14.1b. The probe current equals the local displacement current and is therefore proportional to the time derivative of the electric field. The displacement current produces magnetic fields that are proportional to the current and perpendicular to the probe.

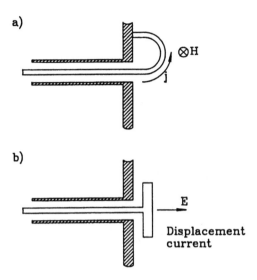

FIGURE 14.1
Connection between the source terms for E and H pulses and the physical devices used to coupled power from a transmission line to a waveguide or cavity. (a) Coupling loop drives E pulses. (b) Capacitive probe drives H pulses.

Extension of the model to cylindrical geometries follows a procedure similar to that of Section 9.4. We shall consider H-type solutions which include the familiar case of pulse propagation in coaxial transmission lines. Taking **H** normal to the direction of propagation, the allowed field components are H_θ, E_r and E_z. We can write the curl of magnetic field intensity in terms of the stream function $\eta = rH_\theta$,

$$\nabla \times \mathbf{H} = \frac{1}{r}\left[-\hat{r}\,\frac{\partial\eta}{\partial z} + \hat{z}\,\frac{\partial\eta}{\partial r}\right]. \tag{14.14}$$

The circuit integral of $\nabla \times \mathbf{H}/\epsilon$, around the path of Figure 2.12 has the form

$$\sum_i W_i\,\eta_i - \eta_o \sum_i W_i\,. \tag{14.15}$$

The coupling constants are

$$W_i = \frac{1}{2}\left[\frac{\cot\theta_{b,i+1}}{\epsilon_{i+1}\,R_{i+1}} + \frac{\cot\theta_{a,i}}{\epsilon_i\,R_i}\right]. \tag{14.6}$$

In Equation 14.16 R_i and R_{i+1} are the average element radii discussed in Section 9.4. Equation 14.10 holds with η replacing H_z if the coefficients have the form

$$A_o = \frac{1}{3\Delta r^2} \sum_i \frac{\mu_i a_i}{R_i}, \tag{14.17}$$

$$C_o = \frac{1}{6\Delta t} \sum_i a_i \frac{\sigma_i \mu_i}{\epsilon_i R_i}. \tag{14.18}$$

The condition $\eta = 0$ always holds at $r = 0$.

By convention, in planar solutions we assign a uniform current density over a region representing a source. We shall modify the form of the source term for convenience in cylindrical solutions. It is best to pick a source function that conserves current in radial flow. For H pulses we assign constant values of the radially weighted flux, $\Gamma = rj$, to source regions. With this choice regions of constant Γ generate TEM-type E_θ or H_θ with a $1/r$ variation in radius. The cylindrical source has the form

$$S_o = \sum_i \frac{\Gamma_{oi}}{2\epsilon_i R_i} \left[\cos\beta_i \left(z_i - z_{i-1} \right) + \sin\beta_i \left(r_i - r_{i-1} \right) \right], \tag{14.19}$$

where β_i is the angle of current density relative to the z axis.

14.2 Electromagnetic Pulse Solutions

Time-domain solutions of Equation 14.4 are relatively easy. Initially E_z equals zero everywhere. Electromagnetic disturbances are initiated with a source current density or with a specified time variation of E_z on a fixed boundary. To determine a time step that satisfies the Courant condition, we must check the entire mesh. A reliable procedure is to calculate the minimum value of the distance between adjacent vertices divided by the local speed of light. There are three types of boundaries. As in previous finite-element treatments, the normal derivative of E_z automatically equals zero on unspecified external boundaries. This condition is equivalent to mirror symmetry or an open-circuit condition. Pulses of E type reflect with positive polarity from an open boundary. The second possibility is fixed field boundary with $E_z = 0$, equivalent to a short-circuit termination. Here, reflected E pulses have inverted polarity. The third option is a drive boundary with a fixed field that follows a specified function of time, $E_z(t)$. A drive boundary can revert to an open or short-circuit condition after initiation of the pulse.

The option to locate termination layers adjacent to open boundaries expands the range of boundary properties. Figure 14.2a illustrates how to implement a matched absorbing boundary. Following the discussion of

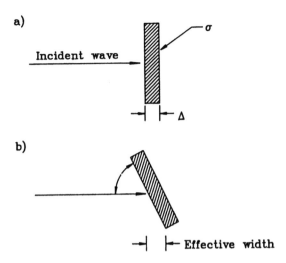

FIGURE 14.2
Absorbing boundary for two-dimensional solutions. (a) Normal incidence; (b) increase in effective layer thickness for $\theta < 90\%$.

Section 13.2, we define a thin element layer of thickness Δ. This element has the ϵ and μ values of the adjacent material and a conductivity,

$$\sigma = \frac{1}{\Delta \sqrt{\mu/\epsilon}}. \tag{14.20}$$

The layer performance closely follows the one-dimensional results (Figure 13.8) for pulses incident at 90°. In two-dimensional solutions we must deal with pulses entering from other angles, reducing the effectiveness of the absorber. Figure 14.2b shows a pulse striking an absorbing layer at an angle $\theta < \pi/2$. For this pulse the effective layer thickness is $\Delta/\sin\theta$, reducing the surface resistance by a factor of $\sin\theta$. In this case the termination is undermatched. Applying Equation 13.18, the predicted reflection coefficient is

$$\frac{E_r}{E_o} = -\frac{1-\sin\theta}{1+\sin\theta}. \tag{14.21}$$

The degradation in performance is not severe. A matched layer absorbs more than 90% of the incident wave energy over the range $30° \le \theta \le 90°$.

Figure 14.3 shows tests of Equation 14.21 for a Gaussian pulse (0.067 m width) incident from vacuum. Figure 14.3a shows results for a 45° corner reflector. The lines are contours of E_z. An application of Equation 13.5 shows that these lines are parallel to the magnetic field in the x–y plane and are separated by equal intervals of magnetic flux derivative. The absorbing layer on the bottom has $\Delta = 0.005$ m and conductivity $\sigma = 0.5301$ mhos/m. The top-left

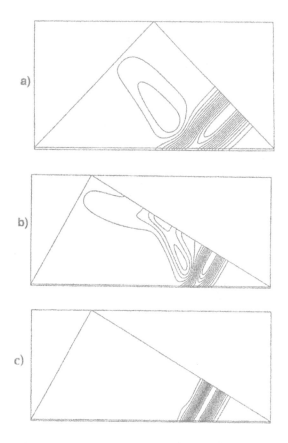

FIGURE 14.3
Performance of matched termination boundary as a function of incidence angle. Gaussian E
pulse in vacuum with 0.067 m full width at half-maximum. (a) Contours of E_z showing incident
and reflected pulses at a 45° corner reflector. Absorbing layer ($\Delta = 0.005$ m, $\sigma = 0.5301$ mhos/m)
at bottom and symmetry boundary at top right. (b) Pulse reflection at $\theta = 30°$. (c) Improvement
of 30° absorption by adjusting the layer conductivity to $\sigma = 0.2650$ mhos/m.

drive boundary creates the pulse, while the top-right edge is an open symmetry boundary. The incident pulse with peak amplitude 1 V/cm is at the bottom right. The reflected pulse traveling at a right angle has an amplitude of
–0.17 V/m, consistent with the prediction of Equation 14.21. Figure 14.3b
shows an example with $\theta = 30°$. The amplitude of the specularly reflected
pulse is –0.37 V/m, close to the predicted value of –0.33 V/m.

One advantage of the termination layer approach is that we can improve
absorption if we know the approximate direction of the pulses. For example,
for pulses that strike the surface at angle θ we can approach ideal absorption
by setting the layer conductivity equal to

$$\sigma = \frac{\sin\theta}{\Delta\sqrt{\mu/\epsilon}}. \tag{14.22}$$

Figure 14.3c shows improvement of the solution of Figure 14.3b by changing the layer conductivity to $\sigma = 0.2650$ mhos/m. In this case, the reflection coefficient is only -0.025.

Figure 14.4 shows applications of Equation 14.4. In Figure 14.4a, the left-hand drive boundary creates a Gaussian E pulse which moves through a vacuum region and strikes a dielectric block ($\epsilon_r = 3.0$). The top and bottom boundaries are open; therefore, the results simulate scattered radiation from an array of blocks. Note the retardation and narrowing of the pulse in the block. Because the impedance of the block is lower than that of free space, the reflected wave has negative polarity. A scan of electric field values along the vertical midplane gives reflected and transmitted pulse amplitudes of -0.268 and 0.732, in agreement with Equation 13.18. Figures 14.4b, c, and d show the propagation of a cylindrical wave on a finite-element mesh. A drive boundary condition on the central rod produces the expanding pulse of Figure 14.4b. The left and right boundaries are shorted and the top and bottom boundaries are open. The E_z contours of Figure 14.4c show the state after reflection. There is interference of positive polarity pulses from the top and bottom with negative pulses from the sides. The negative polarity pulse fronts in the corners result from double reflections from short and open boundaries. The behavior is more easily visualized in the wireframe plot of the first quadrant in Figure 14.4d.

To illustrate cylindrical H solutions we shall consider TEM pulse propagation in a coaxial transmission line. In a uniform line we can make a connection between the field quantity η and the macroscopic quantities of voltage and current used in transmission line theory. The current I at a point in the line equals the total current flow along the central conductor. Equation 9.12 implies that

$$I = \frac{2\pi r B_\theta}{\mu} = 2\pi r H_\theta = 2\pi\eta . \tag{14.23}$$

Equation 13.13 implies that the radial electric field is related to the magnetic intensity by the characteristic impedance of the medium,

$$E_r = H_\theta \sqrt{\frac{\mu}{\epsilon}} = \frac{\eta}{r}\sqrt{\frac{\mu}{\epsilon}} . \tag{14.24}$$

The voltage of the center conductor equals the integral of $-E_r$ from r_o to r_i,

$$V = \left[\frac{1}{2\pi}\sqrt{\frac{\mu}{\epsilon}}\ln\left(\frac{r_o}{r_i}\right)\right] I = \left[\frac{1}{2\pi}\sqrt{\frac{\mu}{\epsilon}}\ln\left(\frac{r_o}{r_i}\right)\right] 2\pi\eta . \tag{14.25}$$

The bracketed quantity in Equation 14.25 is the characteristic impedance of the coaxial line,

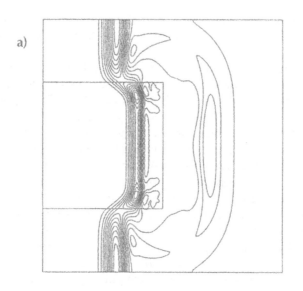

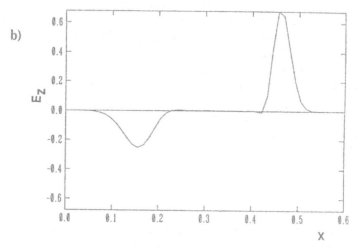

FIGURE 14.4
Examples of time-domain pulse propagation, Gaussian E pulses. (a) Contours of E_z for a pulse striking a dielectric block with $\epsilon = 3.0\epsilon_o$. (b) Scan of E_z along the midplane of part a. (c) Contours of E_z for a cylindrically symmetric expanding pulse. (d) Electric field contours for the pulse of part c after reflection from open boundaries at the left and right and shorted boundaries at the bottom and top. (e) Wireframe plot of the data of part d.

$$Z_o = \frac{1}{2\pi} \sqrt{\frac{\mu}{\epsilon}} \ln\left(\frac{r_o}{r_i}\right). \tag{14.26}$$

Because the fields of TEM pulses vary as $1/r$, the constant value of η uniquely identifies V and I.

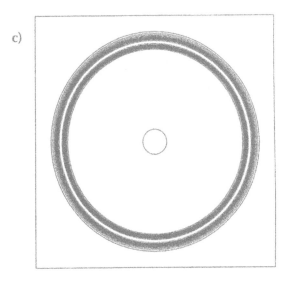

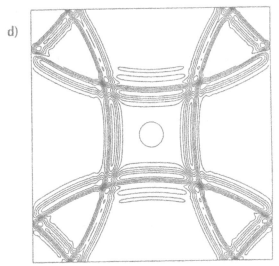

FIGURE 14.4 (continued)

Figure 14.5a shows the geometry for a benchmark test of an ideal impedance transition. The line of length 1 m has uniform radii $r_i = 0.05$ m and $r_o = 0.15$ m. There is a transition from $\mu = \mu_o$ to $\mu = 2.5\mu_o$ at the midpoint, giving an impedance change from 65.97 Ω to 104.31 Ω. The metal walls at the inner and outer radii correspond to an open-circuit condition for H pulses. A TEM pulse is initiated by setting a variation of η on the left-hand boundary. Figure 14.5a shows the incident pulse — the contours of H_θ lie along electric field lines. Figure 14.5b shows reflected and transmitted pulses after striking the interface. The relative amplitudes of −0.225 and 0.775 are consistent with Equation 13.19. Figure 14.5c shows a more interesting example, a nonideal

e)

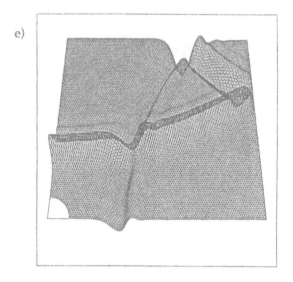

FIGURE 14.4 (continued)

impedance change arising from a sharp transition of the inner radius. The downstream and upstream impedances are 65.97 and 23.25 Ω. A one-dimensional model predicts transmitted and reflected H_θ amplitudes of 1.46 and 0.46. The contour and wireframe plots of Figures 14.5d and e show pulses that are only roughly in agreement. Two-dimensional effects at the transition cause a complex local radiation pattern.

14.3 Frequency-Domain Equations

We shall next extend the material of Section 13.4 to two-dimensional frequency-domain solutions. As in the previous section we divide waves into two classes: E and H waves. Waves of the E type are polarized with electric field along z and magnetic intensity components H_x and H_y. Here, we solve finite element equations for E_z with appropriate boundary conditions and then calculate the magnetic intensity components from Equation 13.5. The field components for H waves are H_z, E_x, and E_y. Solutions in cylindrical coordinates give azimuthally symmetric waves that propagate in the r and z directions. The primary field component of E waves is E_θ with subsidiary fields H_r and H_z. The field components for cylindrical H waves are H_θ, E_r, and E_z.

To begin, consider frequency-domain E wave solutions in planar geometry. The governing equation is,

$$-\nabla \times \left(\frac{1}{\mu} \nabla \times \mathbf{E} \right) = -\epsilon \omega^2 \mathbf{E} + j\omega \mathbf{J}_o.$$

(14.27)

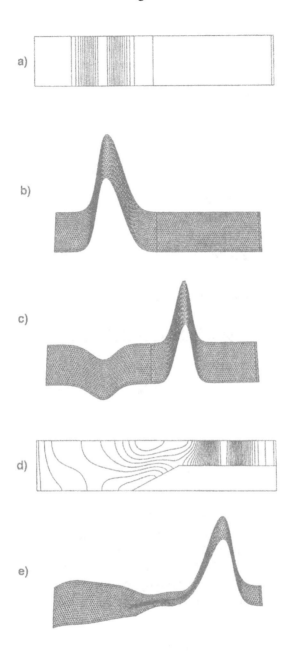

FIGURE 14.5
Benchmark calculations, cylindrical H pulses in coaxial transmission lines. (a) Propagation of a TEM pulse and reflection from an ideal impedance change. $L = 1$ m, $r_i = 0.05$ m, $r_o = 0.15$ m. Transition from $\mu = \mu_o$ to $\mu = 2.5_o$ at $z = 0.5$ m. $Z_1 = 65.97$ Ω, $Z_2 = 104.31$ Ω Contours of rH_θ for the incident pulse. (b) Wireframe plot of the pulse of part a. (c) Pulse of part *a* after striking boundary. (d) Nonideal impedance change associated with a change in line geometry. $Z_1 = 65.97$ Ω, $Z_2 = 23.25$ Ω Contours of rH_θ after reflection. (e) Wireframe plot of the data of part d.

In Equation 14.27 μ and ϵ may be complex quantities to represent resistive losses or nonideal material response. The complex number current source J_o contains information on the amplitude and phase of drives. To derive finite-element equations, we take integrals of Equation 14.27 over the standard area surrounding a vertex that encloses one third of the surrounding elements. The result is

$$-E_{zo} \sum_i W_i + \sum_i W_i E_{zi} = -E_{zo}\, \omega^2 \sum_i \frac{\epsilon_i a_i}{3} + \sum_i \frac{J_i a_i}{3}. \qquad (14.28)$$

Equation 14.28 can be written as

$$\sum_i E_{zi} W_i - E_{zo}\left[\sum_i W_i - A_o\right] = S_o. \qquad (14.29)$$

The coefficients in Equation 14.29 are

$$W_i = \frac{\cot\theta_{bi+1}/\mu_{i+1} + \cot\theta_{bi}/\mu_i}{2}, \qquad (14.30)$$

$$A_i = \omega^2 \sum_i \frac{\epsilon_i a_i}{3}, \qquad (14.31)$$

and

$$S_o = j\omega \sum_i \frac{J_i a_i}{3}. \qquad (14.32)$$

The set of relationships represented by Equation 14.29 can be solved with adaptations of the matrix inversion methods of Chapter 6. Conversion of the routines simply requires replacing real number variables with complex variables. In the search for a pivot element in Gauss-Jordan reduction, the absolute magnitude operation is replaced by the complex magnitude.

The components of magnetic intensity are equal to

$$H_x = \frac{j}{\mu\omega} \frac{\partial E_z}{\partial y},$$

$$H_y = -\frac{j}{\mu\omega} \frac{\partial E_z}{\partial x}. \qquad (14.33)$$

As discussed in Section 11.4, we determine spatial derivatives of the real and imaginary parts of E_z. Note that μ may complex if there are losses in ferrites or other magnetic materials. Equation 14.33 implies that lines of **H** lie along contours of consant E_z.

The equation for H wave propagation in a planar system is

$$\sum_i H_{zi} W_i - H_{zo}\left[\sum_i W_i - A_o\right] = -S_o.$$ (14.34)

Expressions for the coefficients W_i and S_o are the same as those of Equations 14.12 and 14.9. The quantity A_i is

$$A_o = \omega^2 \sum_i \frac{\mu_i a_i}{3}.$$ (14.35)

The electric field components in the x–y plane are equal to

$$E_x = \frac{j}{\omega\epsilon}\left[-\frac{\partial H_z}{\partial y} + j_{ox}\right],$$

$$E_y = \frac{j}{\omega\epsilon}\left[\frac{\partial H_z}{\partial x} + j_{oy}\right].$$ (14.36)

Note that electric field lines follow contours of constant H_z except within source regions.

Modifications for cylindrical geometry follow the discussion of Section 14.1. Solutions for H waves are expressed in terms of $\eta = rH_\theta$. The expressions for W_i and A_i contain factors of $1/R_i$ (where R_i is an average element radius). The source term follows from Equation 14.19. The electric field components are

$$E_r = \frac{j}{\omega\epsilon r}\left[\frac{\partial \eta}{\partial z} + \Gamma_{or}\right],$$

$$E_z = \frac{j}{\omega\epsilon r}\left[\frac{\partial \eta}{\partial r} + \Gamma_{oz}\right].$$ (14.37)

14.4 Methods for Scattering Solutions

Figure 14.6 shows an application of Equation 14.29, simulation of a dual slit diffraction pattern. A metal boundary on the left-hand side has two small openings that admit a plane electromagnetic wave. We model the surface by

FIGURE 14.6
Dual slit diffraction pattern. Distance between slits, 1 μm; wavelength of radiation, λ = 0.472 μm.

taking $E_z = [0,0]$ on the boundary and $E_z = [1,0]$ over the slits. The other three sides of the solution region are surrounded by matched termination layers to simulate open space. The distance between slits is 1 μm and the radiation wavelength is λ = 0.472 μm. The contour plot of Figure 14.6 shows diffraction and interference of the emerging waves. The dashed line is the theoretical prediction of the first interference null.

One important application for frequency-domain solutions is the scattering of electromagnetic radiation from objects in free space. There are two requirements to effect such solutions with a numerical method that uses a bounded computational volume.

- A perfectly absorbing boundary around the volume
- A source inside the volume that creates pure plane waves but does not interfere with the propagation of scattered waves

The first requirement is easy. We construct an anechoic chamber by surrounding the solution volume with a matched termination layer. The second is more challenging. Fixed field drive boundaries inside the solution region cannot be used because they reflect scattered waves. Therefore, internal current sources are necessary. We can determine the spatial distribution of current with the *distributed source method*. To appreciate the method, it is informative to examine first an unsuccessful approach. Figure 14.7a shows the geometry. The intention is to use a planar current layer to generate E waves traveling to the right. Waves moving to the left are immediately

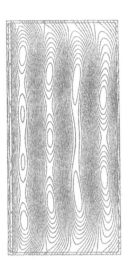

FIGURE 14.7
Approach to the generation of plane waves in an anechoic chamber. Vacuum region surrounded by matched termination layer with a planar drive current layer at the left. Contours of the real part of E_z; waves travel from left to right.

absorbed in the adjacent termination layer. Figure 14.7b shows the resulting solution for a radiation wavelength equal to half the box width. The waves approximate traveling plane waves but are clearly far from ideal. The upper and lower boundaries cause the problem. The discontinuities of the current sheet at the top and bottom give small transverse field components that reflect at low incidence angle from the termination layers. The pattern of Figure 14.7b results from the interference of these components with the propagating plane waves.

To achieve perfect plane waves in the presence of an absorbing boundary we must throw some thought at the problem. Rather than try to guess the correct current density distribution, we will work backward starting from the desired waveform. To begin, consider an anechoic chamber with a termination layer but no scattering objects. The desired field variation of an E wave is

$$E_z(x, y) = \xi \exp\left[-j\left(k_x x + k_y y\right)\right]. \tag{14.38}$$

If $k_x > 0$ and $k_y = 0$, the expression of Equation 14.38 represents a traveling wave with amplitude ξ and wavelength $\lambda = 2\pi/k_x$ moving in the $+x$ direction. Suppose we seek an E wave solution with no sources and a field of the form $E_z - \xi \exp[-j(k_x x + k_y y)]$. Substitution into Equation 14.29 gives

$$\sum_i E_{zi} W_i - \xi \sum_i \exp\left[-j\left(k_x x_i + k_y y_i\right)\right] W_i$$

$$-E_{zo}\left(\sum_i W_i - A_o\right) + \xi \exp\left[-j\left(k_x x_o + k_y y_o\right)\right]\left(\sum_i W_i - A_o\right) = 0. \tag{14.39}$$

Moving the known terms in Equation 14.39 to the right-hand side gives

$$\sum_i E_{zi} W_i - E_{zo} \left(\sum_i W_i - A_o \right) =$$

$$\xi \left[\sum_i \exp\left[-j(k_x x_i + k_y y_i)\right] W_i - \exp\left[-j(k_x x_o + k_y y_o)\right] \left(\sum_i W_i - A_o \right) \right]. \tag{14.40}$$

Equation 14.40 has the same form as Equation 14.29 if we view the right-hand side as a source function. Noting that the total field must be zero, a numerical solution of Equation 14.40 gives the desired plane wave. The important feature is that solutions inside the chamber follow Equation 14.38 *with absorbing wall effects included*.

The above discussion suggests the following steps for an ideal scattering solution:

- Set up an anechoic chamber with no scattering objects and calculate the source terms on the right-hand side of Equation 14.40 using the standard subroutines to evaluate W_i and A_i.
- Introduce the scattering objects and include their contributions to ϵ and μ when calculating values of W_i and A_i for the left-hand side of Equation 14.40.
- Apply standard matrix inversion to find the field solution with the sources and scattering objects. To isolate the contribution of the scattering objects, subtract the right-hand side of Equation 14.40 from the total field.

The procedure is easy to implement in a finite-element code if we make a list of regions corresponding to scattering objects. In the initial calculation of W_i and A_i to find the source terms, the region numbers of the scattering objects are replaced by that of the uniform background medium. Note that the method is not limited to incident plane waves. We could also construct attenuated, cylindrical, or spherical drive waves by changing the form of Equation 14.38.

Figure 14.8 illustrates the procedure for E wave scattering from dielectric and metal bodies in vacuum. To optimize the performance of the termination layer, we construct a cylindrical anechoic chamber with an axis centered on the object (Figure 14.8a). Figure 14.8b shows the ideal plane wave solution with no object ($\epsilon = \epsilon_o$). Figures 14.8c and d show E_z contours of the total solution for a metal body ($\epsilon = 10^5 \epsilon_o$) and a dielectric ($\epsilon = 1.5\epsilon_o$). Finally, Figure 14.8e plots scattering fields isolated from the solution of Figure 14.8d.

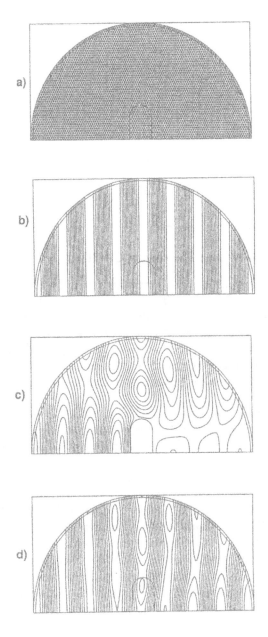

FIGURE 14.8

Scattering solutions using the distributed source method. (a) Anechoic chamber with matched termination layer centered on the scattering object. (b) Ideal plane E waves created by a distributed source. (c) Total field in the presence of a conducting body ($\epsilon = 10^5\epsilon_o$). (d) Total field in the presence of a dielectric body ($\epsilon = 1.5\epsilon_o$). (e) Isolated scattering fields from the dielectric body of part c.

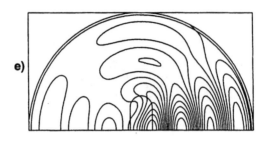

e)

FIGURE 14.8 (continued)

14.5 Waveguides and Resonant Cavities

This section covers resonance calculations in two-dimensional structures. The theory applies to the cut-off modes of uniform waveguides or axisymmetric modes of cylindrical resonators. The procedure is similar to that of Section 13.5. We excite a cavity at a set of frequencies with a drive current and sense the response with a probe at the expected position of maximum primary field. Depending on the mode and the phase of the drive, we search for a zero crossing of the real or imaginary part of the probe signal with the correct polarity change.

To test the method we shall solve for the pillbox cavity of Figure 14.9, the simplest cylindrical resonator. The resonant modes are field solutions consistent with the wall conditions that $E_\parallel = 0$ and $H_\perp = 0$. We divide the modes into two classes, depending on the disposition of field components relative to the z axis. Solutions of E-wave type for rE_θ with subsidiary field components B_r and B_z are called TE modes. Solutions of H-wave type for rH_θ yielding E_r and E_z are called TM modes. The modes have names of the form TM_{mnp}, where the indices denote the field characteristics. The index m designates an azimuthal variation of the form $\exp(jm\theta)$. We assume azimuthal symmetry so that m = 0. The indices n and p indicate the complexity of field variations in the r and z directions, respectively. Low numbers correspond to simple variations and generally have low frequencies. The predicted primary field variations for TM_{0np} modes in a pillbox cavity of radius R and length D are given by

$$H_\theta = H_o\, J_1\left(\chi_n r/R\right)\, \cos(p\pi z/D) \qquad (14.41)$$

with frequencies

$$f_{mp} = \frac{1}{2\pi\sqrt{\mu\epsilon}}\left[\frac{\chi_n^2}{R^2} + \frac{p^2\pi^2}{D^2}\right] \qquad (14.42)$$

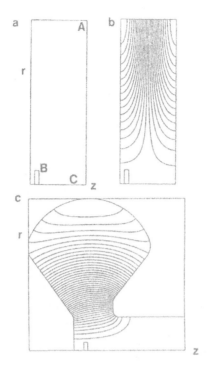

FIGURE 14.9
Resonant modes in cylindrical cavities. (a) Pillbox geometry. For H wave solutions metal walls are open circuit boundaries and $\eta = 0$ on axis. Location of drive current and sensing probe shown. Cavity radius, 0.70 m; Cavity length, 0.25 m. (b) Contours of η (electric field lines) for the TM_{011} mode at 621.329 MHz. (c) Contours of η for the TM_{010} mode in a klystron cavity.

In Equation 14.41, the quantity J_1 is a Bessel function and χ_i represents zeroes of the J_0 Bessel function: $\chi_1 = 2.4048$, $\chi_2 = 5.5201$, $\chi_3 = 8.6537$, ...

The benchmark geometry of Figure 14.9 has $R = 0.70$ m and $D = 0.25$ m with an element size of about 0.01 m. As in the one-dimensional solutions, the accuracy increases with smaller elements and decreases for higher mode numbers. For H wave solutions the metal walls are represented by open-circuit boundaries on the left, right, and top. The Dirichlet condition $\eta = 0$ holds on the axis (bottom boundary). For modes of type TM_{0n0}, we expect an electric field E_z concentrated near the axis. These modes are excited by a small drive region with an axial current ($\beta = 0.0$). The probe is located near the outer radius where we expect the maximum value of η according to Equation 14.41. The numerical result for the TM_{010} mode is 163.919 MHz (accurate to 0.002%). The values are 589.366 MHz for the TM_{030} mode (0.08% accuracy) and 621.329 MHz for the TM_{011} mode (0.04% accuracy). Figure 14.9b shows contours of rH_θ for the TM_{011} mode which lie along electric field lines.

Figure 14.10 illustrates a practical calculation for a superconducting proton accelerator. The accelerating structure consists of five coupled cavities — the simulation of Figure 14.10a represents half of the structure with a symmetry boundary on the right-hand side. The cavities are excited by an axial current at 0.0° phase in the small disk-shaped region at the bottom right. In the lossless structure, the drive creates purely real values of the quantity rH_θ. A probe at the position marked with a circle senses the reciprocal of the real part of the primary field, $V_p = 1/Re(rH_\theta)$. The resonance search involves location of the zero crossings of V_p from positive to negative polarity. The theory of coupled cavity arrays (see, for instance, Humphries, 1986) shows that a five-cavity system has five resonant modes with TM_{010}-type fields at different frequencies. The one with the highest frequency is called the *π-mode* because the phase of the axial electrical field reverses by 180° between adjacent cavities. The other modes have different values of phase shift: $0, \pi/4, \pi/2$, and $3\pi/4$. In the simulation of Figure 14.10 we expect to detect only three resonances because the $\pi/4$ and $3\pi/4$ modes are excluded by the symmetry boundary. Figure 14.10b shows a scan of probe output as a function of frequency. Arrows show the locations of the $0, \pi/2$, and π modes. The calculation gives values of 682.60 MHz for the 0 mode, 694.09 MHz for the $\pi/2$ mode, and 701.62 MHz for the π mode. Figure 14.10c shows the corresponding distributions of $E_z(0,z)$. Finally, Figure 14.10d shows electric field lines for the π-mode, the desired field pattern for particle acceleration. Here, a synchronized proton travels the length of one cavity in one half an RF period. At a frequency of 701.62 MHz and a cavity length of 0.1374 m, the assembly is matched to protons traveling at velocity 0.643c.

14.6   Power Losses and Q Factors

In this section we shall discuss how to apply Equation 13.63 to calculate RF power deposition and the Q factors of resonant structures. The procedures to calculate stored electromagnetic energy and volume power dissipation on a triangular mesh are similar to those for static field energy (Section 10.2). They involve scans through elements of the solution volume. As an illustration, consider an E wave solution in a planar geometry. For each element, the first step is to find the field components. The amplitude of the primary field component is the average of values at the corners of the triangle. If i_1, i_2, and i_3 are the indices of the element vertices, the average values of the real and imaginary parts of the primary field are

$$E_{zr} = \frac{E_{zr}(i_1) + E_{zr}(i_2) + E_{zr}(i_3)}{3},$$

$$E_{zi} = \frac{E_{zi}(i_1) + E_{zi}(i_2) + E_{zi}(i_3)}{3}.$$

(14.43)

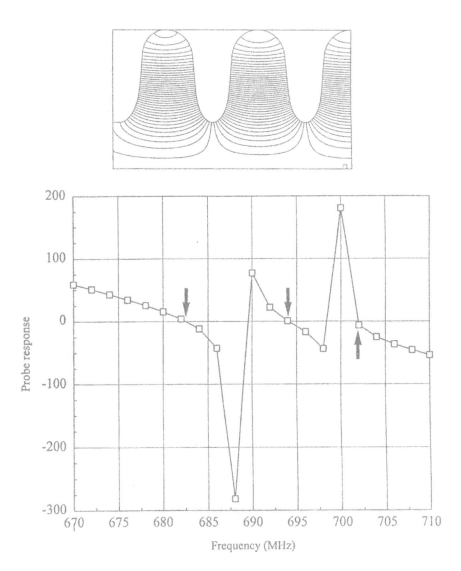

FIGURE 14.10
Resonant modes in a proton accelerator structure, coupled superconducting cavities.
(top) Simulation geometry for half of a five-cavity assembly. Symmetry boundary on the right-
hand side, drive current region at the bottom right. Electric field lines for the π-mode.
(bottom) Scan of probe output as a function of frequency. Arrows show the locations of the 0,
$\pi/2$, and π modes. (next page) Distributions of $E_z(0,z)$ for the 0 mode (682.60 MHz), $\pi/2$ mode
(694.09 MHz), and the π mode (701.62 MHz).

The subsidiary field components are given by derivatives of the primary
components. For example, the three-point formula of Section 7.3 gives the
following approximation for the real part of H_x,

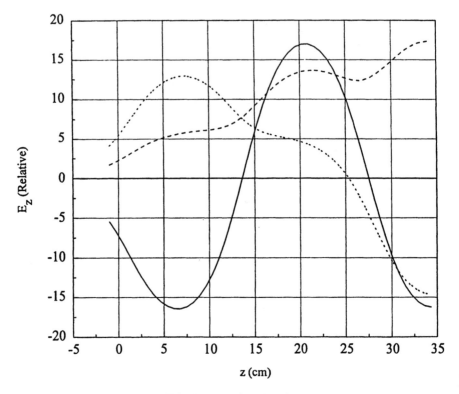

FIGURE 14.10 (continued)

$$H_{xr} = \frac{\left[E_{zr}(i_2) - E_{zr}(i_1)\right](x_3 - x_1) - \left[E_{zr}(i_3) - E_{zr}(i_1)\right](x_2 - x_1)}{(y_2 - y_1)(x_3 - x_1) - (y_3 - y_1)(x_2 - x_1)}. \quad (14.44)$$

Following Equation 13.63, the total field energy in the element per unit length in z is

$$dU = \left[\frac{\epsilon_r\left(E_{zr}^2 + E_{zi}^2\right) + \mu_r\left(H_{xr}^2 + H_{xi}^2 + H_{yr}^2 + H_{yi}^2\right)}{4}\right]a. \quad (14.45)$$

In Equation 14.45, a is the element area in the x–y plane and ϵ_r and μ_r are the real parts of the dielectric constant and magnetic permeability. The quantity in brackets is the field energy density. The additional factor of 2 in the denominator follows from time averaging of the harmonic functions. Similarly, the power dissipated in the element per unit length in z is

$$dP = \left[\frac{\epsilon_i\left(E_{zr}^2 + E_{zi}^2\right) + \mu_i\left(H_{xr}^2 + H_{xi}^2 + H_{yr}^2 + H_{yi}^2\right)}{2}\right]a\omega. \quad (14.46)$$

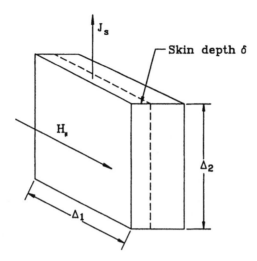

FIGURE 14.11
RF power loss by currents driven in a metal wall.

The quantities ϵ_i and μ_i are the imaginary parts of the element materials properties and ω is the angular frequency of the electromagnetic fields. The quantities in Equations 14.45 and 14.46 can be summed over elements to give total energy deposition in regions or used to generate plots of energy and power distributions.

In vacuum cavities for charged-particle acceleration power losses result mainly from resistive dissipation from currents driven in the metal walls. At the field levels required in accelerators power losses may be high even for good conductors. Currents driven by high-frequency electromagnetic fields are confined to a layer on the surface of metals with thickness equal to the skin depth (Equation 11.57). For copper-lined cavities in the range 250 to 1000 MHz the skin depth is only a few microns. Therefore, it is impractical to apply the volume methods of Equation 14.46. Surface integrals give more accurate results.

Consider a small segment of a metal surface shown in Figure 14.11. The boundary condition on the surface of a good conductor is that the magnetic intensity is parallel to the surface. We denote the field amplitude as $H_\parallel$. Exclusion of the field from the volume of the material implies that the surface carries a linear current density of amplitude $J_s = H_\parallel$. The surface segment has dimensions Δ_1 along the direction of magnetic intensity and Δ_2 along the direction of current. If the metal has volume resistivity ρ, the total resistance of the segment is $R = \rho\Delta_2/\Delta_1\delta$. The time-averaged power deposited in the segment is $\frac{1}{2}R(J_s\Delta_1)^2$. Dividing by $\Delta_1\Delta_2$ gives the time-average power per unit area of the surface,

$$p = \frac{\rho H_\parallel^2}{2\delta} = \frac{R_s H_\parallel^2}{2}. \tag{14.47}$$

The quantity R_s in Equation 14.47 with units of Ω is called the *surface resistivity*. It is given by the expression

$$R_s = \sqrt{\pi f / \mu_o \rho} = \frac{1}{6\delta} \qquad (14.48)$$

where f is the RF frequency.

The following procedure is used to find wall losses. Consider first H-type solutions where metal walls are open boundaries. During the mesh generation process we define one or more line regions with unspecified boundary conditions where we want to evaluate power deposition. After completing the solution, we identify all line segments of the mesh where both vertices have the target region number. In a planar geometry, suppose the vertex coordinates of one such segment are (x_1, y_1) and (x_2, y_2). The power loss on the segment (per unit length in z) is

$$dp = \frac{1}{2} \left[(H_1 + H_2)/2 \right]^2 \sqrt{(x_2 - x_1)^2 + (y_2 - y_1)^2} . \qquad (14.49)$$

The expression for a segment with coordinates (z_1, r_1) and (z_2, r_2) in a cylindrical system is

$$dp = \frac{1}{2} \left[(H_1 + H_2)/2 \right]^2 \frac{2\pi (r_1 + r_2)}{2} \sqrt{(z_2 - z_1)^2 + (r_2 - r_1)^2} . \qquad (14.50)$$

The multiple factors of 2 are written out to clarify the physical content of Equation 14.50. In a program redundant calculations should be eliminated to reduce floating point operations.

The procedure is more involved for E wave solutions. After finding vectors that lie on lines of $E_z = 0$ or $rE_\theta = 0$, we identify the two elements adjacent to each vector. Because the vector lies on a metal boundary, the routine to calculate magnetic intensity returns valid values only for the material element that lies inside the solution volume. It is sufficient to take the magnitude of H in this element because we know that the magnetic intensity must lie parallel to the conducting surface. This value is used in place of the averages in Equations 14.49 or 14.50.

To illustrate the method, we shall calculate Q factors for TM_{010} modes in accelerator cavities. Figure 14.12 shows two 303 MHz cavities. The first is a standard pillbox of length d = 0.2 m and radius R = 0.3789 m. The second has the same axial length and a rounded outer boundary. We shall find that this cavity has a higher Q factor; therefore, it consumes less power to achieve the same accelerating gradient. In the solution, both cavities are driven by a small current source near the axis. Resonant calculations in lossless structures give relative field levels — we can adjust the solution to represent any excitation by scaling all values of the primary field. For the pillbox cavity the value of

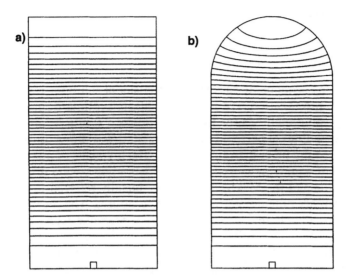

FIGURE 14.12
Comparison of the Q factors for two 303-MHz accelerator cavities with the same axial length cavities. Contours of rH_θ, lower boundary is the z axis. (a) Pillbox cavity, outer radius 0.3789 m. (b) Cavity with contoured outer wall, outer radius 0.4000 m.

the electric field on axis is $E_z(0,z) = E_o = 3.6103 \times 10^4$ V/m. The predicted stored energy (see, for instance, Jackson, 1975) is

$$U = \pi R^2 d \left[\frac{\epsilon E_o^2}{2} \right] J_1^2 (2.405). \tag{14.51}$$

Equation 14.51 predicts a value $U = 1.410 \times 10^{-4}$ J, close to the numerical prediction of $U = 1.369 \times 10^{-4}$ J from Equation 14.45. The integral of $H_\parallel^2/2$ over the metal surface of the cavity using Equation 14.50 equals 1684 A². Suppose the inside of the cavity is lined with high-purity, polished copper with $\rho = 1.712 \times 10^{-8}$ Ω-m. At 303 MHz, the surface resistance is $R_s = 4.531 \times 10^{-3}$ Ω and the skin depth is $\delta = 3.778$ μm. The product of the surface resistivity and surface current integral gives a total power loss of P = 7.630 W. The quality factor is therefore $Q = 2\pi f U/P = 34{,}165$. The analytic formula for the TM_{010} mode in a cylindrical cavity is

$$Q = \frac{d}{\delta} \frac{1}{1 + d/R}. \tag{14.52}$$

Inserting values in Equation 14.52 gives Q = 35,290, close to the numerical value. Similar calculations for the smooth cavity give $U = 1.080 \times 10^{-4}$ J, $\iint dA$ $H_\parallel^2/2 = 1211$ A², and P = 3.750 W. These figures imply an improved quality factor of Q = 37497.

14.7   Finite-Difference Time-Domain Method in Three Dimensions

The following sections discuss three-dimensional solutions for electromagnetic wave propagation. We shall concentrate on time-domain methods. Boundary value methods for frequency-domain solutions are beyond the present capabilities of personal computers. Models of three-dimensional systems may require more than 10^6 elements. The inversion of matrices for such a large set of equations is a challenge. This section covers finite-difference representations of the Maxwell equations, usually referred to as the FDTD (finite-difference time-domain) method. This approach has been widely used and is informative to review. The difference representation has a close correspondence to the differential form of the Maxwell equations. In the next section, we shall study alternate methods based on an element viewpoint.

The differential Maxwell equations for a Cartesian coordinate system are

$$\frac{\partial H_x}{\partial t} = \frac{1}{\mu}\left(\frac{\partial E_y}{\partial z} - \frac{\partial E_z}{\partial y}\right),$$

$$\frac{\partial H_y}{\partial t} = \frac{1}{\mu}\left(\frac{\partial E_z}{\partial x} - \frac{\partial E_x}{\partial z}\right),$$

$$\frac{\partial H_z}{\partial t} = \frac{1}{\mu}\left(\frac{\partial E_x}{\partial y} - \frac{\partial E_y}{\partial x}\right),$$

$$\frac{\partial E_x}{\partial t} = \frac{1}{\epsilon}\left(\frac{\partial E_z}{\partial y} - \frac{\partial E_y}{\partial z} - \sigma E_x\right), \tag{14.53}$$

$$\frac{\partial E_y}{\partial t} = \frac{1}{\epsilon}\left(\frac{\partial H_x}{\partial z} - \frac{\partial H_z}{\partial x} - \sigma E_y\right),$$

$$\frac{\partial E_z}{\partial t} = \frac{1}{\epsilon}\left(\frac{\partial H_y}{\partial x} - \frac{\partial H_x}{\partial y} - \sigma E_z\right).$$

To simplify the discussion, internal current sources have been omitted from Equation 14.53. Other limiting conditions are similar to the treatment of Section 14.1: (1) there is no free space charge, (2) the quantities ϵ and μ are real numbers, and (3) an isotropic conductivity represents losses in materials.

The most widely used differencing scheme for Equation 14.53 (Taflove and Brodwin, 1975) follows from the work of Yee (Yee, 1966). The goal is to determine time- and space-centered approximations for derivatives of Equation 14.53 to achieve second-order accuracy. The method applies to a regular mesh with uniform spacing along each axis: Δx, Δy, and Δz. We shall

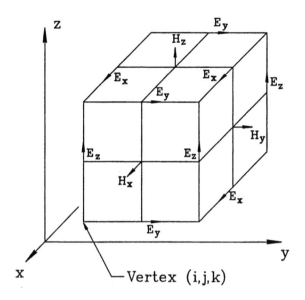

FIGURE 14.13
Mesh for three-dimensional finite-difference time-domain electromagnetic simulations. Definitions of field components on a Yee cell.

concentrate on the special case where $\Delta x = \Delta y = \Delta z = \delta$ and take a uniform time step Δt. Values of E and H are assumed to be offset by $\Delta t/2$ to center the equations in time. For space-centering field components are defined at the offset positions shown in Figure 14.13. The Yee cell illustrated leads to a symmetric difference form for the Maxwell equations.

The differencing procedure is involved to write out but conceptually simple. The equations to advance values of magnetic intensity from time $(n-\frac{1}{2})\Delta t$ to $(n+\frac{1}{2})\Delta t$ are

$$H_x^{n+\frac{1}{2}}\left[i,j+\frac{1}{2},k+\frac{1}{2}\right] = H_x^{n-\frac{1}{2}}\left[i,j+\frac{1}{2},k+\frac{1}{2}\right] + \frac{\Delta t}{\mu\left[i,j+\frac{1}{2},k+\frac{1}{2}\right]\delta}$$

$$\times \left(E_y^n\left[i,j+\frac{1}{2},k+1\right] - E_y^n\left[i,j+\frac{1}{2},k\right] - E_z^n\left[i,j,k+\frac{1}{2}\right]\right.$$

$$\left. + E_z^n\left[i,j+1,k+\frac{1}{2}\right]\right),$$

$$(14.54)$$

$$H_y^{n+\frac{1}{2}}\left[i+\frac{1}{2},j,k+\frac{1}{2}\right] = H_y^{n-\frac{1}{2}}\left[i+\frac{1}{2},j,k+\frac{1}{2}\right] + \frac{\Delta t}{\mu\left[i+\frac{1}{2},j,k+\frac{1}{2}\right]\delta}$$

$$\times \left(E_z^n\left[i+1,j,k+\frac{1}{2}\right] - E_z^n\left[i,j,k+\frac{1}{2}\right] - E_x^n\left[i+\frac{1}{2},j,k\right]\right.$$

$$\left. + E_x^n\left[i+\frac{1}{2},j,k+1\right]\right),$$

$$H_z^{n+\frac{1}{2}}\left[i+\tfrac{1}{2},j+\tfrac{1}{2},k\right]=H_z^{n-\frac{1}{2}}\left[i+\tfrac{1}{2},j\tfrac{1}{2},k\right]+\frac{\Delta t}{\mu\left[i+\tfrac{1}{2},j+\tfrac{1}{2},k\right]\delta}$$

$$\times\left(E_x^n\left[i+\tfrac{1}{2},j+1,k\right]-E_x^n\left[i+\tfrac{1}{2},j,k\right]-E_y^n\left[i,j+\tfrac{1}{2},k\right]\right.$$

$$\left.+E_y^n\left[i+1,j+\tfrac{1}{2},k\right]\right).$$

A comparison of Equation 14.54 with Figure 14.13 shows that the spatial derivations at a point depend on field components with symmetric displacements. The following equations advance the electric fields.

$$E_x^{n+1}\left[i+\tfrac{1}{2},j,k\right]=\left[1+\frac{\sigma\left[i+\tfrac{1}{2},j,k\right]\Delta t}{2\epsilon\left[i+\tfrac{1}{2},j,k\right]}\right]$$

$$\left[E_x^n\left[i+\tfrac{1}{2},j,k\right]\left(1-\frac{\sigma\left[i+\tfrac{1}{2},j,k\right]\Delta t}{2\epsilon\left[i+\tfrac{1}{2},j,k\right]}\right)+\frac{\Delta t}{\epsilon\left[i+\tfrac{1}{2},j,k\right]\delta}\right.$$

$$\times\left(H_z^{n+\frac{1}{2}}\left[i+\tfrac{1}{2},j+\tfrac{1}{2},k\right]-H_z^{n+\frac{1}{2}}\left[i+\tfrac{1}{2},j-\tfrac{1}{2},k\right]\right.$$

$$\left.\left.+H_y^{n+\frac{1}{2}}\left[i+\tfrac{1}{2},j,k-\tfrac{1}{2}\right]-H_y^{n+\frac{1}{2}}\left[i+\tfrac{1}{2},j,k+\tfrac{1}{2}\right]\right)\right],$$

$$E_y^{n+1}\left[i,j+\tfrac{1}{2},k\right]=\left[1+\frac{\sigma\left[i,j+\tfrac{1}{2},k\right]\Delta t}{2\epsilon\left[i,j+\tfrac{1}{2},k\right]}\right]$$

$$\left[E_y^n\left[i,j+\tfrac{1}{2},k\right]\left(1-\frac{\sigma\left[i,j+\tfrac{1}{2},k\right]\Delta t}{2\epsilon\left[i,j+\tfrac{1}{2},k\right]}\right)+\frac{\Delta t}{\epsilon\left[i,j+\tfrac{1}{2},k\right]\delta}\right.\qquad(14.55)$$

$$\times\left(H_x^{n+\frac{1}{2}}\left[i,j+\tfrac{1}{2},k+\tfrac{1}{2}\right]-H_x^{n+\frac{1}{2}}\left[i,j+\tfrac{1}{2},k-\tfrac{1}{2}\right]\right.$$

$$\left.\left.+H_z^{n+\frac{1}{2}}\left[i-\tfrac{1}{2},j+\tfrac{1}{2},k\right]-H_z^{n+\frac{1}{2}}\left[i+\tfrac{1}{2},j+\tfrac{1}{2},k\right]\right)\right],$$

$$E_z^{n+1}\left[i,j,k+\tfrac{1}{2}\right]=\left[1+\frac{\sigma\left[i,j,k+\tfrac{1}{2}\right]}{2\epsilon\left[i,j,k+\tfrac{1}{2}\right]}\right]$$

$$\left[E_z^n\left[i,j,k+\tfrac{1}{2}\right]\left(1-\frac{\sigma\left[i,j,k+\tfrac{1}{2}\right]\Delta t}{2\epsilon\left[i,j,k+\tfrac{1}{2}\right]}\right)+\frac{\Delta t}{\epsilon\left[i,j,k+\tfrac{1}{2}\right]\delta}\right.$$

$$\times\left(H_y^{n+\frac{1}{2}}\left[i+\tfrac{1}{2},j,k+\tfrac{1}{2}\right]-H_y^{n+\frac{1}{2}}\left[i-\tfrac{1}{2},j,k+\tfrac{1}{2}\right]\right.$$

$$\left.\left.+H_x^{n+\frac{1}{2}}\left[i,j-\tfrac{1}{2},k+\tfrac{1}{2}\right]-H_x^{n+\frac{1}{2}}\left[i,j+\tfrac{1}{2},k+\tfrac{1}{2}\right]\right)\right].$$

The solution procedure for Equations 14.54 and 14.55 is straightforward. We need storage for six field quantities in each unit cell. Ideally, we would also store values of ϵ, μ, and σ on three edges of each cell. Values on the remaining edges can be determined from neighboring cells. This count implies a total of 15 real numbers per cell. An alternative is to store material properties at vertices and take averages to find edge values during the solution. This strategy reduces the storage requirement to nine real numbers per cell at the expense of a longer run time. Field quantities are initially zero. Pulses are created by current sources or specified time-dependent boundary fields. The procedure at each time step is to loop through all cells advancing **E** components from $(n-1)\Delta t$ to $n\Delta t$ and then to advance **H** components from $(n-\frac{1}{2})\Delta t$ to $(n+\frac{1}{2})\Delta t$. The time step is governed by the following form of the Courant stability condition:

$$\Delta t \preceq \frac{\sqrt{\epsilon[i,j,k]\,\mu[i,j,k]}}{\sqrt{\dfrac{1}{\Delta x^2} + \dfrac{1}{\Delta y^2} + \dfrac{1}{\Delta z^2} +}} = \frac{\sqrt{\epsilon[i,j,k]\,\mu[i,j,k]}\,\delta}{\sqrt{3}}. \tag{14.56}$$

The second form of Equation 14.56 holds for a cubic mesh.

The Yee formulation is simple to program and runs quickly on meshes with uniform values of Δx, Δy, and Δz. Equations that preserve second-order accuracy on a nonuniform mesh are much more complex (Section 4.1). Like other finite-difference approaches we have discussed the formulation represents material properties as a continuum of values. The method is therefore best suited to problems like plasma simulations where dielectric properties vary smoothly in space. As we saw in Section 4.3, the method gives a poor representation of boundaries between materials with different ϵ or μ. The spatial offsets of field quantities exacerbate the problem. Furthermore, it would be extremely challenging to include materials with field-dependent properties.

The definition of planar conducting or open-circuit boundaries is easily accomplished with the Yee formulation. Consider a conducting surface in the plane j of Figure 14.13. We set quantities in the plane equal to zero: $E_z[i,j,k+\frac{1}{2}] = 0$, $E_x[i+\frac{1}{2},j,k] = 0$, and $H_y[i+\frac{1}{2},j,k\frac{1}{2}] = 0$. These relations imply that the parallel component of electric field and normal component of magnetic intensity are zero in the plane, equivalent to the condition on the surface of a perfect conductor. Conversely, we could field quantities equal to zero in the plane $(j+\frac{1}{2})$: $E_y[i,j+\frac{1}{2},k] = 0$, $H_x[i,j+\frac{1}{2},k\frac{1}{2}] = 0$, and $H_z[i,j+\frac{1}{2},k] = 0$. Here, the normal component of **E** and the parallel component of **H** are zero, equivalent to the open-circuit condition.

Ideal absorbing boundaries are more of a problem. Termination layers generally give poor results in FDTD calculations because of the ambiguity in the spatial distributions of material properties. An alternative approach developed by Mur (Mur, 1981) is to code special relationships for points on the boundaries of the solution volume. These relationships simulate the propagation of waves to infinity. To understand the *lookback technique*, consider setting up a free-space boundary on plane j in Figure 14.13. Let points with $y > y_j$ be inside the solution volume. We make the following provisions.

- The mesh has uniform spacing δ and cells adjacent to the boundary have uniform values of ϵ and μ with $\sigma = 0$
- The time step has the value $\Delta t = \delta(\epsilon\mu)^{\frac{1}{2}}/2$ (consistent with the Courant condition)
- The values of field components at the position j+½ adjacent to the boundary are stored at two previous time steps

To advance fields at the boundary, we must know quantities at locations j–½ and j+½. The essence of the Mur technique is to substitute values of quantities at the position j+½ and the retarded time $t - 2\Delta t$ for the corresponding quantities at j–½ outside the solution volume. These values would have applied at the points if pulses propagated unimpeded through the boundary.

The Mur prescription simulates an ideal absorbing boundary for normally incident pulses under the conditions listed. In practice, there are several difficulties in applying the lookback technique. The necessity of storing past field values at adjacent points leads to complex codes for boundaries other than simple planes parallel to the axes. A major disadvantage is that the time step is locked to the properties of a particular surface. It is impossible to apply the method when different materials abut free-space boundaries. Furthermore, optimization of absorbing properties for pulses incident at different angles is largely an empirical process. In contrast, the behavior of termination layers has a strong theoretical foundation. We can predict the performance of boundary layers for oblique incidence and maximize absorption for known pulse properties.

14.8   Three-Dimensional Element-Based Time-Domain Equations

This section describes an alternative to the Yee formulation based on finite-element methods applied on a regular mesh. The solutions are easier to interpret and the method correctly represents material discontinuities. The main advantage is that we can use termination layers to represent free-space boundaries. Therefore, techniques for one- and two-dimensional simulations from previous sections can be utilized.

We shall use the regular meshes that we developed for electrostatics (Section 4.6). The mesh has box elements with arbitrary variations of Δx_i, Δy_j, and Δz_k along the axes. Figure 14.14 shows projections of the mesh along the three coordinate planes. The material properties $\epsilon[i,j,k]$, $\mu[i,j,k]$, and $\sigma[i,j,k]$ are uniform over an element volume. In contrast to previous treatments, we shall associate the primary field quantities with elements rather than vertices. Because the electric field plays an important role in many applications, we shall use **E** as the primary quantity and treat **H** as a subsidiary field quantity. The electrical field components $E_x[i,j,k]$, $E_y[i,j,k]$, and $E_z[i,j,k]$ are constant over an element volume. To develop space-centered difference equations, we assume that magnetic intensity components are shifted in space to occupy volumes between the elements. Figure 14.14 shows that $H_x[i,j,k]$, $H_y[i,j,k]$, and $H_z[i,j,k]$ are centered near vertex $[i,j,k]$ and extend over a region that includes one eighth of the volumes of elements $[i,j,k]$, $[i+1,j,k]$, $[i,j+1,k]$,$[i+1,j+1,k]$, $[i,j,k+1]$, $[i+1,j,k+1]$, $[i,j+1,k+1]$, and $[i+1,j+1,k+1]$. For time-centering, we adopt the Yee method. Electric fields are defined at times $t = n\Delta t$ and magnetic intensity at $t = (n+\frac{1}{2})\Delta t$.

As in previous treatments, we shall work from integral forms of the Maxwell curl equations,

$$\oint \mathbf{E} \cdot d\mathbf{l} = -\iint dA\, \mu\, \frac{\partial}{\partial t}(\mathbf{H} \cdot \hat{n}) \tag{14.57}$$

and

$$\oint \mathbf{H} \cdot d\mathbf{l} = \iint dA \left[\epsilon \frac{\partial}{\partial t}(\mathbf{E} \cdot \hat{n}) + J_o \cdot \hat{n} + \sigma\, \mathbf{E} \cdot \hat{n} \right]. \tag{14.58}$$

To begin, we can apply Equation 14.58 to derive linear equations for E_x, E_y, and E_z in each element. The following discussion concentrates on the equation for E_x. The circuit integral of magnetic intensity on the left-hand side of the equation is applied with positive rotation about the path shown in Figure 14.14a. Components of **H** that overlap the element are averaged in the x-direction. Adding all components gives the following expression for the circuit integral around element $[i,j,k]$,

$$\text{IntH}_x^{n+\frac{1}{2}}[i,j,k] = \tag{14.59}$$

$$\left(H_y^{n+\frac{1}{2}}[i-1,j-1,k-1] + H_y^{n+\frac{1}{2}}[i-1,j,k-1] + H_y^{n+\frac{1}{2}}[i-1,j,k-1] + H_y^{n+\frac{1}{2}}[i,j-1,k-1] \right.$$

$$\left. -H_y^{n+\frac{1}{2}}[i-1,j,k] - H_y^{n+\frac{1}{2}}[i,j,k] - H_y^{n+\frac{1}{2}}[i-1,j-1,k] - H_y^{n+\frac{1}{2}}[i,j-1,k-1] \right)$$

$$\Delta y[j]/4$$

$$+\left(H_z^{n+\frac{1}{2}}[i-1,j,k-1] + H_z^{n+\frac{1}{2}}[i,j,k-1] + H_z^{n+\frac{1}{2}}[i-1,j,k] \right.$$

$$+H_z^{n+\frac{1}{2}}[i,j,k] - H_z^{n+\frac{1}{2}}[i-1,j-1,k] - H_z^{n+\frac{1}{2}}[i,j-1,k]$$

$$\left. -H_z^{n+\frac{1}{2}}[i-1,j-1,k-1] - H_z^{n+\frac{1}{2}}[i,j-1,k-1] \right) \Delta z[k]/4.$$

The x-component of the area integral on the right-hand side of Equation 14.58 is

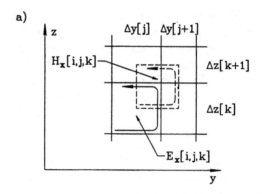

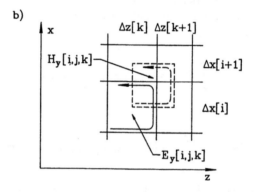

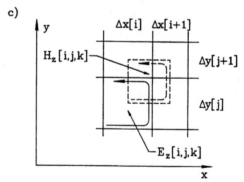

FIGURE 14.14
Regular mesh for three-dimensional finite-element electromagnetic simulations on a regular mesh. Projections in a Cartesian coordinate system show circuit integral paths around elements and H cells (dashed line). (a) y–z plane; (b) z–x plane; (c) x–y plane.

$$\left(\epsilon[i,j,k] \frac{E_x^{n+1}[i,j,k] - E_x^n[i,j,k]}{\Delta t} + J_{ox}^{n+\frac{1}{2}} \right.$$

$$\left. + \sigma[i,j,k] \frac{E_x^{n+1}[i,j,k] + E_x^n[i,j,k]}{2} \right) \Delta y[j]\, \Delta z[k]. \tag{14.60}$$

In Equation 14.60, note that terms involving the electric field are written as expressions centered at time $t = (n+\frac{1}{2})\Delta t$. Combining Equations 14.59 and 14.60 gives an equation to advance the x-component of the electric field,

$$E_x^{n+1}[i,j,k] = \left(\cfrac{1}{1 + \cfrac{\sigma[i,j,k]\Delta t}{2\epsilon[i,j,k]}} \right) \tag{14.61}$$

$$\times \left[\left(1 - \frac{\sigma[i,j,k]\Delta t}{2\epsilon[i,j,k]} \right) E_x^n[i,j,k] - \frac{J_{ox}^{n+\frac{1}{2}}[i,j,k]\Delta t}{\epsilon[i,j,k]} + \frac{IntH_x^{n+\frac{1}{2}}\Delta t}{\epsilon[i,j,k]\Delta y[j]\Delta z[k]} \right].$$

We can derive an equation to advance H from Equation 14.57. For the component $H_x[i,j,k]$ we take a circuit integral of electric field around the dashed line in Figure 14.14a, again averaging in x. The left-hand side of Equation 14.57 has the form

$$IntE_x^n[i,j,k] = \left(E_y^n[i,j,k] + E_y^n[i+1,j,k] - E_y^n[i,j,k+1] - E_y^n[i+1,j,k+1] \right) \frac{\Delta y[j]}{4} \tag{14.62}$$

$$+ \left(E_y^n[i,j+1,k] + E_y^n[i+1,j+1,k] - E_y^n[i,j+1,k+1] - E_y^n[i+1,j+1,k+1] \right) \frac{\Delta y[j+1]}{4}$$

$$+ \left(E_z^n[i,j+1,k] + E_z^n[i+1,j+1,k] - E_z^n[i,j,k] - E_z^n[i+1,j,k] \right) \frac{\Delta z[k]}{4}$$

$$+ \left(E_z^n[i,j+1,k+1] + E_z^n[i+1,j+1,k+1] - E_z^n[i,j,k+1] - E_z^n[i+1,j,k+1] \right) \frac{\Delta z[k+1]}{4}.$$

The area integral on the right-hand side involves a weighted average of magnetic permeability over eight cells. Using time-centered derivatives, the expression becomes,

$$\left[\left(\mu[i,j,k] + \mu[i+1,j,k] \right) \Delta y[j] \Delta z[k] + \left(\mu[i,j+1,k] + \mu[i+1,j+1,k] \right) \Delta y[j+1] \Delta z[k] \right.$$

$$+ \left(\mu[i,j,k+1] + \mu[i+1,j,k+1] \right) \Delta y[j] \Delta z[k+1]$$

$$+ \left(\mu[i,j+1,k+1] + \mu[i+1,j+1,k+1] \right) \Delta y[j+1] \Delta y[k+1] \right] \tag{14.63}$$

$$\left(\frac{H_x^{n+\frac{1}{2}}[i,j,k] - H_x^{n-\frac{1}{2}}[i,j,k]}{\Delta t} \right).$$

Combining Equations 14.62 and 14.63 gives an equation to advance H_x,

$$H_x^{n+\frac{1}{2}}[i,j,k] = H_x^{n-\frac{1}{2}}[i,j,k] - \frac{\Delta t}{MuA_x^n[i,j,k]} \ IntE_x^n[i,j,k], \qquad (14.64)$$

where

$$Mu\,A_x[i,j,k] = \left[\left(\mu[i,j,k] + \mu[i+1,j,k]\right) \ \Delta y[j] \ \Delta z[k]\right.$$

$$+ \left(\mu[i,j+1,k] + \mu[i+1,j+1,k]\right) \ \Delta y[j+1] \ \Delta z[k]$$

$$\qquad\qquad\qquad\qquad\qquad\qquad\qquad\qquad\qquad\qquad (14.65)$$

$$+ \left(\mu[i,j,k+1] + \mu[i+1,j,k+1]\right) \ \Delta y[j] \ \Delta z[k+1]$$

$$+ \left(\mu[i,j+1,k+1] + \mu[i+1,j+1,k+1]\right) \ \Delta y(j+1) \ \Delta z[k+1].$$

A similar procedure leads to difference equations for the y and z components of **E** and **H**.

The solution process for internal elements is straightforward. The first operation in a time step is to advance electric field components for all elements, replacing variables *in situ* because they do not depend on neighboring values of **E**. The second operation is to advance components of **H**. The stability condition is less stringent than that for FDTD calculations (Equation 14.56). The time step must be shorter than the minimum propagation time for radiation across an element. If v_i is the speed of light in an element, then we can write the Courant condition as

$$\Delta t \le \min\left[\frac{\Delta x[i]}{v[i]}, \frac{\Delta y[j]}{v[i]}, \frac{\Delta z[k]}{v[i]}\right]. \qquad (14.66)$$

To specify conditions on solution boundaries, we shall set properties of boundary layers rather than apply conditions at vertices. To avoid index errors implementing the technique, we assume that there is a peripheral layer of H cells with value zero,

$$H_x[0,j,k] = H_x[i,0,k] = H_x[i,j,0] = H_x[I,j,k] \ H_x[i,J,k] = H_x[i,j,K] = 0,$$

$$H_y[0,j,k] = H_y[i,0,k] = H_y[i,j,0] = H_y[I,j,k] \ H_x[i,J,k] = H_y[i,j,K] = 0, \quad (14.67)$$

$$H_z[0,j,k] = H_z[i,0,k] = H_z[i,j,0] = H_z[I,j,k] \ H_x[i,J,k] = H_z[i,j,K] = 0.$$

A boundary consists of a single layer of elements on a side of the solution volume. Table 14.1 shows material properties in the layers to implement common boundary conditions.

We can illustrate the validity of the method with benchmark calculation comparing results to those generated with the two-dimensional method of Section 14.1. The solution region is an axially extended box with dimensions

TABLE 14.1

Properties of Boundary Layers of Thickness Δ

Condition	Vacuum Adjacent Material			Non-Vacuum Adjacent Material		
	ϵ	μ	σ	ϵ	μ	σ
Conductive wall	∞	μ_o	0	∞	μ	0
Open circuit	ϵ_o	∞	0	ϵ	∞	0
Perfect absorber	ϵ_o	μ_o	$1/\Delta(\mu_o/\epsilon_o)^{1/2}$	ϵ	μ	$1/\Delta(\mu/\epsilon)^{1/2}$

(in meters) of $0.0 \leq x \leq 1.0$, $0.0 \leq y \leq 1.0$, and $0.0 \leq z \leq 10.0$. Elements in the half volume from $x = 0.0$ to $x = 0.5$ have $\epsilon_r = 1.0$ while those in the other half have the dielectric constant of polyethylene ($\epsilon_r = 2.7$). The solution box is surrounded by absorbing layers of thickness $\Delta = 0.01$. In locations adjacent to the air region the layer has the properties $\epsilon_r = 1.0$, $\mu_r = 1.0$, and $\sigma = 0.2650$ mhos/m. The layer adjacent to polyethylene has $\epsilon_r = 2.7$, $\mu_r = 1.0$, and $\sigma = 0.4354$ mhos/m. A pulse is initiated by a field boundary condition on a plate at $x = 0$. The plate extends over an area $0.25 \leq y \leq 0.75$ and $2.5 \leq z \leq 7.5$. The field conditions are that $E_x = 0.0$, $E_y = 0.0$, and that E_z follows the Gaussian variation of Equation 13.47 with an amplitude of 1 V/m. During the initial pulse transit time across the x dimension, the fields on the center plane ($z = 5.0$) of the three-dimensional solution should agree exactly with those of a two-dimensional solution for a system with infinite length in z.

Figure 14.15a shows results from a two-dimensional calculation using a conformal triangular mesh. There are approximately 100 elements on a side. For the planar geometry, the primary field component E_z is defined at vertices. The plot shows contours of E_z at 3.0 nsec. The drive condition applies along the vertices marked with a dashed line in the figure. The solution exhibits some features of one-dimensional pulse propagation. A transmitted pulse of reduced amplitude and spatial width moves forward in the dielectric, and there is a negative reflected pulse. Fields propagating inward from the edges of the drive plate increase the complexity of the field pattern. There is a region of negative E_z behind the pulse front. Figure 14.16a shows $E_z(t)$ detected by probes located in the air region ($x = 0.25$, $y = 0.50$) and in the polyethylene ($x = 0.75$, $y = 0.50$). The front of the transmitted pulse has reduced amplitude consistent with Equation 13.18 and the same temporal width as the incident pulse. Figures 14.15b and 14.16b show corresponding results from a three-dimensional calculation. The uniform mesh has 1,061,208 elements with $\Delta x = \Delta y = 0.01$ and $\Delta z = 0.1$. The absorbing layer extends around the periphery and the drive field condition applies over the thin rectangular region on the left-hand side. Figure 14.15b is a contour plot of E_z in a solution slice at $z = 5.0$ m. The results are almost identical to those of the two-dimensional calculation using an entirely different finite-element approach. The plot is more coarse because there are half as many rectangular elements in a cross section of the three-dimensional solution. The probe results of Figure 14.16b are also consistent. Finally, Figure 14.15c is a contour plot of a slice of the three-dimensional solution at y = 0.5. The figure has been expanded by a factor of 10 along the x direction to display the fields more clearly. There is a well-defined transmitted pulse with finite axial extent. Fields propagating from the axial edges create an interesting interference pattern along z in the reflected pulse.

As a final note, an open-circuit layer gives a good representation of a boundary with mirror symmetry when drive surfaces or current densities are well removed. On the other hand, interference effects give an imperfect solution when sources are adjacent to the layer. In the example of Figure 14.17 a pulsed plate contacts the upper open-circuit layer. The resulting solution does not exhibit ideal reflection symmetry. In this case, it is better to enforce symmetry along the boundary explicitly. In the example, the values of magnetic intensity on the boundary should be given by

$$H_x[i,J,k] = 0$$

$$H_y[i,J,k] = H_y[i,J-1,k] \qquad (14.68)$$

$$H_z[i,J,k] = 0$$

Chapter 14   Exercises

14.1. Develop time-domain equations for E-type waves on a two-dimensional square mesh with spacing Δ. The values $E_z(i,j)$ are defined at vertices. Consider a homogeneous medium with no losses or sources.

(a) Find the finite-difference form of Equation 13.7.

(b) Apply Equation 14.1 around the square elements. Take the average value of E_z along a side as the average of values at the connected vertices.

14.2. Figure E14.1 is a cross section of a parallel plate transmission line with a uniform fill medium with ε and μ.

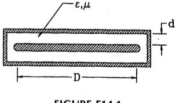

FIGURE E14.1

(a) In the limit that d « D, derive expressions for the capacitance per unit length, inductance per unit length, and characteristic impedance.

(b) Derive expressions that relate the field quantities E_x and H_y to V and I following the discussion of Section 14.2.

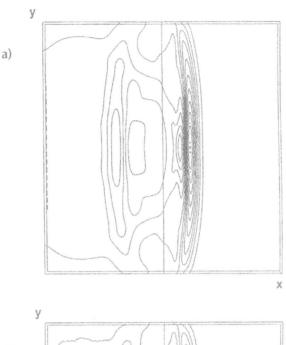

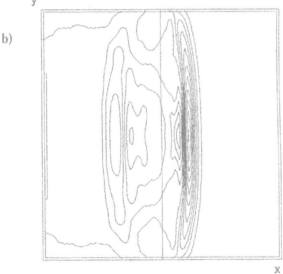

FIGURE 14.15
Pulse propagation in a box-shaped anechoic chamber. (a) Two-dimensional calculation using a conformal triangular mesh. Left-hand region, $\epsilon r = 1.0$; right-hand region; $\epsilon r = 2.7$. Boundary condition on line region at the left is a 0.22 nsec Gaussian pulse of uniform Ez. Contours of Ez at t = 3.0 nsec. Boundaries: $x_{min} = -0.01$; $x_{max} = 1.01$; $y_{min} = -0.01$; $y_{max} = 1.01$. (b) Slice of a three-dimensional calculation at z = 5.0 with the same conditions as Part a. Boundaries: $x_{min} = -0.01$; $x_{max} = 1.01$; $y_{min} = -0.01$; $y_{max} = 1.01$. (c) Slice of a three-dimensional calculation at y = 0.5 with the same conditions as Part a. Boundaries: $x_{min} = -0.01$; $x_{max} = 1.01$; $z_{min} = -0.10$; $z_{max} = 10.1$.

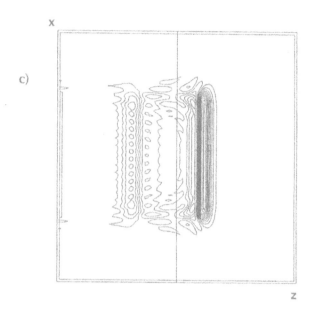

FIGURE 14.15 (continued)

14.3.   Write expressions for the subsidiary field components H_r and H_z for a cylindrical solution with primary field component E_θ.

14.4.   Consider scattering solutions where we place dielectric objects in a rectangular solution space with dimensions $x_{min} = 10.0$, $x_{max} = 20.0$, $y_{min} = -5.0$, and $y_{max} = 5.0$. A cylindrical source of radius r_o centered at $(0,0)$ produces an electric field of the form

$$E(r,t) = \frac{E_o r_o}{r} \exp\left[-jkr + j\omega t\right]$$

where $r = (x^2 + y^2)^{1/2}$. Applying the distributed source method of Section 14.4, give an expression for source current density as a function of position that would generate the wave.

14.5.   A pillbox cavity has radius $R = 0.15$ m and length $D = 0.05$ m.

   (a) What is the resonant frequency for TM_{021} mode?

   (b) Sketch field variations for the mode.

14.6.   A half-wave transmission line resonator shorted at each end is excited at the center. The coaxial vacuum line has inner radius $R_i = 0.01$ m, outer radius $R_o = 0.05$ m, and length $L = 2.0$ m.

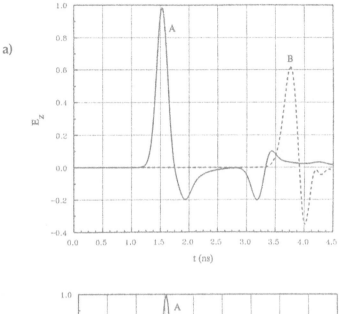

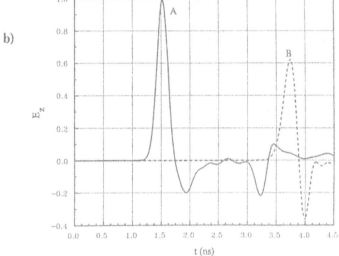

FIGURE 14.16
Probe signals of $E_z(t)$ for calculations of FIGURE 14.15. Probe A at x = 0.25, y = 0.50, and probe B at x = 0.75 and y = 0.50. (a) Two-dimensional calculation; (b) three-dimensional calculation.

(a) What is the resonant frequency?

(b) If the voltage amplitude at the center is V_o, give expressions for the variation of voltage and current over the length of the line.

(c) What is the total stored electromagnetic energy of the mode?

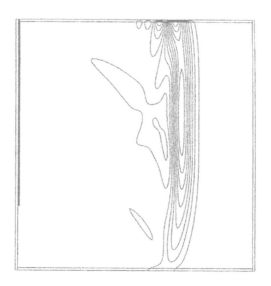

FIGURE 14.17
Imperfect implementation of a mirror reflection boundary when a drive field or current density region abuts an open-circuit layer. Absorbing layers on bottom, left, and right boundaries. Top boundary is an open-circuit layer. Drive plate on top left-hand side. Contours of E_z.

    (d) Assuming copper electrodes with volume resistivity $\rho = 1.7 \times 10^{-8}$ Ω-m, find the total resistive power loss. (Use the results of Exercise 11.12.)

    (e) What is the Q factor of the resonator?

14.7.   Write one-dimensional time-domain coupled electromagnetic equations for E_x and H_y using the element approach of Section 14.6.

Bibliography

Akin, J.E., *Application and Implementation of Finite-element Methods*, Academic Press, New York, 1982.

Alder, B., Fernbach, S., and Rotenburg, M., Eds., *Methods in Computational Physics — Plasma Physics*, Academic Press, New York, 1970.

Allaire, P.E., *Basics of the Finite Element Method*, Wm. C. Brown, Dubuque, IA, 1985.

Arfken, G., *Mathematical Methods for Physicists*, Academic Press, New York, 1985.

Balanis, C.A., *Antenna Theory: Analysis and Design*, Harper and Row, New York, 1982.

Bathe, K.J. and Wilson, E.L., *Numerical Methods in Finite-element Analysis*, Prentice-Hall, Englewood Cliffs, NJ, 1976.

Becker, E.B., Carey, G.F., and Oden, J.T., *Finite Elements — An Introduction*, Prentice-Hall, Englewood Cliffs, NJ, 1981.

Bewley, I.V., *Two-dimensional Fields in Electrical Engineering*, Dover, New York, 1963.

Bickford, W.B., *A First Course in Finite-element Methods*, Richard D. Irwin, Homewood, IL, 1990.

Binns, K.J., Lawrence, P.J., and Trowbridge, *The Analytic and Numerical Solution of Electric and Magnetic Fields*, Wiley, New York, 1992.

Birdsall, C.K. and Langdon, A.B., *Plasma Physics Via Computer Simulation*, McGraw-Hill, New York, 1985.

Booton, R.C., Jr., *Computational Methods for Electromagnetics and Microwaves*, Wiley, New York, 1992.

Brebbia, C.A., Ed., *Applied Numerical Modeling*, Wiley, New York, 1978.

Brebbia, C.A., *Finite-element Systems: A Handbook*, Springer-Verlag, New York, 1985.

Brebbia, C.A. and Connor, J.J., *Fundamentals of Finite-element Technique*, Butterworth, London, 1973.

Brebbia, C.A. et al., *Boundary Elements*, CML Publishers, Southhampton, 1983.

Cairo, L. and Kahan, T., *Variational Techniques in Electromagnetism*, Gordon and Breach, New York, 1965.

Chari, M.V. and Silvester, P.P., Eds., *Finite Elements for Electrical and Magnetic Field Problems*, Wiley, New York, 1980.

Cheng, D.K., *Field and Wave Electromagnetics, Second Edition*, Addison-Wesley, Reading, MA, 1992, 38.

Chung, T.J., *Finite-element Analysis in Fluid Dynamics*, McGraw-Hill, New York, 1978.

Csendes, Z.J., Ed., *Computational Electromagnetism*, North-Holland, New York, 1986.

Davies, A.J., *The Finite-element Method: A First Approach*, Clarendon, Oxford, 1980.

Davis, P.J. and Rabinowitz, P., *Methods of Numerical Integration*, Academic Press, New York, 1975.

Duff, I.S., Erisman, A.M., and Reid, J.K., *Direct Methods for Sparse Matrices*, Oxford University Press, Oxford, 1986.

Eikichi, Y., Ed., *Analysis Methods for Electromagnetic Wave Problems*, Artech House, Norwood, MA, 1990.

Ferziger, J.H., *Numerical Methods for Engineering Applications*, Wiley, New York, 1981.

George, P.L., *Automatic Mesh Generation — Applications to Finite-element Methods*, Wiley, New York, 1992. .

Goldberg, S., *Introduction to Difference Equations*, Dover, New York, 1986.

Golub, G.H. and Van Loan, C.F., *Matrix Computations*, Johns Hopkins University Press, Baltimore, 1983.

Green, C.D., *Integral Equation Methods*, Barnes and Noble, New York, 1969.

Guenther, R.B. and Lee, J.W., *Partial Differential Equations of Mathematical Physics and Integral Equations*, Prentice-Hall, Englewood Cliffs, NJ, 1988.

Gupta, K.C. et al., *Computer-aided Design of Microwave Circuits*, Artech House, Dedham, MA, 1981.

Halstead, J.B., Liquid Water — Dielectric Properties, in Franks, F., Ed., *Water — A Comprehensive Treatise*, Plenum Press, New York, 1972.

Hansen, R.C., Ed., *Moment Methods in Antennas and Scattering*, Artech House, Norwood, MA, 1990.

Harrington, R.F., *Time Harmonic Electromagnetic Fields*, McGraw-Hill, New York, 1961.

Harrington, R.F., *Field Computation by Moment Methods*, Macmillan, New York, 1968.

Henrici, P., *Introduction to Numerical Analysis*, Wiley, New York, 1964.

Hildebrand, F.B., *Introduction to Numerical Analysis*, Dover, New York, 1974.

Hildebrand, F.B., *Methods of Applied Mathematics*, Prentice-Hall, Englewood Cliffs, NJ, 1965, 137.

Hinton, E. and Owen, D.R., *Finite-element Programming*, Academic Press, New York, 1977.

Hinton, E. and Owen, D.R., *An Introduction to Finite-element Computations*, Pineridge Press, Swansea, 1979.

Hoole, S.R., *Computer-aided Analysis and Design of Electromagnetic Devices*, Elsevier, New York, 1989.

Huebner, K.H., *The Finite-element Method for Engineers*, Wiley, New York, 1975.

Hughes, T.J., *The Finite-element Method: Linear, Static and Dynamics Finite-element Analysis*, Prentice-Hall, Englewood Cliffs, NJ, 1987.

Humphries, S., *Principles of Charged Particle Acceleration*, Wiley, New York, 1986, 103.

Humphries, S., *Charged Particle Beams*, Wiley, New York, 1990, chaps. 2 and 7.

Isaacson, E. and Keller, H.B., *Analysis of Numerical Methods*, Dover, New York, 1966.

Itoh, T., Ed., *Numerical Techniques for Microwave and Millimeter-wave Passive Structures*, Wiley, New York, 1989.

Jackson, J.D., *Classical Electrodynamics*, Second Edition, Wiley, New York, 1975, 174 and 336.

Jin, J., *The Finite-element Method in Electromagnetics*, Wiley, New York, 1993.

Kantorovich, L.V. and Krylov, V.I., *Approximate Methods of High Analysis*, Wiley, New York, 1964.

Kardestuncer, H., Ed., *Finite-element Handbook*, McGraw-Hill, New York, 1987.

Kelly, L.G., *Handbook of Numerical Methods and Applications*, Addison-Wesley, Reading, MA, 1967.

Kong, J.A., Ed., *Research Topics in Electromagnetic Theory*, Wiley, New York, 1981.

Kunz, K.S. and Luebbers, R.J., *The Finite-difference Time-domain Method for Electromagnetics*, CRC Press, Boca Raton, FL, 1993, chap. 18.

Lapidus, L. and Pinder, G.F., *Numerical Solution of Partial Differential Equations in Science and Engineering*, Wiley, New York, 1982.

Ley, B.J., *Computer Aided Analysis and Design for Electrical Engineers*, Holt, Rinehart and Winston, New York, 1989.

Marchuk, G.I. and Shaidurov, V.V., *Difference Methods and Their Extrapolations*, Springer-Verlag, New York, 1983.

Martin, C. and Carey, G.F., *Introduction to Finite-element Analysis: Theory and Application*, McGraw-Hill, New York, 1978.

McCaig, M. and Clegg, A.G., *Permanent Magnets in Theory and Practice, Second Edition*, Wiley, New York, 1987.

Mikhlin, S.G., *Variational Methods in Mathematical Physics*, Macmillan, New York, 1964.

Mikhlin, S.G. and Smolitskiy, K.I., *Approximate Methods for Solution of Differential and Integral Equations*, Elsevier, New York, 1967.

Miller, E.K., Medgyesi-Mitschang, L., and Newman, E.H., *Computational Electromagnetics*, IEEE Press, Piscataway, NJ, 1992.

Milne, W.E., *Numerical Solutions of Differential Equations*, Dover, New York, 1970.

Mirshekar-Syahkal, D., *Spectral Domain Method for Microwave Integrated Circuits*, Wiley, New York, 1990.

Mittra, R., Ed., *Computer Techniques for Electromagnetics*, Pergamon Press, Oxford, 1973.

Mittra, R., Ed., *Numerical and Asymptotic Techniques in Electromagnetics*, Springer-Verlag, New York, 1975.

Moiseiwitch, B.L., *Variational Principles*, Interscience, London, 1966.

Morgan, M.A., Ed., *Finite-element and Finite-difference Methods in Electromagnetic Scattering*, Elsevier, New York, 1990.

Morse, P.M. and Feshback, *Methods of Theoretical Physics*, McGraw-Hill, New York, 1953.

Mur, G., IEEE Transactions of Electromagnetic Compatibility, 23, 377, 1981.

Nakamura, S., *Computational Methods in Engineering and Science*, Wiley, New York, 1977.

Nering, E.D., *Linear Algebra and Matrix Theory*, Wiley, New York, 1965.

Norris, D.H. and de Vries, G., *An Introduction to Finite-element Analysis*, Academic Press, New York, 1978.

Oden, J.T. and Reddy, J.N., *An Introduction to the Mathematical Theory of Finite Elements*, Wiley, New York, 1976.

Potter, D., *Computational Physics*, Wiley, New York, 1973, 78.

Press, W.H., Teukolsky, S.A., Vettering, W.T., and Flannery, B.P., *Numerical Recipes in FORTRAN, Second Edition*, Cambridge University Press, Cambridge, 1992, chaps. 3 and 16.

Reddy, J.N. and Rasmussen, *Advanced Engineering Analysis*, Wiley, New York, 1982.

Reddy, J.N., *An Introduction to the Finite-element Method*, McGraw-Hill, New York, 1984.

Richtmyer, R.D. and Morton, K.W., *Difference Methods for Initial-value Problems, Second Edition*, Interscience, New York, 1976.

Sabonnadiere, J.C. and Coulomb, J.L., *Finite-element Methods in CAD: Electrical and Magnetic Fields*, Springer-Verlag, New York, 1987.

Sadiku, M.N., *Numerical Techniques in Electromagnetics*, CRC Press, Boca Raton, FL, 1992, chap. 6.

Scott, C., *The Spectral Domain Method in Electromagnetics*, Artech House, Norwood, MA, 1989.

Silvester, P.P. and Ferrari, R.L., *Finite Elements for Electrical Engineers*, Cambridge University Press, New York, 1983.

Silvester, P.P. and Pelosi, G., Eds., *Finite Elements for Wave Electromagnetics*, IEEE Press, Piscataway, NJ, 1994.

Smith, G.D., *Numerical Solution of Partial Differential Equations — Finite-difference Methods*, Clarendon Press, Oxford, 1978.

Stakgold, I., *Boundary Value Problems of Mathematical Physics*, Macmillan, New York, 1968.

Steele, C.W., *Numerical Computation of Electric and Magnetic Fields*, Van Nostrand Reinhold, New York, 1987.

Strait, B.J., *Applications of Moment Methods to Electromagnetics*, SCEEE Press, St. Cloud, FL, 1980.

Strang, G. and Fix, G.J., *An Analysis of the Finite-element Method*, Prentice-Hall, Englewood Cliffs, NJ, 1973.

Strikwerda, J.C., *Finite Difference Schemes and Partial Differential Equations*, Wadsworth and Brooks, Belmont, CA, 1989.

Taflove, A. and Brodwin, M.E., *IEEE Transactions on Microwave Theory and Techniques*, MTT-23, 623, 1975.

Tai, C.T., *Dyadic Green's Functions in Electromagnetic Theory*, Intext, Scranton, PA, 1971.

Thom, A. and Apelt, C.J., *Field Computations in Engineering and Physics*, D. Van Nostrand, London, 1961.

Uslenghi, P.L., Ed., *Electromagnetic Scattering*, Academic Press, New York, 1978.

Vemuri, V. and Karplus, W.J., *Digital Computer Treatment of Partial Differential Equations*, Prentice-Hall, Englewood Cliffs, NJ, 1981.

Vichevetsky, R., *Computer Methods for Partial Differential Equations*, Prentice-Hall, Englewood Cliffs, NJ, 1981.

Vorobev, Y.U., *Method of Moments in Applied Mathematics*, Gordon and Breach, New York, 1965.

Wachspress, E.L., *A Rational Finite-element Basis*, Academic Press, New York, 1975.

Wait, R. and Mitchell, A.R., *Finite-element Analysis and Applications*, Wiley, New York, 1985.

Warren, J.L. et al., *Reference Manual for the Poisson/Superfish Group of Codes*, Los Alamos National Laboratory, LA-UR-87-126, 1987.

Weinstock, R., *Calculus of Variations*, Dover, New York, 1974.

Yee, K.S., *IEEE Transactions on Antennas and Propagation* AP-14, 302, 1966.

Young, D.M. and Gregory, R.T., A Survey of Numerical Mathematics, Dover, New York, 1973.

Zienkiewicz, O.C., *The Finite-element Method in Engineering Science*, McGraw-Hill, New York, 1971.

Index

A

Adiabatic exponent, 174
Ampere's force law 183
Ampere's law
  differential, 187
  estimating magnet drive current, 202
  finite-element form, 198
  integral, 185–186
  with anisotropic materials, 211
  with displacement current, 255
Amplification factor, diffusion
  solutions, 278, 279
Anechoic chamber, 346
Arbitrary mesh, 52
Attenuation, electromagnetic wave, 303

B

Back-substitution method, 75, 316
Biot and Savart law, 183
Bisection method, 134, 324
Block tridiagonal form, 126
Boundary conditions
  absorbing, lookback technique, 359
  Dirichlet, 33
  electric field on a dielectric surface, 22
  electric field on a metal surface, 18, 19
  electrostatic finite-element, 33
  electrostatic potential, 17
  finite-difference time-domain on Yee
    cells, 359
  finite-element (3D), 365
  for vector potential, magnetic
    diffusion, 290, 293
  free-space, 344
  ideal gas flow, 176
  ideal absorbing boundaries,
    electromagnetics, 302, 335
  magnetostatic solutions, 191, 202

Neumann, 34–35, 154
  on magnetic material, 197
  potential on a metal surface, 18, 33
  reactive boundaries, electromagnetic
    solutions, 313
  short-circuit and open-circuit,
    electromagnetic waves, 302
  symmetry, electrostatics, 35
  symmetry, finite-element
    electromagnetics (3D), 366
  time-domain electromagnetics (2D),
    334
Boundary plot, 146
Boundary value solutions
  definition, 2
  iterative for non-linear materials, 160
  successive over-relaxation, 104
Boyle's law, 173

C

C Magnet, 201
Calculus of variations, 47–49
  Euler equation, 49
  functionals, 48
Capacitance
  definition, 221
  mutual, 221
Capacitive probe, source term in
  electromagnetic solution, 332–333
Capacitor
  boundary in electromagnetic
    solutions, 314
  current through, 221
  displacement current in, 254
  energy, parallel plate, 47
Centered difference operators, 67, 69
Charge density
  as element property, 29
  definition, 13–14

O

Ohm's law, 103, 240, 288
Open-circuit condition, electromagnetic
    waves, 302
Operating point, permanent magnet,
    206
Orthogonal functions, 65

P

Paramagnetism, 194
Particle-in-cell method, 233
Permanent magnet
    demagnetization curve, 204
    design procedure, 206
    easy axis, 204
    energy product, 208
    intrinsic demagnetization curve,
        205
    properties, 204–205
Permeability, magnetic, 196–197
Pillbox cavity resonator, 348
Pinch, magnetic, 291
Pivoting, matrix operations, 113–114
Plot
    contour, 148
    element, color-coded, 150
    elevation or wireframe, 151
    hidden surface, 153
    vector sorting in, 149
Poisson equation, 17
    condition of minimum energy,
        49–50
    difference form (1D), 74
    electric fields with complex dielectric
        constant (AC), 260
    finite-element (1D), 74
    solution by back-substitution, 75
    solution by tridiagonal matrix
        inversion, 119–123
    with beam charge, 236
    with isotropic dielectrics, 22
Polynomial interpolation, 165
Potential, complex number (AC), 260,
    262
Power density
    electromagnetic, 318
    magnetic field (AC), 266
    resistive media, 263

Power loss, resonator wall, 353
Poynting's theorem, 317
Pyramid functions, 54

Q

Quality factor, 322, 324
    pillbox cavity, 355
    resonators (2D), 355

R

Reflection coefficient, electromagnetic
    wave at a boundary, 302
Relativistic dynamics, 230
Relaxation method, 104
    boundary value solutions, 104
    complex vector potential (AC),
        265
    residual, 105
    spectral radius, 106
    stability, 105
    with Chebyshev acceleration, 106
    with nonlinear materials, 160, 202
Remanence field, 195, 205
Residual
    complex for magnetic field solutions
        (AC), 265
    in relaxation solutions, 105
Resistive medium
    electric fields (AC), 258, 261
    electrostatics, 103
    power density, in 263
    with Hall effect, 239
Resolution, mesh, 25
Resonant solution, definition, 6
Resonator
    lumped element, 321
    mode frequencies, pillbox cavity,
        348
    mode search procedures, 319,
        323–324, 348
    one-dimensional, 321
    quality factor, 322
    RF accelerator, 350
    two-dimensional, 348
    wall power losses, 353
Rotation matrices, 91, 152
Roundoff errors, 25
Runge-Kutta procedure, 71